MOLECULAR AND CELLULAR THERAPIES FOR MOTOR NEURON DISEASES

MOLECULAR AND CELLULAR THERAPIES FOR MOTOR NEURON DISEASES

Edited by

NICHOLAS BOULIS, DEIRDRE O'CONNOR
AND ANTHONY DONSANTE
Emory University, Atlanta, GA, United States

ACADEMIC PRESS

An imprint of Elsevier
elsevier.com

Academic Press is an imprint of Elsevier
125 London Wall, London EC2Y 5AS, United Kingdom
525 B Street, Suite 1800, San Diego, CA 92101-4495, United States
50 Hampshire Street, 5th Floor, Cambridge, MA 02139, United States
The Boulevard, Langford Lane, Kidlington, Oxford OX5 1GB, United Kingdom

Notices
Knowledge and best practice in this field are constantly changing. As new research
and experience broaden our understanding, changes in research methods, professional
practices, or medical treatment may become necessary.

Practitioners and researchers must always rely on their own experience and knowledge
in evaluating and using any information, methods, compounds, or experiments described
herein. In using such information or methods they should be mindful of their own
safety and the safety of others, including parties for whom they have a professional
responsibility.

To the fullest extent of the law, neither the Publisher nor the authors, contributors, or
editors, assume any liability for any injury and/or damage to persons or property as a
matter of products liability, negligence or otherwise, or from any use or operation of any
methods, products, instructions, or ideas contained in the material herein.

British Library Cataloguing-in-Publication Data
A catalogue record for this book is available from the British Library

Library of Congress Cataloging-in-Publication Data
A catalog record for this book is available from the Library of Congress

ISBN: 978-0-12-802257-3

For Information on all Academic Press publications
visit our website at https://www.elsevier.com

  Working together
to grow libraries in
developing countries

www.elsevier.com • www.bookaid.org

Publisher: Mara Conner
Acquisitions Editor: Melanie Tucker
Editorial Project Manager: Kristi Anderson
Production Project Manager: Chris Wortley
Designer: Matthew Limbert

Typeset by MPS Limited, Chennai, India

Cover image: "Spinal Cord." 12K gold, ink, and dye on stainless steel panel.
By Greg Dunn, 2014. www.gregadunn.com

Contents

4. Molecular Mechanisms of Amyotrophic Lateral Sclerosis

M. COLLINS AND R. BOWSER

5. An Introduction to the Natural History, Genetic Mapping, and Clinical Spectrum of Spinal Muscular Atrophy

A. McDONOUGH, L. URQUIA AND N. BOULIS

6. Genetics of Spinal Muscular Atrophy

A.H.M. BURGHES AND V.L. McGOVERN

7. Introduction to Gene and Stem-Cell Therapy

D.M. O'CONNOR

8. Gene Therapy for Amyotrophic Lateral Sclerosis: Therapeutic Transgenes

A. DONSANTE

12. Clinical Trials to Date
B.J. MADER AND N. BOULIS

List of Contributors

N. Boulis Emory University, Atlanta, GA, United States

R. Bowser St Joseph's Hospital and Medical Center, Phoenix, AZ, United States

A.H.M. Burghes The Ohio State University, Columbus, OH, United States

K.S. Chen University of Michigan, Ann Arbor, MI, United States

M. Collins St Joseph's Hospital and Medical Center, Phoenix, AZ, United States

S. Corti University of Milan, Milan, Italy; IRCCS Foundation Ca' Granda Ospedale Maggiore Policlinico, Milan, Italy

A. Donsante Emory University, Atlanta, GA, United States

I. Faravelli University of Milan, Milan, Italy; IRCCS Foundation Ca' Granda Ospedale Maggiore Policlinico, Milan, Italy

E.L. Feldman University of Michigan, Ann Arbor, MI, United States

J.D. Glass Emory University, Atlanta, GA, United States

L. Karumbaiah The University of Georgia, Athens, GA, United States

K.P. Kenna University of Massachusetts Medical School, Worcester, MA, United States

J.E. Landers University of Massachusetts Medical School, Worcester, MA, United States

C.L. Lorson University of Missouri, Columbia, MO, United States

B.J. Mader Emory University, Atlanta, GA, United States

M.J. Magnussen Emory University, Atlanta, GA, United States

A. McDonough Emory University, Atlanta, GA, United States

V.L. McGovern The Ohio State University, Columbus, OH, United States

M.R. Miller University of Missouri, Columbia, MO, United States

D.M. O'Connor Emory University, Atlanta, GA, United States

E.Y. Osman University of Missouri, Columbia, MO, United States

S.L. Stice The University of Georgia, Athens, GA, United States

R.L. Swetenburg The University of Georgia, Athens, GA, United States

N. Ticozzi IRCCS Istituto Auxologico Italiano, Milan, Italy; 'Dino Ferrari' Center – Università degli Studi di Milano, Milan, Italy

L. Urquia Emory University, Atlanta, GA, United States

Acknowledgment

This book is dedicated to the people diagnosed with this disease and their families, who have refused to take no for an answer and have dedicated themselves to raising funding for research and volunteering for risky groundbreaking clinical trials, including Josh Thompson and his family, Ed Tesoro, Ted Harada, and Christina Clark.

Molecular and Extracellular Cues in Motor Neuron Specification and Differentiation

R.L. Swetenburg, S.L. Stice and L. Karumbaiah

The University of Georgia, Athens, GA, United States

OUTLINE

Molecular and Cellular Therapies for Motor Neuron Diseases.
DOI: http://dx.doi.org/10.1016/B978-0-12-802257-3.00001-8   1

INTRODUCTION

Motor neurons (MN) are a diverse group of cells without which complex life would not be possible. MNs are responsible for integrating signals from the brain and the sensory systems to control voluntary and involuntary movements. Though MNs can be split into cranial and spinal subsets, this chapter will focus on spinal MNs, as they are a key target of disease and injury. As such, MNs are the focus of regenerative efforts to alleviate these public health burdens. During late gastrulation and neurulation, the developing spinal cord, termed the neural tube, is patterned into distinct progenitor domains. MNs are specified from progenitors in the ventral neural tube. Once specified, newly born MNs are further specified into columns, pools, and subtypes, forming a unique topography. From these columns and pools, axons reach out to their targets under varying guidance cues. All MNs are cholinergic cells which integrate with the motor control circuit, the sensory system, and their outlying targets to control movement. MNs are unique in that their targets lie outside the central nervous system (CNS), meaning that they require novel methods for seeking out and synapsing on them. Here, we present an overview of MN differentiation and development. We will focus mainly on signaling events, transcription factor markers, and the extracellular matrix (ECM) as they pertain to MN development. These cells are targets of permanent and often deadly diseases including amyotrophic lateral sclerosis, spinal muscular atrophy, multiple sclerosis, and injuries such as spinal cord injury. Only by understanding how these cells progress through development can we understand how to treat these maladies which currently have little hope of a cure. Further, by decoding the major events and players in development, we can better recapitulate them in vitro for cell replacement therapy, or harness the underlying principles for regeneration in the adult. Given the growing importance of the MN–glia interaction in a number of neurodegenerative diseases, we will also discuss the initial specification of oligodendrocyte precursor cells (OPCs) in detail, as they share a common progenitor with MNs.

SPECIFICATION OF NEUROECTODERM

Vertebrate embryos specify the ectoderm in late gastrulation. This germ layer will become the epidermis and the nervous system. The anterior neural ectoderm is distinguished from the epidermis by its inability to bind bone morphogenic proteins (BMPs) due to the inhibitors secreted from the Spemann–Mangold organizer region of the gastrula.[1] These inhibitors—noggin, chordin, and follistatin—bind and neutralize the effects of BMPs, creating a permissive transcriptional environment for neural progression.[2-5] Posteriorly, the neural plate is specified by fibroblast growth factors (FGFs) and Wingless-related integration site (Wnt) proteins that also suppress BMP activity.[6] Additionally, retinoid signaling from the paraxial mesoderm specifies the cells of the future spinal cord.[7] This newly specified neural plate then thickens as cells proliferate and invaginates through the convergent extension, forming the neural groove. The neural groove forms hinge points which will ultimately close to form the neural tube – the precursor for the entire CNS.[8] For an in-depth review, see Massarwa et al.[9]

SPINAL CORD PATTERNING

The spinal cord is a two-way information conduit that connects the brain with the sensory and motor systems. To do this, it must generate a highly diverse set of neurons during development. The neural tube provides a three-dimensional template which is patterned by gradients of morphogens to generate this diversity. The early neural tube is composed of multipotent neural stem cells expressing Sex determining region Y box 1 (Sox1).[10] The dorsal neural tube will generate cells linking the

CNS to the sensory peripheral nervous system (PNS). The ventral neural tube will ultimately give rise to the motor control circuit responsible for controlling MNs. Bone morphogenetic proteins specify the dorsal portion of the neural tube, including neuronal subtypes involved in integration of the peripheral sensory nervous system. Ventrally, an initial wave of Sonic hedgehog (Shh) from the notochord patterns the cells into distinct progenitor domains.[11] These domains arise due to cross-repressive actions of two types of transcription factors downstream of Shh signaling: Type I transcription factors are repressed at threshold Shh concentrations, while Type II are expressed below threshold Shh concentrations[11] (Fig. 1.1A). The type I transcription factor paired box protein 6 (Pax6) represses the activity of type II homeobox protein Nkx2.1. Similarly, type II homeobox Nkx6.1 cross-represses developing brain homeobox 2 (Dbx2).[11] The most ventral progenitor domain is the floor plate, which is induced to secrete Shh in a second wave of patterning, followed by the progenitor domains p3, pMN, p2, p1, and p0 (Fig. 1.1B). The combinatory actions of these two classes of proteins yield the five spatially distinct ventral progenitor domains.

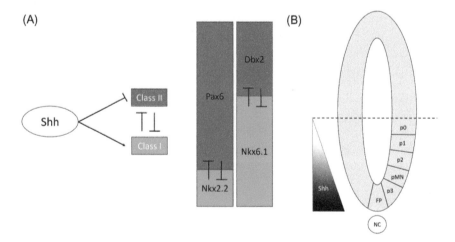

FIGURE 1.1 **A gradient of sonic hedgehog drives progenitor domain formation in the ventral neural tube.** (A) Sonic hedgehog (Shh) signaling from the notochord (NC) drives Class I transcription factors and represses Class II transcription factors at threshold levels. These transcription factors cross-repress to form sharply delineated boundaries. The class I Nkx2.2 represses the class II Pax6, while the Class I Nx6.1 represses Dbx2. (B) This cross-repression leads to five distinct domains: FP (floor plate), p3, pMN (motor neuron progenitor), p2, p1, and p0. p, progenitor.

THE MOTOR NEURON PROGENITOR DOMAIN AND INITIAL NEUROGENESIS

The MN progenitor (pMN) domain is responsible for generating MNs. In mice and chick models, this domain is identified by the expression of the homeobox transcription factors Nkx6.1 and Pax6 and the basic helix–loop–helix (bHLH) oligodendrocyte transcription factor 2 (Olig2).[12] Olig2 expression is obligate for MN specification, as Olig2^{null} mice fail to generate MNs.[13] Initially, Olig2 plays a key role in progenitor proliferation; however, it also drives the expression of neurogenin 2 (Ngn2),[14] a key neural determinant. The first murine MNs are born around E9.5.[13] The homeobox transcription factor Nkx2.2, important for the glial switch and a marker for p0 cells, shows variable expression in humans compared to mouse and chick models: the human pMN domain appears to include both Olig2+/Nkx2.2− as well as Olig2+/Nkx2.2+ cells.[15] This could potentially add to the diversity of human MNs.

MOLECULAR PROGRAMS IN NEWBORN MOTOR NEURONS

As mentioned above, Olig2 drives Ngn2 expression. However, Ngn2 is ultimately responsible for cell cycle exit and neurogenesis,[16,17] in direct contrast to the role of Olig2. Once Ngn2 protein levels surpass those of Olig2, cell cycle exit occurs and cells commit to the neuronal lineage. Olig2 binds and sequesters the MN transcription factor homeobox gene 9 (Hb9, also called MNX1), which is necessary for MN development.[17] LIM homeobox gene Isl1 and LIM homeobox 3 (Lhx3) form a complex with the nuclear LIM interactor which suppresses interneuron fate and specifies MN.[18] Along with Ngn2, this complex stimulates Hb9, which self-stimulates its own expression,[19] while forming a positive feedback loop with Isl1. Isl1 and 2 work in concert to further specify MN cell fate.[20] Lhx3 and Isl1 expressions are necessary for MN generation and the expression of cholinergic genes common to all MNs.[21] However, little is known about potential negative feedback mechanisms in this differentiation process that would limit MN number and organ size. We will discuss this further in the Glial Switch section.

In summary, Shh secreted from notochord drives the expression of Pax6 and Nxk6.1, which in turn drive Olig2 expression. Olig2 expression delineates a mitotic pMN progenitor. Olig2 induces the expression of Ngn2, which is responsible for cell cycle exit, of Lhx3/Isl1 transcription factors, as well as MN-specific Hb9 in newborn, postmitotic MNs (Fig. 1.2).

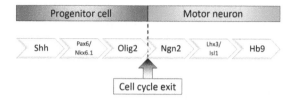

FIGURE 1.2   **Sonic hedgehog (Shh)-driven transcriptional progression of initial motor neuron specification.** Shh drives transcription of Pax6 and Nkx6.1, which drive Olig2, identifying mitotic progenitor cells. Ngn2 is expressed and causes cells to exit the cell cycle. Lhx3 and Isl1 activate Hb9 transcription which identifies a committed motor neuron.

MIGRATION

The topography of MNs is largely correlated with their function. MNs cluster in columns with similar transcription factor expression and like targets. Within muscle-innervating columns, there are MN pools which innervate specific muscle groups. In order to specify this topography, MNs must migrate away from the ventricular progenitor cells to their final destination in the ventral horn of the spinal cord. Newly born neurons detach from the epithelium and migrate radially to the medial and lateral areas of the neural tube. Critical to this migration is cadherin expression driven by beta and gamma catenin signaling in a Wnt-independent manner. In knockout models of either cadherin, MNs fail to properly align to their proper column, although their ultimate muscular targets are not disrupted.[22] This implies that the stereotypic and highly organized topology of MNs is not a modulator of identity or function. The role of this highly specific organization has yet to be elucidated. Transcriptionally, the forkhead box P (Foxp) genes regulate cadherin expression for migration to occur. Specifically, the Foxp2/4 genes allow for the detachment of MN from the neuroepithelium by downregulating cadherin 2 and allow them to migrate toward their final location by further modulation.[23] Although cell bodies migrate within the spinal cord, all MN soma are exclusively contained within the spinal cord. Recently, Isl1/2 has been shown to play an integral part in preventing the cell body from exiting the spinal cord.[24] In knockout animals, MN soma successfully exited the spinal cord into the periphery.[24] One potential mechanism is through the semaphorin–neuropilin repulsive signaling pathway common to axon guidance, as neuropilin was found to be regulated by Isl1/2.[24] For an in-depth review of migration and topography, see Kania (2005).[25]

MOTOR NEURON SUBTYPES AND TARGETS

MNs that innervate similar regions of the body group together in columns with identical molecular properties. The rostrocaudal axis is

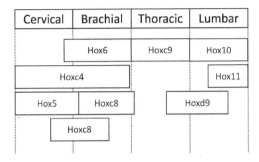

Cervical	Brachial	Thoracic	Lumbar

(Hox6 spanning Brachial–Thoracic; Hoxc9 Thoracic; Hox10 Lumbar; Hoxc4 spanning Cervical–Brachial; Hox11 Lumbar; Hox5 Cervical; Hoxc8 Brachial; Hoxd9 Thoracic–Lumbar; Hoxc8 Brachial)

FIGURE 1.3   Hox gene expression throughout the developing spinal cord. Rostrocaudal identity is conferred through Hox genes expression throughout the neural tube. The repressive and combinatory actions of the factors play a crucial role in generating the wide diversity of MNs.

specified externally by retinoic acid (RA) at the cervical and brachial regions and by growth differentiation factor (GDF11) and FGF8 at the thoracic and lumbar regions.[26] Like much of the developing embryo, this rostrocaudal positional identity is specified internally by Hox gene activation in response to these external cues to delineate the types of MNs in a given region.[27] Hox expression is summarized in Fig. 1.3. Hoxc9 plays a critical role in the organization of the spinal cord by repressing limb-specific Hox genes, thus specifying the thoracic column.[28] Within each column, MNs organize into pools which innervate distinct muscles. Hox genes work in concert with the Hox accessory factor Foxp1 to specify many of the columns and establish motor pools within columns.[29] Notably, in the absence of Foxp1 in knockout animals, there is a lack of defined rostrocaudal motor columns,[29] further strengthening the case for its role in the positional identity of motor columns.

Somatic Motor Neurons

MNs can be separated into somatic and visceral motor columns based on their targets. Somatic MNs synapse directly onto muscles, including the body wall, limbs, and diaphragm. These include the phrenic motor column, medial motor column (MMC), hypaxial motor column, spinal accessory column, and lateral motor column (LMC). The variation in transcription factors expressed in these different columns most likely correlate to function. However, much work is needed to fully understand what the initial cues are that trigger these regulators and how they can account for such a diverse set of cells.

The phrenic motor column, located cervically, controls the diaphragm and respiratory actions. This column is specified by Isl1/2, Hb9, and homeobox gene Pou3F1 under the control of Hox5.[30]

The spinal accessory motor column is located cervically and controls the muscles of the jaw and neck. These neurons are specified under the homeobox transcription factors Nkx9.2 and Nkx2.2[31] and require expression of paired like homeobox 2b (Phox2b).[32]

MMC neurons innervate the muscles of the back and are unique in that they are present throughout the spinal cord, not being restricted to one region or under the influence of Hox genes. These cells in this column all express Hb9, Isl1/2 and are the only neurons to retain Lhx3 expression after specification.[33] Lhx3 is thought to play an important role in overcoming Hox regionalization.[29] These neurons also do not express Foxp1,[29] which likely aids in escaping Hox control.

Body wall and abdominal muscles are innervated by the hypaxial motor column, located only in the thoracic region. These cells express Hb9, Isl1, Ets variant 1 (Etv1), and low levels of Isl2, while being negative for Foxp1.[29] Like the MMC, the lack of Foxp1 expression likely aids in their ability to escape Hox regionalization in order to span a large part of the spinal cord.

Limb-innervating neurons are found in either the brachial or lumbar region in the LMC. This can be further subdivided into the medial (LMCm) and lateral (LMCl) regions that innervate ventral and dorsal muscles, respectively, during development. Further specification includes the type of muscle innervated. These cells express Isl2, Foxp1, and the aldehyde dehydrogenase family gene Aldh1A2, a protein involved in RA synthesis, while downregulating Lhx3.[29] Recently, the Onecut transcription factor family has been implicated in LMCm and LMCl fate decisons.[34] Conditional dicer knockouts result in a severe reduction in LMC MNs through apoptosis, implicating miRNA as a key determinant in their survival.[35]

There are three types of MNs located within a pool synapsing on a specific muscle group: alpha, beta, and gamma. Alpha MNs innervate extrafusal muscles and cause muscle contraction, including reflex responses,[36] synapsing on fast-twitch fatigable, slow-twitch fatigue-resistant, and fast-twitch fatigue-resistant fibers.[37] Beta MNs innervate both extra- and intrafusal muscle fibers, though their identity and role are poorly understood.[38] Gamma MNs synapse on intrafusal fibers to modulate muscle sensitivity.[36] Gamma MNs are a distinct subset and can be identified by the secretion of Wnt7a.[39]

Visceral Motor Neurons

As opposed to somatic MNs, visceral MNs regulate the autonomic nervous system and are found in the thoracic level of the spinal cord in the preganglionic column. These are unique in that they do not synapse on muscles, but rather the ganglia of the PNS. Through this, they are able to regulate smooth muscle, cardiac muscle, and glandular activity. These cells express mothers against decapentaplegic homolog 1 (Smad1),[29] nitric oxide synthase 1,[40] Smad interacting protein 1 (Sip1),[34] and low levels of Foxp1.[29] The presence of Sip1 and Smad1 imply an important role for BMPs, possibly by distinguishing them from the somatic MNs during specification. Isl2 is downregulated to specify this cell type.[20] Recently, the Onecut family of factors which drive Isl1 were revealed to play an important role in visceral/somatic fate decisions.[34] In the more caudal regions, preganglionic motor column neurons regulate the parasympathetic system by projecting to ganglia near their target organs, while in the thoracic regions, they synapse on the ganglia in proximity to the spine to regulate the sympathetic systems of the body.

Axon Targeting

MN axons must exit the CNS in order to project onto their targets. A small subset of MNs, specifically the spinal accessory column, exit dorsally through the lateral exit point.[41] These neurons express the netrin receptor, and are repelled from the midline of the spinal cord which expresses netrin-1.[41] However, most MN axons exit ventrally, driven by the chemokine C-X-C motif receptor 4 (Cxcr4) expressed on the neurons and its ligand chemokine C-X-C ligand 12 (Cxcl12) located in the ventral paraxial mesoderm.[42] Once the axons have exited, they are directed toward the body wall, the limbs, the muscles of the back, or the autonomic ganglia, which will ultimately determine which motor column they will become. MMC MNs express ephrinA 3 and 4 receptors (EphA3/4) which repel them from the ephrin-A1 (Efna1)-secreting dorsal root ganglia to continue to the back muscles.[43] These neurons express FGF receptor 1 , which is also thought to direct them.[44] LMCs, the limb-innervating MNs, use a semaphorin–neuropilin complex in timing the invasion of the limb.[45] Once they have reached the limb bud, LMCl neurons express EphA4 receptor.[46] Ephrins in the ventral limb repel these neurons and they extend to the dorsal limb. LMCm neurons express EphB1 and are similarly repelled from the dorsal limb which expresses ephrinB.[47] Though many of the mechanisms are poorly understood, understanding how axons migrate to their targets will be important for cell replacement therapies.

Once axons reach their target muscle, extrinsic factors play a critical role in finalizing the connections. Muscle-derived glial cell line-derived neurotrophic factor (GDNF), specifically, drives Ets variant 4 (Etv4) which induces MNs to arborize and innervate muscle fibers.[48] The Onecut gene family involved in MN specification has additionally been identified as a transcriptional regulator of neuromuscular junction formation.[49] For a review on axon guidance, see Stifani.[37]

EXTRACELLULAR MATRIX
AND THE NERVOUS SYSTEM

The ECM of the nervous system plays an important role in facilitating neuronal migration and axonal pathfinding during CNS and PNS development. The nervous system ECM is compositionally distinct, consisting of markedly lesser quantities of fibrous proteins such as collagens, and fibronectins, and displaying a higher concentration of glycoproteins and proteoglycans when compared to the ECM of other systemic tissues. The contribution of CNS and PNS ECM components to neuronal function and homeostasis, emphasizing MNs, are presented in this section.

Collagen

In contrast with other tissue-specific ECMs, collagen type IV is the only collagen found in the CNS ECM. A majority of the collagen IV in the CNS ECM self-polymerizes along with fibronectin and laminin to form the basement membrane, which serves as a protective barrier between the vascular endothelial cells and the parenchyma of the brain.[50,51] Collagen IV is also associated with the neural interstitial matrix, where it is found in small amounts along with small amounts of other such adhesive glycoproteins such as fibronectin and laminin, and sulfated glycoproteins such as entactin.[51] These minor components are interlinked within a major network of chondroitin sulfate proteoglycans (CSPGs), hyaluronic acid (HA), and tenascin, which make up the bulk of the neural interstitial matrix.

Collagens in the PNS are broadly classified into the fibrillary type I, III, and V collagens and the basement membrane associated type IV collagen.[52] The fibrillary collagens play an important role in supporting myelinating Schwann cell function. Long-term (14 to 28 days) in vitro cultures of MNs in 3D type I and III collagen scaffolds produced elongated neurites of ~850 μm in length with thick myelin sheaths, indicating that trophic support provided by fibrillary collagens is essential for MN growth.[53] On the contrary, dysregulated fibrillary collagen associated with scar tissue formed after PNS injury is considered to be a major barrier to nerve regeneration.[52]

Laminin

Laminins are large (~800 kDa) ECM-associated glycoproteins thought to be essential for neuronal migration and axonal pathfinding during development. They are heterotrimeric molecules consisting of α, β, and γ chains. These chains present themselves in a variety of different combinations to result in as many as 18 distinct laminin isoforms, a majority of which are present in the nervous system ECM.[54] S-laminin is a homolog of the B1 subunit of laminin. It is known to facilitate MN adhesion, and plays an important role in the formation of neuromuscular junctions. In the developing CNS, S-laminin and laminin are both found to be present in the subplate in the cerebral cortex.[55] This situation changes dramatically in the adult CNS, which is marked by the disappearance of S-laminin, and the restricted expression of laminin to the basement membrane, where it is found in close association with collagen and heparan sulfate proteoglycans (HSPGs).[51]

Laminins are important constituents of the PNS basement membrane found surrounding Schwann cells. As integral adhesive and growth-promoting constituents of the basement membrane, they are believed to organize together the meshwork of structural proteins to form the substratum upon which neuronal migration and axonal pathfinding can take place.[56] The diversity of laminin isoforms can be advantageous in regulating the selective attachment of MNs, which is reportedly mediated by the S-laminin-specific LRE (leucine–arginine–glutamic acid) amino acid sequence.[57]

Fibronectin

Fibronectins are ECM proteins known to play specific roles in neuronal migration during development and after injury.[54] Like laminin, they are large glycoproteins consisting of functional domains that facilitate cellular and ECM interactions.[58] In the CNS they are associated with the basement membrane where they are known to interact with collagen, HSPGs, and tenascin.[59] Fibronectin in the PNS is specifically upregulated after injuries. This is also accompanied by the upregulation of $\alpha_5\beta_1$ integrin on the cell membranes of regenerating MNs and Schwann cells,[54] which is necessary for blood vessel development, pointing to the role of fibronectin in facilitating PNS repair.

Proteoglycans and Glycosaminoglycans

Proteoglycans such as CSPGs and HSPGs, and their associated sulfated glycosaminoglycans (GAGs), as well as unsulfated GAGs such as HA constitute the majority of the CNS ECM. High-molecular-weight HA forms the diffuse meshwork to which CSPGs, HSPGs, tenascins and other fibrous ECMs proteins are linked. This network is more condensed around presynaptic terminals, nodes of Ranvier, and around the perineuronal nets that surround inhibitory interneurons.[60]

Chondroitin Sulfate Proteoglycans

Sulfated CSPGs belonging to the lecitican family (aggrecan, versican, neurocan, and brevican) are the most abundant CSPGs associated with the CNS ECM and play a major role in regulating neuronal plasticity. They are large proteins consisting of multiple domains capable of specifically binding HA, tenascin, and other ECM components. The associated CS-GAG sidechains can be sulfated at various positions, resulting in a complex configuration of sulfated GAGs that are capable of mediating a variety of different functions, some of which include receptor and growth factor binding.[61,62] The oversulfated CS-GAG CS-4,6 sulfate (CS-E) is a potent neurite repellant that is specifically upregulated after CNS injury.[63,64] Nerve repulsive CSPGs are also believed to regulate axonal regeneration following peripheral neuron crush injury, where they are found to bind and inhibit the growth-promoting activity of laminin.[65]

Heparan Sulfate Proteoglycans

In contrast to the predominantly nerve-repulsive role played by CSPGs, large basement-membrane-associated HSPGs such as perlecan are known to complex with laminin to induce neurite outgrowth.[66] However, in the context of neurodegenerative diseases such as Alzheimer's disease, HSPGs are thought to exacerbate the formation of amyloid-β and prevent its proteolytic degradation.[67] In the PNS, HSPGs such as glypican-1 are upregulated in the dorsal root ganglia after sciatic nerve transection, where it is believed to promote target MN reinnervation.[68]

Hyaluronic Acid

HA is an unsulfated GAG that interacts with CD44 and other cell surface receptors in the brain ECM.[69] HA is abundantly found in the neural interstitial matrix where it is complexed with CSPGs. The upregulation of HA in response to CNS injuries is believed to inhibit remyelination of neurons. However, the role of HA in directly influencing these outcomes remains to be elucidated owing to the fact that it is often bound to CSPGs.[51] Recent studies have indeed demonstrated that the enzymatic degradation of HA by the hyaluronidase PH20, but not other hyaluronidases, results in digestion products that can further impede remyelination.[70] In the PNS, HA is found to be densely deposited at the nodes of Ranvier in peripheral neuron axons.[71] Previous studies evaluating the regenerative potential and function of transected sciatic nerves treated with HA report significant improvements in nerve conduction velocity and muscle mass, and a significant reduction in scarring when compared to saline-treated animals.[72] These seemingly contradictory functions point to the diverse roles played by HA in the CNS and PNS, and also highlight the importance of the role of enzymatic degradation of these GAGs in regulating these end outcomes.

MOTOR NEURON CELL DEATH

The developing spinal cord makes MNs in excess, with about 40% undergoing cell death.[73] Almost all MNs are specified before this death begins. This phenomenon was observed when the removal of a limb bud led to increased death in MNs,[74] leading to the neurotrophic theory: that pro-survival molecules exist in limited quantities during development allowing for selected survival or programmed cell death of neurons.[75] During development, different MN subtypes require different cocktails of factors to survive. The best-studied of these is GDNF. GDNF is expressed by skeletal muscles and can be transported retro- and anterogradely.[76,77] GDNF binds its receptor, the tyrosine kinase Ret, as well as the GFRalpha1 receptor.[78] Exogenous brain-derived neurotrophic factor (BDNF) also enhances MN survival but is not required.[79] Other factors include ciliary neurotrophic factor , neurotrophic factor, leukemia inhibitory factor, hepatocyte growth factor), insulin-like growth factor, and vascular endothelial growth factor.[80]

THE GLIAL SWITCH

The pMN domain which generates MNs also generates OPCs. During progenitor and MN specification, Olig2 and Nkx2.2 cross-repress each other to establish the pMN-p3 border.[81] However, Olig2 and Nkx2.2 are coexpressed during the oligodendrogliogenic phase of development with Olig2 being necessary for oligodendrocyte production[82] and Nkx2.2 knockouts yielding extremely depleted oligodendrocyte pools.[83] Olig2 overexpression alone is insufficient for OPC generation; however, dual overexpression with Nkx2.2 leads to an increase in OPCs from this domain.[84] This switch from cross-repression to coexpression suggests that there is a common convergence of extracellular signaling focusing on the simultaneous activation of these two transcription factors. Interestingly, dicer inactivation leads to a high loss of OPCs, implicating miRNA as a key determinant in OPC specification.[35] One hallmark of the neurogenic to gliogenic switch is a downregulation of pro-neural genes, specifically Ngn2.[85] Olig2 is a bHLH repressor and its phosphorylation state affects protein–protein binding kinetics. Olig2 is typically phosphorylated at Ser147 during MN differentiation, while the site is dephosphorylated during OPC specification. Abolishing the phosphorylation site blocks MN differentiation, while having no effect on OPCs. The proposed mechanism of action involves homodimerization of Olig2 during neurogenesis and heterodimerization during oligodendrogliogenesis.[86]

Floor plate cells play a critical role in OPC specification. In zebrafish, the floor plate secretes Indian hedgehog b, which is necessary for OPC specification in a Shh-independent manner.[87] Another role for the floor plate in mouse and chick models is secretion of sulfatase 1, which positively regulates Shh signaling through HSPG modifications, potentially by triggering a dorsal shift in perceived Shh levels, thereby increasing Nkx2.2 (expressed under high concentrations of Shh) in cells already expressing Olig2 (e.g., in the pMN domain).[88] Notch-Delta is a form of cell–cell juxtacrine signaling and is additionally implicated in progenitor fate determination. However, findings here have been contradictory. In zebrafish, Notch signaling has been shown to increase OPCs,[89] while in the chick, attenuation of Notch signaling through its ligand Jagged2 enhanced OPC generation.[90] Notch has been shown to induce cell cycle exit,[91] which implies a permissive, rather than instructive role for Notch-Delta in OPC specification.

Elegant modeling work in the olfactory bulb has shown a negative feedback system in which committed neuronal cells influence both progenitor cell renewal and neuronal/glial fate decisions.[92] Gokoffski and coworkers show that committed neuronal cells secrete transforming growth factor-β (TGF-β) family proteins (Activin β-B and GDF11) which inhibit progenitor populations from expanding, while this is countered by production of TGF-β inhibitors (follistatin) from the progenitor cells themselves. Therefore, when neuronal signals outnumber progenitor signals (e.g., after neuronal populations and signaling reach a threshold), the progenitor cells exit the cell cycle and differentiate toward glia. The study also showed that TGF-β signaling could upregulate Achaete–Scute family bHLH transcription factor 1 (Ascl1), which is necessary for OPC specification in the ventral neural tube.[93] While this study examined olfactory bulb, similar feedback mechanisms may be at play in the neural tube.[94] Recently, TGF-β has been implicated in the glial switch in the spinal cord making it a strong candidate for further study.[95]

GENERATING MOTOR NEURONS FROM PLURIPOTENT STEM CELLS

Regenerative medicine requires either deriving cell types in vitro for transplantation or understanding the cues governing development to generate lost or injured cell types from endogenous stores of progenitors. The study and optimization of pluripotent stem cell (PSC) differentiation into MN progenitors, MNs, and oligodendrocytes is of the utmost importance in realizing these cures, while serving the dual purpose of acting as developmentally relevant models which are easier to interrogate than embryos in utero. Much of the groundbreaking work that led to the initial decoding of in vivo MN specification also led to the first derivation of MNs from PSCs. Murine MNs were induced from PSCs using RA and Shh. When they were transplanted into mice, they were able to synapse onto muscles.[96] In vitro, these cells are electrophysiologically active and form neuromuscular junctions with myocytes.[97] Human MNs have also been derived from embryonic stem cells (ESCs).[98,99] Shh mimics including purmorphamine[100,101] and HH-ag1.3[97] have been used to increase the efficiency and decrease the cost of MN generation from PSCs in vitro. More recently, Smad inhibition greatly increased the yield of neural stem cells from ESCs and has been used in concert with RA and Shh cues to generate MN.[102]

With the advent of induced pluripotent stem cells (iPSCs) offering patient-specific treatment hopes,[103] human iPSCs were also shown to generate MNs in vitro under developmental cues including RA and Shh.[101] Murine iPSCs were shown to mature and integrate into chick spinal cord by histological analysis.[104] Optogenetic neuronal control[105] has provided further proof that transplanted PSC-derived MNs can directly restore function to limb muscles.[106]

Exogenous gene expression has also been used to increase the efficiency of MN differentiation in addition to, or instead of, external patterning factors. Exogenous expression of Isl1 and Lhx3 has been shown to more efficiently transform ESCs to MNs than Shh alone.[18] ESCs can be directly converted to MNs with as little as three factors: Ngn2, Isl1, and Lhx3.[107] MNs can be transdifferentiated from fibroblasts, as well, with the input of three groups of factors: fibroblast to neuronal factors, pMN progenitor factors, and MN transcription factors.[108]

However, once MNs are initially specified, further differentiation into specific columns or pools has proved challenging, though some gains have been made in this area. Alterations to protocols can drastically alter MN subtype. RA addition to murine PSC cultures results in cervical MNs.[96] Though RA addition is extremely efficient in producing cervical MNs from PSCs,[96] it is interesting to note that MNs can be specified in vitro in the absence of retinoid signaling, leading to greater malleability in later specification and maturation toward a wider variety of subtypes.[109] For instance, FGF and GDF11 addition push MNs to a more caudal identity.[26] In human MN derivation, the use of purmorphamine and smoothened agonist, when compared to recombined Shh protein, results in the shift in columnar identity from MMC to LMC neurons.[110] Importantly, transplanted MNs can develop into appropriate thoracic and lumbar fates when transplanted into the corresponding location in the embryonic spinal cord.[111] Recently, efficient generation of LMC and MMC MNs was achieved through exogenous Foxp1 expression,[112] likely by escaping Hox gene regionalization.

When there is no precedent for generating a specific MN subtype from PSCs, there are combined in vivo and in vitro methods for delineating transcriptional regulation. For instance, in the difficult-to-study phrenic motor column, Machado and coworkers[113] isolated these MNs in vivo and compared their gene expression with control MNs using microarray analysis. Putative determinants based on this analysis were then exogenously expressed in PSCs followed by another round of comparative analysis. Through this systematic approach, Pou3f1 and Hoxa5, as well as Notch signaling, were all found to be key determinants of phrenic motor column identity. Future studies will further elucidate methods for generating the vast diversity of MNs for cell transplantation studies and disease and developmental modeling.

PSC-derived MNs are a valuable tool for MN studies outside the context of an animal. Using microfluidics and stripe assays, Nedelec and coworkers were able to show growth cone collapse occurs through two separate mechanisms, one dependent and one independent of local protein synthesis, in response to semaphorins in both human and murine cell lines.[114] PSC-derived MNs make excellent candidates for drug screening as well, particularly in human-derived cells where in vivo tests are inappropriate and iPSCs can be generated from patients. For instance, in a large-scale screen of small molecules, kenpaullone, an inhibitor of glycogen synthase kinase 3 and the mitogen-activated kinase pathway (MAP4K4), was recently shown to greatly increase the survival of iPSC-derived MNs from an amyotrophic lateral sclerosis patient.[115]

Other developmental phenomena have been recapitulated using PSCs, including the fate decisions in pMN progenitors. A negative feedback mechanism appears to control this switch from neurogenesis to gliogenesis, while progenitor mitosis is triggered externally, rather than being an inherent quality.[94] Understanding these rules opens the door for regenerative medicine to regrow lost populations of progenitors, MNs, or OPCs within the adult body with simple cues.

Differences between human and mouse ESC differentiation has also provided insights into species differences in developmental properties. For instance, during neurogenesis human ESCs coexpress Olig2+ and Nkx2.2+ in a subset of MN progenitors. In contrast, murine ESCs segregate these transcription factors until the gliogenic phase.[15] Differentiating stem cells into MNs has even given insights into how Shh signaling occurs. For instance, exosomes containing integrins are necessary for Olig2 expression.[116] For a review of generating MNs from PSCs, see Davis-Dusenbery et al.[117]

GENERATING OLIGODENDROCYTE PRECURSOR CELLS FROM PLURIPOTENT STEM CELLS

Understanding oligodendrocyte differentiation from PSCs is important in light of diseases that affect myelin, including multiple sclerosis and demyelination following spinal cord injury. Additionally, oligodendrocytes were the first human-ESC-derived cell type to be approved for clinical trial.[118] In fact, oligodendrocytes were generated from murine ES cells before MNs.[119] These ESC-derived oligodendrocytes myelinate axons both in vitro and in vivo. Human ESC-derived oligodendrocytes similarly remyelinated axons when transplanted.[120] Recently, there has been an emphasis on transplanting neural stem and progenitor cells to an injured or diseased site. It is interesting to note that most of the daughter cells from these transplants develop into glia,[121–123] which means that understanding the glial switch will better inform the ways in which these cells are transplanted. Compared to murine PSCs, oligodendrocytes derived from human ESCc show divergence in their response to FGF signaling,[124] implying that these processes are not wholly conserved between species. Most oligodendroglial specification protocols are similar to MN derivation protocols using RA and Shh to specify pMN cells, however they often require long periods of time.[125,126] This approach likely works by eliminating neural progenitors through division and neurons through unfavorable culture conditions. It is interesting to note that OPCs can be coerced to generate type I and II astrocytes in vitro through Notch signaling.[127] In addition to remyelination, OPC transplantation has been shown to decrease neuropathic pain after spinal cord contusions, a beneficial side effect for further transplantation studies.[128] OPCs have been generated from fibroblasts through the expression of transcription factors Sox10, Olig2, and Zfp536.[129] For an in-depth review of OPC development into mature oligodendrocytes, see Alsanie et al.[130]

CONCLUSION

Our current understanding of the journey from neuroectodermal specification to fully mature MNs has provided an excellent road map for the diversification of neuronal identity, as well as presented many clues for potential cell replacement and regenerative therapies. However, much work remains, both in vitro and in vivo, to fully understand the mechanisms governing these essential cells and to make these treatments a reality.

References

1. Hamburger V. Ontogeny of neuroembryology. *J Neurosci Off J Soc Neurosci* 1988;**8**(10):3535–40.
2. Smith WC, Knecht AK, Wu M, Harland RM. Secreted noggin protein mimics the Spemann organizer in dorsalizing Xenopus mesoderm. *Nature* 1993;**361**(6412):547–9.
3. Hemmati-Brivanlou A, Kelly OG, Melton DA. Follistatin, an antagonist of activin, is expressed in the Spemann organizer and displays direct neuralizing activity. *Cell* 1994;**77**(2):283–95.
4. Hemmati-Brivanlou A, Melton DA. Inhibition of activin receptor signaling promotes neuralization in Xenopus. *Cell* 1994;**77**(2):273–81.
5. Sasai Y, Lu B, Steinbeisser H, Geissert D, Gont LK, De Robertis EM. Xenopus chordin: a novel dorsalizing factor activated by organizer-specific homeobox genes. *Cell* 1994;**79**(5):779–90.
6. Stern CD. Neural induction: old problem, new findings, yet more questions. *Development* 2005;**132**(9):2007–21.
7. Novitch BG, Wichterle H, Jessell TM, Sockanathan S. A requirement for retinoic acid-mediated transcriptional activation in ventral neural patterning and motor neuron specification. *Neuron* 2003;**40**(1):81–95.
8. Colas JF, Schoenwolf GC. Towards a cellular and molecular understanding of neurulation. *Dev Dynam Off Publ Am Assoc Anat* 2001;**221**(2):117–45.
9. Massarwa R, Ray HJ, Niswander L. Morphogenetic movements in the neural plate and neural tube: mouse. Wiley interdisciplinary reviews. *Dev Biol* 2014;**3**(1):59–68.
10. Zhang X, Huang CT, Chen J, et al. Pax6 is a human neuroectoderm cell fate determinant. *Cell Stem Cell* 2010;**7**(1):90–100.
11. Jessell TM. Neuronal specification in the spinal cord: inductive signals and transcriptional codes. *Nat Rev Genet* 2000;**1**(1):20–9.
12. Novitch BG, Chen AI, Jessell TM. Coordinate regulation of motor neuron subtype identity and pan-neuronal properties by the bHLH repressor Olig2. *Neuron* 2001;**31**(5):773–89.
13. Lu QR, Sun T, Zhu Z, et al. Common developmental requirement for Olig function indicates a motor neuron/oligodendrocyte connection. *Cell* 2002;**109**(1):75–86.
14. Ma YC, Song MR, Park JP, et al. Regulation of motor neuron specification by phosphorylation of neurogenin 2. *Neuron* 2008;**58**(1):65–77.
15. Marklund U, Alekseenko Z, Andersson E, et al. Detailed expression analysis of regulatory genes in the early developing human neural tube. *Stem Cells Dev* 2014;**23**(1):5–15.
16. Mizuguchi R, Sugimori M, Takebayashi H, et al. Combinatorial roles of olig2 and neurogenin2 in the coordinated induction of pan-neuronal and subtype-specific properties of motoneurons. *Neuron* 2001;**31**(5):757–71.
17. Lee SK, Lee B, Ruiz EC, Pfaff SL. Olig2 and Ngn2 function in opposition to modulate gene expression in motor neuron progenitor cells. *Genes Dev* 2005;**19**(2):282–94.
18. Lee S, Cuvillier JM, Lee B, Shen R, Lee JW, Lee SK. Fusion protein Isl1-Lhx3 specifies motor neuron fate by inducing motor neuron genes and concomitantly suppressing the interneuron programs. *Proc Natl Acad Sci USA* 2012;**109**(9):3383–8.
19. Tanabe Y, William C, Jessell TM. Specification of motor neuron identity by the MNR2 homeodomain protein. *Cell* 1998;**95**(1):67–80.
20. Thaler JP, Koo SJ, Kania A, et al. A postmitotic role for Isl-class LIM homeodomain proteins in the assignment of visceral spinal motor neuron identity. *Neuron* 2004;**41**(3):337–50.
21. Cho HH, Cargnin F, Kim Y, et al. Isl1 directly controls a cholinergic neuronal identity in the developing forebrain and spinal cord by forming cell type-specific complexes. *PLoS Genet* 2014;**10**(4):e1004280.

22. Demireva EY, Shapiro LS, Jessell TM, Zampieri N. Motor neuron position and topographic order imposed by beta- and gamma-catenin activities. *Cell* 2011;**147**(3):641–52.
23. Rousso DL, Pearson CA, Gaber ZB, et al. Foxp-mediated suppression of N-cadherin regulates neuroepithelial character and progenitor maintenance in the CNS. *Neuron* 2012;**74**(2):314–30.
24. Lee H, Kim M, Kim N, et al. Slit and Semaphorin signaling governed by Islet transcription factors positions motor neuron somata within the neural tube. *Exp Neurol* 2015;**269**:17–27.
25. Kania A. Spinal motor neuron migration and the significance of topographic organization in the nervous system. *Adv Exp Med Biol* 2014;**800**:133–48.
26. Liu JP, Laufer E, Jessell TM. Assigning the positional identity of spinal motor neurons: rostrocaudal patterning of Hox-c expression by FGFs, Gdf11, and retinoids. *Neuron* 2001;**32**(6):997–1012.
27. Dasen JS, Jessell TM. Hox networks and the origins of motor neuron diversity. *Curr Top Dev Biol* 2009;**88**:169–200.
28. Jung H, Lacombe J, Mazzoni EO, et al. Global control of motor neuron topography mediated by the repressive actions of a single hox gene. *Neuron* 2010;**67**(5):781–96.
29. Dasen JS, De Camilli A, Wang B, Tucker PW, Jessell TM. Hox repertoires for motor neuron diversity and connectivity gated by a single accessory factor, FoxP1. *Cell* 2008;**134**(2):304–16.
30. Philippidou P, Walsh CM, Aubin J, Jeannotte L, Dasen JS. Sustained Hox5 gene activity is required for respiratory motor neuron development. *Nat Neurosci* 2012;**15**(12):1636–44.
31. Pabst O, Rummelies J, Winter B, Arnold HH. Targeted disruption of the homeobox gene Nkx2.9 reveals a role in development of the spinal accessory nerve. *Development* 2003;**130**(6):1193–202.
32. Hirsch MR, Glover JC, Dufour HD, Brunet JF, Goridis C. Forced expression of Phox2 homeodomain transcription factors induces a branchio-visceromotor axonal phenotype. *Dev Biol* 2007;**303**(2):687–702.
33. Tsuchida T, Ensini M, Morton SB, et al. Topographic organization of embryonic motor neurons defined by expression of LIM homeobox genes. *Cell* 1994;**79**(6):957–70.
34. Roy A, Francius C, Rousso DL, et al. Onecut transcription factors act upstream of Isl1 to regulate spinal motoneuron diversification. *Development* 2012;**139**(17):3109–19.
35. Chen JA, Wichterle H. Apoptosis of limb innervating motor neurons and erosion of motor pool identity upon lineage specific dicer inactivation. *Front Neurosci* 2012;**6**:69.
36. Eccles JC, Eccles RM, Iggo A, Lundberg A. Electrophysiological studies on gamma motoneurones. *Acta Physiol Scand* 1960;**50**:32–40.
37. Stifani N. Motor neurons and the generation of spinal motor neuron diversity. *Front Cell Neurosci* 2014;**8**:293.
38. Bessou P, Emonet-Denand F, Laporte Y. Motor fibres innervating extrafusal and intrafusal muscle fibres in the cat. *J Physiol* 1965;**180**(3):649–72.
39. Ashrafi S, Lalancette-Hebert M, Friese A, et al. Wnt7A identifies embryonic gamma-motor neurons and reveals early postnatal dependence of gamma-motor neurons on a muscle spindle-derived signal. *J Neurosci Off J Soc Neurosci* 2012;**32**(25):8725–31.
40. Dasen JS, Liu JP, Jessell TM. Motor neuron columnar fate imposed by sequential phases of Hox-c activity. *Nature* 2003;**425**(6961):926–33.
41. Dillon AK, Fujita SC, Matise MP, et al. Molecular control of spinal accessory motor neuron/axon development in the mouse spinal cord. *J Neurosci Off J Soc Neurosci* 2005;**25**(44):10119–30.
42. Lieberam I, Agalliu D, Nagasawa T, Ericson J, Jessell TM. A Cxcl12-CXCR4 chemokine signaling pathway defines the initial trajectory of mammalian motor axons. *Neuron* 2005;**47**(5):667–79.

43. Gallarda BW, Bonanomi D, Muller D, et al. Segregation of axial motor and sensory pathways via heterotypic trans-axonal signaling. *Science* 2008;**320**(5873):233–6.
44. Shirasaki R, Lewcock JW, Lettieri K, Pfaff SL. FGF as a target-derived chemoattractant for developing motor axons genetically programmed by the LIM code. *Neuron* 2006;**50**(6):841–53.
45. Huber AB, Kania A, Tran TS, et al. Distinct roles for secreted semaphorin signaling in spinal motor axon guidance. *Neuron* 2005;**48**(6):949–64.
46. Helmbacher F, Schneider-Maunoury S, Topilko P, Tiret L, Charnay P. Targeting of the EphA4 tyrosine kinase receptor affects dorsal/ventral pathfinding of limb motor axons. *Development* 2000;**127**(15):3313–24.
47. Luria V, Krawchuk D, Jessell TM, Laufer E, Kania A. Specification of motor axon trajectory by ephrin-B:EphB signaling: symmetrical control of axonal patterning in the developing limb. *Neuron* 2008;**60**(6):1039–53.
48. Helmbacher F, Dessaud E, Arber S, et al. Met signaling is required for recruitment of motor neurons to PEA3-positive motor pools. *Neuron* 2003;**39**(5):767–77.
49. Audouard E, Schakman O, Rene F, et al. The Onecut transcription factor HNF-6 regulates in motor neurons the formation of the neuromuscular junctions. *PloS One* 2012;**7**(12):e50509.
50. Burnside ER, Bradbury EJ. Manipulating the extracellular matrix and its role in brain and spinal cord plasticity and repair. *Neuropathol Appl Neurobiol* 2014;**40**(1):26–59.
51. Lau LW, Cua R, Keough MB, Haylock-Jacobs S, Yong VW. Pathophysiology of the brain extracellular matrix: a new target for remyelination. *Nat Rev Neurosci* 2013;**14**(10):722–9.
52. Koopmans G, Hasse B, Sinis N. Chapter 19: The role of collagen in peripheral nerve repair. *Int Rev Neurobiol* 2009;**87**:363–79.
53. Gingras M, Beaulieu MM, Gagnon V, Durham HD, Berthod F. In vitro study of axonal migration and myelination of motor neurons in a three-dimensional tissue-engineered model. *Glia* 2008;**56**(3):354–64.
54. Venstrom KA, Reichardt LF. Extracellular-matrix.2. Role of extracellular-matrix molecules and their receptors in the nervous-system. *FASEB J* 1993;**7**(11):996–1003.
55. Hunter DD, Llinas R, Ard M, Merlie JP, Sanes JR. Expression of s-laminin and laminin in the developing rat central nervous system. *J Comp Neurol* 1992;**323**(2):238–51.
56. Hohenester E, Yurchenco PD. Laminins in basement membrane assembly. *Cell Adhes Migr* 2013;**7**(1):56–63.
57. Hunter DD, Cashman N, Morris-Valero R, Bulock JW, Adams SP, Sanes JR. An LRE (leucine-arginine-glutamate)-dependent mechanism for adhesion of neurons to S-laminin. *J Neurosci* 1991;**11**(12):3960–71.
58. Rutka JT, Apodaca G, Stern R, Rosenblum M. The extracellular matrix of the central and peripheral nervous systems: structure and function. *J Neurosurg* 1988;**69**(2):155–70.
59. Ingham KC, Brew SA, Erickson HP. Localization of a cryptic binding site for tenascin on fibronectin. *J Biol Chem* 2004;**279**(27):28132–5.
60. Bruckner G, Brauer K, Hartig W, et al. Perineuronal nets provide a polyanionic, glia-associated form of microenvironment around certain neurons in many parts of the rat brain. *Glia* 1993;**8**(3):183–200.
61. Sugahara K, Mikami T. Chondroitin/dermatan sulfate in the central nervous system. *Curr Opin Struct Biol* 2007;**17**(5):536–45.
62. Deepa SS, Umehara Y, Higashiyama S, Itoh N, Sugahara K. Specific molecular interactions of oversulfated chondroitin sulfate E with various heparin-binding growth factors. Implications as a physiological binding partner in the brain and other tissues. *J Biol Chem* 2002;**277**(46):43707–16.
63. Brown JM, Xia J, Zhuang B, et al. A sulfated carbohydrate epitope inhibits axon regeneration after injury. *Proc Natl Acad Sci U S A* 2012;**109**(13):4768–73.

64. Karumbaiah L, Anand S, Thazhath R, Zhong Y, McKeon RJ, Bellamkonda RV. Targeted downregulation of N-acetylgalactosamine 4-sulfate 6-O-sulfotransferase significantly mitigates chondroitin sulfate proteoglycan-mediated inhibition. *Glia* 2011;**59**(6): 981–96.

65. Zuo J, Hernandez YJ, Muir D. Chondroitin sulfate proteoglycan with neurite-inhibiting activity is up-regulated following peripheral nerve injury. *J Neurobiol* 1998;**34**(1): 41–54.

66. Hantaz-Ambroise D, Vigny M, Koenig J. Heparan sulfate proteoglycan and laminin mediate two different types of neurite outgrowth. *J Neurosci* 1987;**7**(8):2293–304.

67. Rosenmann H, Meiner Z, Kahana E, et al. An association study of a polymorphism in the heparan sulfate proteoglycan gene (perlecan, HSPG2) and Alzheimer's disease. *Am J Med Genet B Neuropsychiatr Genet* 2004;**128B**(1):123–5.

68. Bloechlinger S, Karchewski LA, Woolf CJ. Dynamic changes in glypican-1 expression in dorsal root ganglion neurons after peripheral and central axonal injury. *Eur J Neurosci* 2004;**19**(5):1119–32.

69. Bignami A, Hosley M, Dahl D. Hyaluronic-acid and hyaluronic acid-binding proteins in brain extracellular-matrix. *Anat Embryol* 1993;**188**(5):419–33.

70. Preston M, Gong X, Su WP, et al. Digestion products of the PH20 hyaluronidase inhibit remyelination. *Ann Neurol* 2013;**73**(2):266–80.

71. Abood LG, Abul-Haj SK. Histochemistry and characterization of hyaluronic acid in axons of peripheral nerve. *J Neurochem* 1956;**1**(2):119–25.

72. Ozgenel GY. Effects of hyaluronic acid on peripheral nerve scarring and regeneration in rats. *Microsurgery* 2003;**23**(6):575–81.

73. Hamburger V. Cell death in the development of the lateral motor column of the chick embryo. *J Comp Neurol* 1975;**160**(4):535–46.

74. Hamburger V, Yip JW. Reduction of experimentally induced neuronal death in spinal ganglia of the chick embryo by nerve growth factor. *J Neurosci Off J Soc Neurosci* 1984;**4**(3):767–74.

75. Oppenheim RW. The neurotrophic theory and naturally occurring motoneuron death. *Trends Neurosci* 1989;**12**(7):252–5.

76. Rind HB, Butowt R, von Bartheld CS. Synaptic targeting of retrogradely transported trophic factors in motoneurons: comparison of glial cell line-derived neurotrophic factor, brain-derived neurotrophic factor, and cardiotrophin-1 with tetanus toxin. *J Neurosci Off J Soc Neurosci* 2005;**25**(3):539–49.

77. Rind HB, von Bartheld CS. Anterograde axonal transport of internalized GDNF in sensory and motor neurons. *Neuroreport* 2002;**13**(5):659–64.

78. Airaksinen MS, Titievsky A, Saarma M. GDNF family neurotrophic factor signaling: four masters, one servant? *Mol Cell Neurosci* 1999;**13**(5):313–25.

79. Henderson CE, Yamamoto Y, Livet J, Arce V, Garces A, deLapeyriere O. Role of neurotrophic factors in motoneuron development. *J Physiol Paris* 1998;**92**(3–4):279–81.

80. Gould TW, Enomoto H. Neurotrophic modulation of motor neuron development. *Neurosci Rev J bringing Neurobiol Neurol Psychiatry* 2009;**15**(1):105–16.

81. Sun T, Dong H, Wu L, Kane M, Rowitch DH, Stiles CD. Cross-repressive interaction of the Olig2 and Nkx2.2 transcription factors in developing neural tube associated with formation of a specific physical complex. *J Neurosci Off J Soc Neurosci* 2003;**23**(29):9547–56.

82. Zhou Q, Anderson DJ. The bHLH transcription factors OLIG2 and OLIG1 couple neuronal and glial subtype specification. *Cell* 2002;**109**(1):61–73.

83. Qi Y, Cai J, Wu Y, et al. Control of oligodendrocyte differentiation by the Nkx2.2 homeodomain transcription factor. *Development* 2001;**128**(14):2723–33.

84. Zhou Q, Choi G, Anderson DJ. The bHLH transcription factor Olig2 promotes oligodendrocyte differentiation in collaboration with Nkx2.2. *Neuron* 2001;**31**(5):791–807.

85. Sugimori M, Nagao M, Bertrand N, Parras CM, Guillemot F, Nakafuku M. Combinatorial actions of patterning and HLH transcription factors in the spatiotemporal control of neurogenesis and gliogenesis in the developing spinal cord. *Development* 2007;**134**(8):1617–29.

86. Li H, de Faria JP, Andrew P, Nitarska J, Richardson WD. Phosphorylation regulates OLIG2 cofactor choice and the motor neuron-oligodendrocyte fate switch. *Neuron* 2011;**69**(5):918–29.

87. Chung AY, Kim S, Kim E, et al. Indian hedgehog B function is required for the specification of oligodendrocyte progenitor cells in the zebrafish CNS. *J Neurosci Off J Soc Neurosci* 2013;**33**(4):1728–33.

88. Touahri Y, Escalas N, Benazeraf B, Cochard P, Danesin C, Soula C. Sulfatase 1 promotes the motor neuron-to-oligodendrocyte fate switch by activating Shh signaling in Olig2 progenitors of the embryonic ventral spinal cord. *J Neurosci Off J Soc Neurosci* 2012;**32**(50):18018–34.

89. Snyder JL, Kearns CA, Appel B. Fbxw7 regulates Notch to control specification of neural precursors for oligodendrocyte fate. *Neural Develop* 2012;**7**:15.

90. Rabadan MA, Cayuso J, Le Dreau G, et al. Jagged2 controls the generation of motor neuron and oligodendrocyte progenitors in the ventral spinal cord. *Cell Death Differ* 2012;**19**(2):209–19.

91. Louvi A, Artavanis-Tsakonas S. Notch signalling in vertebrate neural development. *Nat Rev Neurosci* 2006;**7**(2):93–102.

92. Gokoffski KK, Wu HH, Beites CL, et al. Activin and GDF11 collaborate in feedback control of neuroepithelial stem cell proliferation and fate. *Development* 2011;**138**(19):4131–42.

93. Sugimori M, Nagao M, Parras CM, et al. Ascl1 is required for oligodendrocyte development in the spinal cord. *Development* 2008;**135**(7):1271–81.

94. Swetenburg R, White DE, Kemp ML, McDevitt TC, Stice SS. In preparation; 2015.

95. Dias JM, Alekseenko Z, Applequist JM, Ericson J. Tgfbeta signaling regulates temporal neurogenesis and potency of neural stem cells in the CNS. *Neuron* 2014;**84**(5):927–39.

96. Wichterle H, Lieberam I, Porter JA, Jessell TM. Directed differentiation of embryonic stem cells into motor neurons. *Cell* 2002;**110**(3):385–97.

97. Miles GB, Yohn DC, Wichterle H, Jessell TM, Rafuse VF, Brownstone RM. Functional properties of motoneurons derived from mouse embryonic stem cells. *J Neurosci Off J Soc Neurosci* 2004;**24**(36):7848–58.

98. Li XJ, Du ZW, Zarnowska ED, et al. Specification of motoneurons from human embryonic stem cells. *Nat Biotechnol* 2005;**23**(2):215–21.

99. Shin S, Dalton S, Stice SL. Human motor neuron differentiation from human embryonic stem cells. *Stem Cells Dev* 2005;**14**(3):266–9.

100. Li XJ, Hu BY, Jones SA, et al. Directed differentiation of ventral spinal progenitors and motor neurons from human embryonic stem cells by small molecules. *Stem Cells* 2008;**26**(4):886–93.

101. Hu BY, Zhang SC. Differentiation of spinal motor neurons from pluripotent human stem cells. *Nat Protoc* 2009;**4**(9):1295–304.

102. Chambers SM, Fasano CA, Papapetrou EP, Tomishima M, Sadelain M, Studer L. Highly efficient neural conversion of human ES and iPS cells by dual inhibition of SMAD signaling. *Nat Biotechnol* 2009;**27**(3):275–80.

103. Takahashi K, Yamanaka S. Induction of pluripotent stem cells from mouse embryonic and adult fibroblast cultures by defined factors. *Cell* 2006;**126**(4):663–76.

104. Toma JS, Shettar BC, Chipman PH, et al. Motoneurons derived from induced pluripotent stem cells develop mature phenotypes typical of endogenous spinal motoneurons. *J Neurosci Off J Soc Neurosci* 2015;**35**(3):1291–306.

105. Boyden ES, Zhang F, Bamberg E, Nagel G, Deisseroth K. Millisecond-timescale, genetically targeted optical control of neural activity. *Nat Neurosci* 2005;**8**(9):1263–8.

106. Bryson JB, Machado CB, Crossley M, et al. Optical control of muscle function by transplantation of stem cell-derived motor neurons in mice. *Science* 2014;**344**(6179):94–7.
107. Mazzoni EO, Mahony S, Closser M, et al. Synergistic binding of transcription factors to cell-specific enhancers programs motor neuron identity. *Nat Neurosci* 2013;**16**(9): 1219–27.
108. Son EY, Ichida JK, Wainger BJ, et al. Conversion of mouse and human fibroblasts into functional spinal motor neurons. *Cell Stem Cell* 2011;**9**(3):205–18.
109. Patani R, Hollins AJ, Wishart TM, et al. Retinoid-independent motor neurogenesis from human embryonic stem cells reveals a medial columnar ground state. *Nat Commun* 2011;**2**:214.
110. Amoroso MW, Croft GF, Williams DJ, et al. Accelerated high-yield generation of limb-innervating motor neurons from human stem cells. *J Neurosci Off J Soc Neurosci* 2013;**33**(2):574–86.
111. Peljto M, Dasen JS, Mazzoni EO, Jessell TM, Wichterle H. Functional diversity of ESC-derived motor neuron subtypes revealed through intraspinal transplantation. *Cell Stem Cell* 2010;**7**(3):355–66.
112. Adams KL, Rousso DL, Umbach JA, Novitch BG. Foxp1-mediated programming of limb-innervating motor neurons from mouse and human embryonic stem cells. *Nat Commun* 2015;**6**:6778.
113. Machado CB, Kanning KC, Kreis P, et al. Reconstruction of phrenic neuron identity in embryonic stem cell-derived motor neurons. *Development* 2014;**141**(4):784–94.
114. Nedelec S, Peljto M, Shi P, Amoroso MW, Kam LC, Wichterle H. Concentration-dependent requirement for local protein synthesis in motor neuron subtype-specific response to axon guidance cues. *J Neurosci Off J Soc Neurosci* 2012;**32**(4):1496–506.
115. Yang YM, Gupta SK, Kim KJ, et al. A small molecule screen in stem-cell-derived motor neurons identifies a kinase inhibitor as a candidate therapeutic for ALS. *Cell Stem Cell* 2013;**12**(6):713–26.
116. Vyas N, Walvekar A, Tate D, et al. Vertebrate Hedgehog is secreted on two types of extracellular vesicles with different signaling properties. *Sci Rep* 2014;**4**:7357.
117. Davis-Dusenbery BN, Williams LA, Klim JR, Eggan K. How to make spinal motor neurons. *Development* 2014;**141**(3):491–501.
118. https://clinicaltrials.gov/ct2/show/NCT01217008.
119. Liu S, Qu Y, Stewart TJ, et al. Embryonic stem cells differentiate into oligodendrocytes and myelinate in culture and after spinal cord transplantation. *Proc Natl Acad Sci USA* 2000;**97**(11):6126–31.
120. Keirstead HS, Nistor G, Bernal G, et al. Human embryonic stem cell-derived oligodendrocyte progenitor cell transplants remyelinate and restore locomotion after spinal cord injury. *J Neurosci Off J Soc Neurosci* 2005;**25**(19):4694–705.
121. Kumagai G, Okada Y, Yamane J, et al. Roles of ES cell-derived gliogenic neural stem/progenitor cells in functional recovery after spinal cord injury. *PloS One* 2009;**4**(11):e7706.
122. Hooshmand MJ, Sontag CJ, Uchida N, Tamaki S, Anderson AJ, Cummings BJ. Analysis of host-mediated repair mechanisms after human CNS-stem cell transplantation for spinal cord injury: correlation of engraftment with recovery. *PloS One* 2009;**4**(6):e5871.
123. Erceg S, Ronaghi M, Oria M, et al. Transplanted oligodendrocytes and motoneuron progenitors generated from human embryonic stem cells promote locomotor recovery after spinal cord transection. *Stem Cells* 2010;**28**(9):1541–9.
124. Hu BY, Du ZW, Li XJ, Ayala M, Zhang SC. Human oligodendrocytes from embryonic stem cells: conserved SHH signaling networks and divergent FGF effects. *Development* 2009;**136**(9):1443–52.
125. Hu Z, Li T, Zhang X, Chen Y. Hepatocyte growth factor enhances the generation of high-purity oligodendrocytes from human embryonic stem cells. *Differ Res Biol Diversity* 2009;**78**(2–3):177–84.

126. Jiang P, Selvaraj V, Deng W. Differentiation of embryonic stem cells into oligodendrocyte precursors. *J Vis Exp* 2010(39).
127. Chen C, Daugherty D, Jiang P, Deng W. Oligodendrocyte progenitor cells derived from
mouse embryonic stem cells give rise to type-1 and type-2 astrocytes in vitro. *Neurosci
Lett* 2012;**523**(2):180–5.
128. Tao F, Li Q, Liu S, et al. Role of neuregulin-1/ErbB signaling in stem cell therapy
for spinal cord injury-induced chronic neuropathic pain. *Stem Cells (Dayton, Ohio)*
2013;**31**(1):83–91.
129. Yang N, Zuchero JB, Ahlenius H, et al. Generation of oligodendroglial cells by direct
lineage conversion. *Nat Biotechnol* 2013;**31**(5):434–9.
130. Alsanie WF, Niclis JC, Petratos S. Human embryonic stem cell-derived oligodendrocytes: protocols and perspectives. *Stem Cells Dev* 2013;**22**(18):2459–76.

Natural History of Amyotrophic Lateral Sclerosis

M.J. Magnussen and J.D. Glass

Emory University, Atlanta, GA, United States

HISTORY

The term "amyotrophic lateral sclerosis" (ALS) was first proposed by Jean-Martin Charcot in 1874, deriving the name from his method of correlating clinical symptoms and signs with anatomical lesions found at autopsy.[1] While working at the Salpetriere Hospital in France, Charcot and colleagues clinically followed patients with exclusively spastic weakness in addition to those with predominantly weakness and atrophy. Upon their death, Charcot then performed autopsies in order to identify anatomical changes that correlated with the patients' clinical deficits. This led him to discover the "brownish-gray streak marks produced by sclerotic changes" in the lateral columns of the spinal cord correlating with clinical spasticity, as well as lesions that were "systematically limited to the anterior horns of the gray matter" in patients with muscle wasting, or "amyotrophy."[2,3]

The findings of Charcot built upon, and further validated, earlier works by Lockhart Clarke and Charles Bland Radcliffe.[1] Although they failed to formally name the disease, Clarke and Radcliffe described the pathological changes of a patient affected with what appeared to be ALS in 1862.[4] In their work, "An important case of paralysis and muscular atrophy, with disease of the nervous centres," they describe autopsy findings of degeneration of the lateral columns in the spinal cord and atrophy of both the anterior roots of the spinal cord and the bulbar motor nerves.[5] This work, performed over a century ago, was essential to the medical community's understanding of both neuroanatomy and the pathological features of ALS. These early discoveries by Clarke, Charcot, and others have stood the test of time, and remain the clinical and pathological framework upon which current research efforts in ALS are built.

Over the next 100 years, different clinical presentations of diseases, primarily affecting the motor system, were further defined both clinically and pathologically. As these progressive motor disorders were further defined, William Richard Gowers was the first to suggest that these processes may represent syndromic variants of a single disease process.[6] This idea was further refined by Walter Russell Brain in 1962 when he coined the term "motor neurone disease" (MND) as a category of disease including ALS, progressive bulbar palsy (PBP), progressive muscular atrophy (PMA), and progressive spastic paraparesis.[1] The term "MND" remains in use today. The major subsets of MND include ALS, PMA, primary lateral sclerosis (PLS), and PBP. Despite the early pathologic and clinical descriptions of ALS, the salient features of ALS/MND remain unchanged from the original clinical descriptions.

EPIDEMIOLOGY

The majority of epidemiological data regarding ALS are derived from observational morbidity and/or mortality studies. These larger studies often fail to separate the different forms of MND, or distinguish sporadic cases of ALS from familial types. The potential analysis of numerous syndromes as a single disorder, ALS, may introduce errors into these studies, as well as prevent the identification of drastically different disease courses that may be present in the more rare types of MND. Other types of analyses, including case control or cohort studies, are also utilized for identifying possible risk factors for ALS. While these may be better suited to the analysis of specific subtypes of MND, they also carry a risk of introducing sampling bias or recall bias. Aside from the possible sources of error from these study types, they represent an important resource in our understanding of the natural history of ALS.

In their text[7,] Mitsumoto and colleagues presented a review of multiple epidemiological studies of ALS that were reported from 1966 to 1994. This analysis combined data from Europe, Canada, the United States, Mexico, and the Middle East. The mean overall incidence of sporadic-onset ALS was 1.1 ± 0.5 cases per 100,000, while the prevalence was 3.6 ± 1.8 per 100,000 in the population.[7] In that report, the peak age of onset was between 55 and 75 years old, with a male to female ratio ranging from 1.4 to 2.5:1.[7] The average survival after onset of symptoms of ALS was calculated at 3 years with a median disease duration ranging from 23 to 52 months.[7,8]

More recent studies have largely confirmed these estimates for ALS incidence, prevalence, and age of onset.[9-11] However, some mortality-based ALS studies suggest that there has been a rise in the incidence of ALS over the past several decades. This hypothesis of an increased incidence is challenged by other natural-history analyses.[9-12] The seemingly aberrant increase in ALS incidence is most likely attributed to an overall improvement in diagnosis and a more accurate attribution of cause of death in the medical record, rather than a true spike in disease incidence.

An interesting and encouraging finding from recent epidemiological studies is the increase in survival of ALS patients over the last two decades. Both retrospective population analyses and cohort analyses suggest that certain patient populations diagnosed in the 1990s had a median survival almost 1 year longer than those diagnosed prior to that time.[11,13,14] Despite the prolonged survival in those diagnosed in or after the 1990s as compared to earlier populations, the two patient populations showed no differences in their rates of functional decline.[14] This pattern of improved survival without a significant change in functional abilities is similar to that seen in therapeutic trials in ALS patients.[15,16] However, studies have shown this same increase in survival even after correcting for more recent additions to the clinical management of ALS patents, including the drug riluzole, noninvasive ventilation (NIV), and percutaneous endoscopic gastrostomy (PEG) tube placement.[13] This suggests that the survival benefit may be derived from other undefined factors related to the institution of multidisciplinary care clinics, or perhaps to improvements in the management of non-ALS-related comorbidities. While increased survival in ALS patients has been noted by several groups, other ALS populations fail to display similar improvements, despite undergoing comparable disease management.[9] Therefore, further evaluation is necessary to confirm the presence of this phenomenon, as well as to elucidate the actions that may be leading to an improved survival.

In an effort to better understand the underlying pathophysiology of ALS, many studies have attempted to identify risk factors for the development of ALS. To date, both age and gender have been consistently identified as risk factors for the development of ALS.[7,9,10,17] The incidence of ALS begins to increase after the age of 40 years, with a peak incidence occurring in the sixth and seventh decades of life. Interestingly, the relationship of increasing age and incidence of ALS is reversed between ages 75 and 80 years.[7,10,17,18] Therefore, advancing age alone is unlikely to be a lone risk factor, suggesting that during this age range, the motor neurons of susceptible individuals may cross a threshold of cumulative damage, which then leads to the appearance of clinical symptoms.

Traditionally, males have shown a higher incidence of sporadic ALS (SALS) than females.[7,9,10,19] This notion has been challenged as some studies have suggested that the previously reported male:female ratio of ALS incidence of 2:1 is equalizing.[17] Possible explanations for this change in the influence of gender on ALS incidence are multiple, but they are simply hypotheses at this time. Theories directed at explaining this finding range from relating this change to an improved physician awareness and diagnosis of MND, to alterations in socioeconomic factors that have increased female exposure to possible environmental risks. However, this change in the effect of gender on disease incidence is not consistent across all recent studies.[9] Definitive evidence as to whether these reported changes in gender incidence of ALS truly represent a shift in the epidemiology of the disease will require continued analyses of MND.

There is a suggestion that race may impact one's risk of developing ALS. Cronin et al.[20] reviewed 61 epidemiological studies of ALS from Europe, Central and South America, North America and Canada, Asia, the Pacific, Africa, the Middle East, and Russia. They noted a lower incidence of ALS among African, Asian, and Hispanic ethnicities when compared to Caucasians, a finding that was corroborated when comparing Hispanic and African populations to Caucasians within the same country.[12,20,21] The interpretation of these studies is complex, as the heterogeneity of ethnicities within a single racial group may be significant, which could skew the reported racial incidence. Regardless of these limitations, differences in incidence among racial groups suggest predisposing genetic factors play a role in the development of SALS. Prospective, population-based analyses of strictly defined racial groups are needed to further define any racial predilections in ALS.

There are multiple other possible risk factors for ALS reported in the literature, ranging from pesticide exposure to exposure to ultraviolet radiation.[22,23] A relatively well-supported risk factor is smoking, which has been identified in multiple studies as an independent risk factor for ALS.[24–26] Military service, irrespective of duration of service or time served during war, has also been associated with an increased risk of development of ALS.[27]

Two of the more controversial possible risk factors in the literature are heightened levels of physical activity and a history of traumatic brain injury. There are several reports suggesting that accomplished athletes, both amateur and professional, have a higher incidence of ALS, though this relationship of physical activity to ALS has not been a consistent finding.[28-32] Head injury in professional athletes has received a lot of attention of late as a potential risk factor for ALS or even as a cause of a disorder that mimics ALS, but this work remains highly controversial.[33-36] Indeed, a recent clinical and pathological evaluation of ALS patients with and without a history of traumatic brain injury failed to identify any difference in disease progression or pathological findings between the two groups.[37] Clearly, both athleticism and head injury will require further evidence in order to definitively link them with an increased risk of developing ALS.

The most clearly defined risk factor for the development of ALS is a family history of ALS and the harboring of a pathogenic mutation. Familial ALS (FALS) is thought to represent 5–10% of cases.[7,38] Through the discovery of several different genes associated with FALS, multiple possible genetic identifiers of increased risk of developing ALS now exist. Significant discoveries have been made in genetic causes for ALS, though disease causing gene mutations remain unknown for about 30% of FALS. Both FALS and the commonly identified genetic mutations in ALS will be further discussed in a later section of this chapter.

CLINICAL PRESENTATION

Patients with ALS may initially present to the clinic with patterns of upper motor neuron (UMN) signs, lower motor neuron (LMN) signs, or both. Amidst the variable manners of clinical presentation in ALS, a focal onset of symptoms is the most common. When categorizing the clinical phenotype of ALS, individual cases are characterized by the region of disease onset, as well as the presence or absence of UMN and LMN signs. The UMN and LMN signs refer to clinical features that are associated with lesions in motor neurons of the central and peripheral nervous systems, respectively (Table 2.1). The region of disease onset is typically defined as bulbar, limb, or the more rare presentation of diaphragmatic onset disease. Through the identification and monitoring of these phenotype qualities, patients with traditional ALS are separated from those with other types of MND. The distinction of subsets of MND allows for improved prognostication and informed decision making during the care of these individuals.

The most common clinical presentation of MND is ALS, that is the presence of both UMN and LMN symptoms and signs. Typically, patients present with a mixture of muscle atrophy, fasciculations, spasticity, and pathologically brisk reflexes. The most common initial symptom is

TABLE 2.1   Upper and Lower Motor Neuron Signs

Upper motor neuron signs	Lower motor neuron signs
Hyperreflexia	Hyporeflexia/areflexia
Spasticity	Flaccidity
Impaired dexterity	Muscle atrophy
Pathologic reflexes (Babinski, Hoffman)	Fasciculations
Muscle weakness	Muscle weakness

painless weakness, which is the chief complaint for approximately 60% of patients.[7] This is typically a focal weakness involving limb or bulbar musculature, with less than 10% of patients presenting with generalized weakness.[7-9] The most common site of onset for focal weakness is in a limb (60–70%), with an equal distribution of initial involvement between arms and legs.[7-10] Weakness or spasticity in bulbar musculature accounts for an additional 20–30% of presenting symptoms in typical ALS.[7-10] Rarely, patients may present with isolated weakness of the muscles of respiration which has been reported to occur in 1–3% of the ALS population.[8,39]

Despite weakness being the most common symptom at the time of presentation, patients with typical ALS may first present with other complaints. Although rare, some patients will pursue medical evaluation due to isolated fasciculations, weight loss, or muscle cramps. While all of these symptoms are commonly seen during the course of ALS, they are less likely to represent the initial symptoms in these patients. Certainly, these symptoms are common to a number of disorders, which may lead to a rather broad differential diagnosis, especially when there are no other signs to support a diagnosis of ALS.

Progression of symptoms, and thus disability, is the norm for people with ALS. The spread of disease may occur in a contiguous manner from an initially focal region of onset.[7,40-44] For example, progression from the limb of onset to the contralateral limb, or progression rostrally or caudally along the neuraxis.[7,40,41,43] However, there are many clinical examples of what appears to be multifocal onset and/or spread of disease, and even examples where neurological dysfunction will skip a contiguous neuroanatomical region, raising the question of whether there truly is "spread" from a focal region of onset. This question remains an ongoing focus of research.

The progression of ALS is quite variable, though it is generally assumed that the rate of symptom progression follows a linear course from the time of onset.[45,46] Indeed, the course of the disease may vary from less than a year to many decades. In an attempt to determine why the rate

TABLE 2.2 Presenting Features Associated With Prolonged Survival in Amyotrophic Lateral Sclerosis[a]

Prolonged time from onset to diagnosis	Nonbulbar region of onset
Younger age of onset	Normal weight/nutritional status at diagnosis
Pure UMN signs	Nonfamilial ALS[b]
Pure LMN signs	Normal cognition at diagnosis
No dyspnea at onset	

ALS, amyotrophic lateral sclerosis; LMN, lower motor neuron; UMN, upper motor neuron.
[a]The presence or absence of these features does not definitively predict prognosis.
[b]Some types of familial ALS are associated with prolonged survival (e.g. select SOD1 gene mutations).

of progression varies so greatly, risk factors for progression and disease survival have been evaluated. Epidemiological studies have identified multiple risk factors that may have a positive or negative influence on ALS progression. Factors described as being associated with slower progression and better prognosis include: prolonged time from symptom onset to diagnosis, younger age at onset (< 35–40 years old), pure UMN or LMN signs, no dyspnea at onset, nonbulbar region of onset, normal weight at the time of diagnosis, nonfamilial ALS, and normal cognition at the time of diagnosis (Table 2.2).[7,8,47–51] This is not an exhaustive list and certain factors listed require further clarification, as is the case with nonfamilial ALS being associated with improved prognosis. While ALS associated with chromosome 9 open reading frame 72 (C9orf72) mutations has been linked with a reduced survival versus nonmutants, there are other forms of FALS that have a very slow clinical course.[38,50] These factors may be associated with a better survival prognosis; however, the absence of one or more of these disease characteristics does not necessarily impart a poor prognosis.

Unlike the combination of both UMN and LMN signs that is seen in typical ALS, patients with PMA present with pure LMN symptoms and signs. PMA represents < 10% of patients with MND in various population analyses.[7,8,52–54] It is often difficult to differentiate patients with PMA from those with typical ALS with very few UMN signs (LMN-predominant ALS). Overall, patients with PMA have an improved survival time from symptom onset than those with clinically typical ALS.[7,52,54] Within the PMA patient population, predictors of a decreased survival time include generalized weakness at onset and a reduced forced vital capacity at the time of diagnosis.

Analogous to PMA, PLS presents as a pure UMN disorder. PLS is also a rare presentation, representing 2–4% of the patients in the MND population.[7,8,54,55] As with PMA, the diagnosis of PLS is complicated by the challenge of distinguishing those with pure PLS from those patients with UMN-predominant ALS. Of course, needle electromyography may identify LMN disease not evident on clinical examination, and many patients will develop more typical ALS over time. However, the continued presence of pure UMN signs 4 years after symptom onset is considered acceptable for making the diagnosis of PLS. The distinction between UMN-predominant ALS and PLS may be important because survival is believed to be longer in pure UMN ALS.[55,56] The question of prognosis in PLS versus UMN-predominant ALS is currently under study.

Another relatively uncommon clinical phenotype of MND is PBP. Clinically, PBP is characterized by the selective, progressive paralysis of the bulbar musculature involving the UMN, LMN, or both. The portion of ALS patients diagnosed with PBP varies across different population analyses from 0% to 2.2%, despite bulbar onset disease occurring in 20–30% of ALS patients.[7–10,57,58] As with PMA and PLS, the diagnosis of PBP requires longitudinal clinical evaluation of patients looking for other systemic signs that constitute a diagnosis of ALS.

FALS constitutes about 10% of the ALS population.[7,38] The clinical presentation of patients from ALS families is largely indistinguishable from those with SALS, though the presence of coincident dementia may suggest an underlying genetic mutation. Certainly, obtaining a thorough family history for any relatives with ALS, dementia, or other neurodegenerative diseases is the best initial screening method to identify those with possible FALS. Numerous genetic mutations have been identified as either directly causing ALS, or increasing the risk for disease (Table 2.3). The majority of disease-causing mutations, however, are four different genes: C9orf72, superoxide dismutase 1 (SOD1), TAR DNA binding protein (TARDBP), and fused in sarcoma (FUS). The known gene mutations that are associated with ALS are ever changing as this area of research progresses. With only 50–70% of the mutations responsible for FALS identified (depending on the population studied), ample room exists for further solidifying the genetics and heritability of ALS.[59,60]

The disease heterogeneity seen in SALS is also present in patients with familial disease. Even in family members carrying the same gene mutation the onset and course of disease may be vastly different. Examples from the author's experience include a mother who presented with bulbar disease in her early 60s and died within 2 years, and her daughter who developed progressive paraplegia in her 30s and survived for > 7 years. Both carried the same SOD1 mutation. Indeed, some disease mutations (i.e., C9orf72, VCP, TARDBP) may present with vastly different diseases (alone or in combination) including ALS, frontotemporal dementia, and even Padget's disease of bone (valosin containing protein (VCP) mutation). Thus, the clinician must maintain a high level of suspicion for familial disease, and must also realize that genetic mutations may be found in some patients with apparently sporadic disease.[59] The identification

TABLE 2.3   Familial Amyotrophic Lateral Sclerosis Genes and Clinical Features

Gene name	Locus name	Inheritance	Locus/clinical features
Cu/Zn superoxide dismutase 1	ALS 1	AD (some AR)	21q22.1/typical ALS
C9orf72	ALS-FTD2	AD/sporadic	Chromosome 9 repeat expansion
ALSIN	ALS 2	AR	2q33.1/juvenile onset UMN, slowly progressive
Senataxin (SETX)	ALS 4	AD	9q34.13/slowly progressive, LMN
Dynactin	–	AD	2p13/slowly progressive, vocal cord paralysis
Tau (MAPT)	–	AD	17q21.1/associated with FTD
–	ALS 5	AR	15q15–21/juvenile onset, slowly progressive
Fused in sarcoma (FUS)	ALS 6	AD	16q12/typical ALS
–	ALS 7	AD	20p13/typical ALS
Vesicle-ass protein B	ALS 8	AD	20q13.33/typical ALS
TDP-43	ALS 10	AD	1p36/typical ALS
Progranulin	–	AD	ALS with dementia
Ubiquilin 2	ALS 15	X-linked	ALS or ALS/FTD
Valosin-containing protein	ALS 14	AD	ALS, FTD, IBM, Padgett's disease
Matrin3 (MATR3)	–	AD	ALS, distal myopathy, vocal cord paralysis

Data from: Byrne S, Elamin M, Bede P, Shatunov A, Walsh C, Hardiman O, et al. Cognitive and clinical characteristics of patients with amyotrophic lateral sclerosis carrying a C9orf72 repeat expansion: a population-based cohort study. Lancet Neurol 2012;**11**:232–40; Sabatelli M, Conti A, Zollino M. Clinical and genetic heterogeneity of amyotrophic lateral sclerosis. Clin Genet 2013;**83**:408–16; Andersen PM, Al-Chalabi A. Clinical genetics of amyotrophic lateral sclerosis: what do we really know? Nat Rev Neurol 2011;**7**:603–15; Renton AE, Majounie E, Waite A, Simon-Sanchez J, Rollinson S, Traynor BJ, et al. A hexanucleotide repeat expansion in C9orf72 is the cause of chromosome 9p-21 linked ALS-FTD. Neuron. 2011; 72, 257-268. Johnson JO, Pioro EP, Boehringer A, Chia R, Feit H, Traynor BJ, et al. Mutations in the Martin 3 gene cause familial amyotrophic lateral sclerosis. Nature Neuroscience. 2014; 17, 664-666. http://dx.doi.org/10.1038/nn.3688
ALS, amyotrophic lateral sclerosis; AD, autosomal dominant; AR, autosomal recessive; FTD, frontotemporal dementia; IBM, inclusion body myopathy; LMN, lower motor neuron; UMN, upper motor neuron.

of patients carrying disease-causing mutations is important, allowing in some cases for confirmation of diagnosis, but also to provide education for patients and their families regarding the expected disease course, as well as counseling vis-à-vis risk to family members.

PATHOGENESIS

The majority of patients with ALS undergo the progressive degeneration of both the UMN and LMN systems. Upon autopsy, there is evidence of atrophy within the motor cortex and the ventral spinal roots due to the selective loss of pyramidal cells, as well as anterior horn cells and brainstem motor neurons. There is also sclerotic change within the lateral columns of the spinal cord and ventral spinal roots due to the loss of myelinated axons within the corticospinal tracts and anterior horn cell destruction. The pathological hallmark of ALS is the presence of ubiquitinated protein accumulations within the cytoplasm of motor neurons and glial cells. In the majority of SALS cases, TAR DNA-binding protein 43 (TDP-43) is the major component of these cytoplasmic inclusions.[38,54]

There is pathological evidence to suggest the presence of a presymptomatic period, during which motor neuron loss is occurring. This theory is suggested by the finding that up to 30% of motor neurons can be lost in ALS patients in the absence of clinical evidence of weakness in corresponding muscles.[61] Similarly, in poliomyelitis, it has been suggested that clinical weakness does not occur until more than 50% of motor neurons were lost.[62] The implication of a presymptomatic period suggests that early therapeutic interventions may modify the clinical progression of disease.

The pathogenesis of ALS remains unknown, and as a result many competing theories exist regarding possible disease mechanisms. The majority of hypotheses attempt to connect pathogenesis in both FALS and SALS populations. In reality though, genetic models of disease outnumber models of sporadic disease, simply because target genes and their mutations are a rich source of focused experimental paradigms. Oxidative stress was an early topic of interest based on the original discovery of pathogenic mutations in the gene for Cu/Zn SOD1. This continues to be a focus of interest for some laboratories, even though it has now been demonstrated that these mutations do not create a loss of function phenotype for the antioxidative function of the SOD1 protein. Even so, a gain in function might also result in increased oxidative stress and oxidative damage to proteins as has been found in both animal models and human neuropathology.[38]

Major areas of current investigation include protein misfolding and degradation, and alterations in RNA processing. The damage and misfolding of proteins can be a normal process within cells that is handled by several systems responsible for refolding, recycling, and degradation. These systems include the ubiquitin proteasome system, autophagy, and the lysosomal degradation system.[63] Normal functioning of these systems is essential for maintaining cellular health, and there is evidence for abnormalities in each of these pathways that may lead to motor neuron dysfunction and death.[38,64] In reality, however, none of the pathway abnormalities identified in cellular and animal models have been reliably identified in patients with ALS. Nevertheless, intervention directed toward correcting these pathways is a major focus for the development of new therapies for ALS.

Genetic discoveries have identified a commonality among genes that are involved in RNA and DNA transport and processing, suggesting a common pathogenesis. These genes include TDP43, FUS, hnRNPA1, and the most common genetic cause of ALS, C9orf72. Mutations impacting the function of these proteins are hypothesized to affect the normal economy of DNA translation and RNA transport in and out of the nucleus (and possibly transport into axons and dendrites).[38,64–66] This might create an imbalance in the production and retention of essential proteins within various cellular compartments, leading to cellular dysfunction and possibly death. For the case of the C9orf72 hexanucleotide expansion mutation, there is evidence that there may be sequestering of essential RNA proteins within the expanded region, or that the expanded DNA may be translated into abnormal peptides that are possibly toxic to the cell.[67] All of these mechanisms are currently under intense study which likely will result in a better understanding of disease pathogenesis and hopefully new, highly focused, therapeutic trials.

Murine and human studies suggest that neuroinflammation is a pivotal factor contributing to disease progression and survival in ALS. This evidence suggests that an increased presence of functional T-regulatory and Th2 lymphocytes in the spinal cord of ALS patients is correlated with a slow disease progression, whereas a peripheral reduction in functional T-regulatory lymphocytes early in the disease has been associated with a rapidly progressive course.[68–70] Extrapolated from these findings is the theory that an uninhibited, motor-neuron-directed, and cellular-mediated immune response is an instrumental aspect of disease progression in ALS.[38,70,71] This theory provides an additional avenue for the development of future therapeutic trials in ALS. Overall, it seems unlikely that a single disease mechanism is responsible for ALS. Instead, the pathogenesis of ALS is likely secondary to a combination of several different processes, which may even differ among individuals with different genetic profiles.

TREATMENTS

The lack of a clearly defined pathogenesis of ALS has led to limited therapeutic options. Currently, there is no cure for ALS. However, some interventions have been identified which lead to improved survival. The only medication that has been proven to prolong survival in the treatment of ALS is riluzole.[15,16] The exact mechanism of action of riluzole in ALS is not known, but it is thought to have a neuroprotective function by blocking glutamatergic neurotransmission and inactivating voltage-dependent sodium channels peripherally.[72,73] Although the survival benefit provided by the addition of riluzole is modest, it is the only available medication that impacts survival in ALS. Therefore, it is the author's opinion that it should be considered in the treatment of all individuals with ALS.

Two additional therapeutic interventions, which have been proven beneficial in individuals with ALS, are the use of non-invasive positive pressure ventilation (NIPPV) and the placement of a PEG tube. Malnutrition has been identified as a poor prognostic factor in ALS, as well as a factor that reduces individuals' quality of life.[48,49] The placement of a PEG tube has been shown to prolong survival as well as improve the quality of life of individuals with ALS.[74,75] Similar to the benefits seen with the placement of a PEG tube, the use of NIV in patients with ALS has also been shown to prolong survival and increase individuals' quality of life.[76,77] The evidence of the ability of these interventions to alter the natural history of ALS argues for their inclusion in the patient care plan whenever appropriate.

The development of therapeutic interventions for ALS is an endeavor that has repeatedly fallen short while attempting to translate developments in basic science to human treatments. Despite numerous successful experimental treatments in rodent disease models, human clinical trials have failed to recapitulate these same results. The discordance between laboratory and clinical results may be secondary to inadequate animal modeling of disease, the significant genetic heterogeneity of the human ALS population, poor selection of trial participants, or the cumulative impact from each of these confounders. The ability to stratify patients based on the genetics of their disease and their expected disease course may provide an opportunity to optimize the patient population included in research trials, thereby improving the odds of uncovering additional therapies. Such an approach requires further expansion of our genetic understanding of ALS, the identification of reliable predictors of disease course, as well as the development of efficient methods of genetic screening.

SUMMARY

ALS is a progressive, neurodegenerative disease that primarily affects the motor neurons throughout the nervous system. It is a relatively rare disease, which typically begins clinically in the sixth or seventh decade of life and is slightly more common in males than females. Despite numerous different clinical presentations, typical ALS often begins with focal limb weakness with evidence of lesions involving both the upper and LMNs. In its typical course, ALS exhibits a linear progression of symptoms at variable rates among patients. The median disease survival is 2–3 years from symptom onset with death most commonly due to a decline in respiratory muscle strength and its sequelae. There are certain clinical features such as diffuse body involvement at onset, early respiratory involvement, and early bulbar involvement that are typically associated with a decreased survival. Genetic mutations have been associated with both familial and sporadic cases of ALS; however, a family history of ALS is typically associated with a more rapid disease progression. Although numerous mutations have been related to the development of ALS, the genetic abnormality responsible for the majority of familial cases of ALS remains unknown.

The pathogenesis of ALS is yet to be defined and is most likely multifactorial with both environmental and genetic components. Riluzole is the only available medication that is proven to prolong survival. Other interventions that improve patient survival include the placement of a PEG tube in response to weight loss, as well as the use of NIPPV in response to respiratory muscle weakness. The discovery of future therapeutic options, hinges upon continued advancements within the genetics and pathogenesis of ALS.

References

1. Turner MR, Swash M. The expanding syndrome of amyotrophic lateral sclerosis: a clinical and molecular odyssey. *J Neurol Neurosurg Psychiatry* 2015. http://dx.doi.org/10.1136/jnnp-2014-308946.
2. Charcot J. Sclerose des cordons lateraux de la moelle epiniere chez une femme hysterique atteinte de contracture permanente des quatre membres. *Bull Soc Med (Paris)* 1865:24–35.
3. Charcot J-M, Joffroy A. Deux cas d'atrophie musculaire progressive avec lesions de la substance grise et de faisceaux anterolateraux de la moelle epiniere. *Arch Physiol Norm Pathol* 1869 1;2;3:354-367;628-49;744-57.
4. Turner MR, Swash M, Ebers GC. Lockhart Clarke's contribution to the description of amyotrophic lateral sclerosis. *Brain* 2010;**133**:3470–9.
5. Radcliffe CB, Lockhart Clarke J. An important case of paralysis and muscular atrophy with disease of the nervous centres. *Brit Foreign Medico-Chirurgical Rev* 1862;**30**:215–25.
6. Gowers W. *A manual of diseases of the nervous system*. London: J & A Churchill; 1886.
7. Mitsumoto H, Chad DA, Pioro EP. *Amyotrophic lateral sclerosis*. Philadelphia, PA: F. A. Davis Company; 1998.
8. Norris F, Shepherd R, Denys E, Kwei U, Mukai E, Elias L, et al. Onset, natural history and outcome in idiopathic adult motor neuron disease. *J Neurol Sci* 1993;**118**:48–55.
9. O'Toole O, Traynor BJ, Brennan P, Sheehan C, Frost E, Corr B, et al. Epidemiology and clinical features of amyotrohpic lateral sclerosis in Ireland between 1995 and 2004. *J Neurol Neurosurg Psychiatry* 2008;**79**:30–2.
10. Logroscino G, Traynor BJ, Hardiman O, Chio A, Mitchell D, Swingler RJ, et al. Incidence of amyotrophic lateral sclerosis in Europe. *J Neurol Neurosurg Psychiatry* 2010;**81**:384–90.

11. Testa D, Lovati R, Ferrarini M, Salmoiraghi F, Filippini G. Survival of 793 patients with amyotrphic lateral sclerosis diagnosed over a 28-year period. *Amyotroph Lateral Scler Other Motor Neuron Disord* 2004;**5**:208–12.
12. Noonan CW, White MC, Thurman D, Wong LY. Temporal and geographic variation in United States motor neuron disease mortality, 1969–1998. *Neurology* 2005;**64**:1215–21.
13. Czaplinski A, Yen AA, Simpson EP, Appel SH. Slower disease progression and prolonged survival in contemporary patients with amyotrophic lateral sclerosis: is the natural history of amyotrophic lateral sclerosis changing? *Arch Neurol* 2006;**63**:1139–43.
14. Qureshi MSD, Schoenfeld DA, Paliwal Y, Shui A, Cudkowicz ME. The natural history of ALS is changing: improved survival. *Amyotrophic Lateral Scler* 2009;**10**:324–31.
15. Bensimon G, Lacombiez L, Meininger V. A controlled trial of riluzole in amyotrophic lateral sclerosis, ALS/Riluzole Study Group. *N Engl J Med* 1994;**330**(9):585–91.
16. Miller RG, Mitchell JD, Moore DH. Riluzole for amyotrophic lateral sclerosis (ALS) / motor neuron disease (MND). *Cochrane Databse Syst Rev* 2012(3):CD001447.
17. Logroscino G, Traynor BJ, Hardiman O, Chio' A, Couratier P, Beghi E, et al. Descriptive epidemiology of amyotrophic lateral sclerosis: new evidence and unsolved issues. *J Neurol Neurosurg Psychiatry* 2008;**79**:6–11.
18. McGuire V, Longstreth Jr WT, Koepsell TD, van Belle G. Incidence of amyotrophic lateral sclerosis in three counties in western Washington state. *Neurology* 1996;**47**:571–3.
19. Traynor B, Codd M, Corr B, Forde C, Frost E, Hardiman O. Incidence and prevalence of ALS in Ireland, 1995-1997: a population-based study. *Neurology* 1999;**52**:504–9.
20. Cronin S, Hardiman O, Traynor BJ. Ethnic variation in the incidence of ALS: a systematic review. *Neurology* 2007;**68**:1002–7.
21. Zaldivar T, Guitierrez J, Lara G, Carbonara M, Logroscino G, Hardiman O. Reduced frequency of ALS in an ethnically mixed population: a population based mortality study. *Neurology* 2009;**72**:1640–5.
22. Kang H, Cha ES, Choi GJ, Lee WJ. Amyotrophic lateral sclerosis and agricultural environments: a systematic review. *J Korean Med Sci* 2014;**29**:1610–7.
23. Steriner I, Birmanns B, Panet A. Sun exposure and amyotrophic lateral sclerosis. *Ann Intern Med* 1994;**120**:893.
24. Sutedja NA, Veldink JH, Fischer K, Kromhout H, Wokke JH, Huisman MH, et al. Lifetime occupation, education, smoking and risk of ALS. *Neurology* 2007;**69**:1508–14.
25. de Jong SW, Huisman MH, Sutedja NA, van der Kooi AJ, de Visser M, Schelhaas HJ, et al. Smoking, alcohol consumption, and the risk of amyotrophic lateral sclerosis: a population-based study. *Am J Epidemiol* 2012;**176**:233–9.
26. Armon C. An evidence-based medicine approach to the evaluation of the role of exogenous risk factors in sporadic amyotrophic lateral sclerosis. *Neuroepidemiology* 2003;**22**:217–28.
27. Weisskopf MG, O'Reilly EJ, McCullough ML, Calle EE, Thun MJ, Ascherio A, et al. Prospective study of military service and mortality from ALS. *Neurology* 2005;**64**:32–7.
28. Huisman MH, Seelen M, de Jong SW, Dorresteijn KR, van Doormaal PT, van der Kooi AJ, et al. Lifetime physical activity and the risk of amyotrophic lateral sclerosis. *J Neurol Neurosurg Psychiatry* 2013
29. Pupillo E, Messina P, Giussani G, Logroscino G, Zoccolella S, Beghi E, et al. Physical activity and amyotrophic lateral sclerosis: a European population-based case-control study. *Ann Neurol* 2014;**75**:708–16.
30. Scarmeas N, Shih T, Stern Y, Ottman R, Rowland LP. Premorbid weight, body mass, and varsity athletics in ALS. *Neurology* 2002;**59**:773–5.
31. Beghi E, Logroscino G, Chio A, Hardiman O, Millul A, Mitchell D, et al. Amyotrophic lateral sclerosis, physical exercise, trauma and sports: results of a population-based pilot case-control study. *Amyotroph Lateral Scler* 2010;**11**:289–92.
32. Chio A, Benzi G, Dossena M, Mutani R, Mora G. Severely increased risk of amyotrophic lateral sclerosis among Italian professional football players. *Brain* 2005;**128**:472–6.

33. Lehman EJ, Hein M, Baron SL, Gersic CM. Neurodegenerative causes of death among retired National Football League players. *Neurology* 2012;**79**:1970–4.

34. Chen H, Richard M, Sandler DP, Umbach DM, Kamel F. Head injury and amyotrophic lateral sclerosis. *Am J Epidemiol* 2007;**166**:810–6.

35. McKee AC, Gavett BE, Stern RA, Nowinski CJ, Cantu RC, Kowall NW, et al. TDP-43 proteinopathy and motor neuron disease in chronic traumatic encephalopathy. *J Neuropathol Exp Neurol* 2010;**69**:918–29.

36. Bedlack RS, Genge A, Amato AA, Shaibani A, Jackson CE, Kissel JT, et al. Correspondence regarding: TDP-43 proteinopathy and motor neuron disease in chronic traumatic encephalopathy. J Neuropathol Exp Neurol 2010;69;918–29. *J Neuropathol Exp Neurol* 2011;**70**:96–7.

37. Fournier CN, Gearing M, Upadhyayula SR, Klein M, Glass JD. Head injury does not alter disease progression or neuropathologic outcomes in ALS. *Neurology* 2015

38. Al-Chalabi A, Jones A, Troakes C, King A, Al-Sarraj S, van den Berg LH. The genetics and neuropathology of amyotrophic lateral sclerosis. *Acta Neuropathol* 2012;**124**:339–52.

39. Turner MR, Wicks P, Brownstein CA, Massagil MP, Toronjo M, Talbot K, et al. Concordance between site of onset and limb dominance in amyotrophic lateral sclerosis. *J Neurol Neurosurg Psychiatry* 2011;**82**:853–4.

40. Brooks BR, Sufit R, DePaul R, Tan YD, Sanjak M, Robbins J, et al. Design of clinical therapeutic trials in amyotrophic lateral sclerosis. In: Rowland L, editor. *Amyotrophic lateral sclerosis and other motor neuron disease adv neurol*. New York, NY: Raven Press; 1991. p. 521–46.

41. Ravits JM, La Spada AR. ALS motor phenotype heterogeneity, focality, and spread: deconstructing motor neuron degeneration. *Neurology* 2009;**73**:805–11.

42. Ravits J, Paul P, Jorg C. Focality of upper and lower motor neuron degeneration at the clinical onset of ALS. *Neurology* 2007;**68**:1571–5.

43. Ravits J, Laurie P, Fan Y, Moore DH. Implications of ALS focality: rostral-caudal distribution of lower motor neuron loss postmortem. *Neurology* 2007;**68**:1576–82.

44. Turner MR, Brockington A, Scaber J, Hollinger H, Marsden R, Shaw PJ, et al. Pattern of spread and prognosis in lower limb-onset alS. *Amyotroph Lateral Scler* 2010;**11**:369–73.

45. Munsat TL, Andres PL, Finison L. The natural history of motoneuron loss in amyotrophic lateral sclerosis. *Neurology* 1988;**38**:409–13.

46. Ringel SP, Murchy J, Alderson MK, et al. The natural history of amyotrophic lateral sclerosis. *Neurology* 1993;**43**:1316–22.

47. Turner MR, Parton MJ, Shaw E, Leigh PN, Al Chalabi A. Prolonged survival in motor neuron disease: a descriptive study of Kings' database 1990-2002. *J Neurol Neurosurg Psychiatry* 2003;**74**:995–7.

48. Desport JC, Preux PM, Truong TC, Vallat JM, Sautereau D, Couratier P. Nutritional status is a prognostic factor for survival in ALS patients. *Neurology* 1999;**53**:1059–63.

49. Marin B, Desport JC, Kajeu P, Jesus P, Nicolaud B, Couratier P, et al. Alteration of nutritional status at diagnosis is a prognostic factor for survival of amyotrophic lateral sclerosis patients. *J Neurol Neurosurg Psychiatry* 2011;**82**:628–34.

50. Byrne S, Elamin M, Bede P, Shatunov A, Walsh C, Hardiman O, et al. Cognitive and clinical characteristics of patients with amyotrophic lateral sclerosis carrying a C9orf72 repeat expansion: a population-based cohort study. *Lancet Neurol* 2012;**11**:232–40.

51. Elamin M, Bede P, Byrne S, Jordan N, Gallagher L, Hardiman O, et al. Cognitive changes predict functional decline in ALS: a population-based longitudinal study. *Neurology* 2013;**80**:1590–7.

52. Kim WK, Liu X, Sander J, Pasmantier M, Andrews J, Mitsumoto H, et al. Study of 962 patients indicates progressive muscular atrophy is a form of ALS. *Neurology* 2009;**73**:1686–92.

53. Van den Berg-Vos RM, Visser J, Kalmijn S, Fischer K, de Visser M, de Jong V, et al. A long-term prospective study of the natural course of sporadic adult-onset lower motor neuron syndromes. *Arch Neurol* 2009;**66**:751–7.

54. Lonergan R, Mitsumoto H, Murray B. Amyotrophic lateral sclerosis. In: Kaminski H, Katirji B, Ruff R, editors. *Neuromuscular disorders in clinical practice*. New York, NY: Springer Science + Business Media; 2014. p. 395–423.
55. Gordon PH, Cheng B, Katz IB, Pinto M, Hays AP, Rowland LP, et al. The natural history of primary lateral sclerosis. *Neurology* 2006;**66**:647–53.
56. Gordon PH, Cheng B, Katz IB, Mitsumoto H, Rowland LP. Clinical features that distinguish PLS, upper motor neuron-dominant ALS, and typical ALS. *Neurology* 2009;**72**:1948–52.
57. Argyriou AA, Polychronopoulos P, Papapetropoulos S, Ellul J, Andriopoulos I, Katsoulas G, et al. Clinical and epidemiological features of motor neuron disease in south-western Greece. *Acta Neurol Scand* 2005;**111**:108–13.
58. Karam C, Scelsa SN, Macgowan DJ. The clinical course of progressive bulbar palsy. *Amyotroph Lateral Scler* 2010;**11**:364–8.
59. Sabatelli M, Conti A, Zollino M. Clinical and genetic heterogeneity of amyotrophic lateral sclerosis. *Clin Genet* 2013;**83**:408–16.
60. Andersen PM, Al-Chalabi A. Clinical genetics of amyotrophic lateral sclerosis: what do we really know? *Nat Rev Neurol* 2011;**7**:603–15.
61. Wohlfart G. Collateral regeneration in partially denervated muscles. *Neurology* 1958;**8**:175–80.
62. Sharrard WJ. The distribution of the permanent paralysis in the lower limb in poliomyelitis; a clinical and pathological study. *J Bone Joint Surg Br* 1955;**37-B**:540–58.
63. Nixon RA. The role of autophagy in neurodegenerative disease. *Nat Med* 2013;**19**:983–97.
64. Al-Chalabi A, Kwak S, Mehler M, Rouleau G, Siddique T, Strong M, et al. Genetic and epigenetic studies of amyotrophic lateral sclerosis. *Amyotroph Lateral Scler Frontotemporal Degener* 2013;**14**:44–52.
65. Kim HJ, Kim NC, Wany YD, Scarborough EA, Moore J, Diaz Z, et al. Mutations in prion-like domains in hnRNPA2B1 and hnRNPA1 cause multisystem proteinopathy and ALS. *Nature* 2013;**495**:467–73.
66. Honda H, Hamasaki H, Wakamiya T, Koyama S, Suzuki SO, Iwaki T, et al. Loss of hnRNPA1 in ALS spinal cord motor neurons with TDP-43-positive inclusions. *Neuropathology* 2015;**35**:37–43.
67. Gendron TF, Belzil VV, Zhang YJ, Petrucelli L. Mechanisms of toxicity in C9FTLD/ALS. *Acta Neuropathol* 2014;**127**:359–76.
68. Butovsky O, Siddiqui S, Gabriely G, Lanser AJ, Dake B, Murugaiyan G, et al. Modulating inflammatory monocytes with a unique microRNA gene signature ameliorates murine ALS. *J Clin Invest* 2012;**122**:3063–87.
69. Beers DR, Henkel JS, Zhao W, Wang J, Huang A, Wen S, et al. Endogenous regulatory T lymphocytes meliorate amyotrophic lateral sclerosis in mice and correlate with disease progression in patients with amyotrophic lateral sclerosis. *Brain* 2011;**134**:1293–314.
70. Henkel JS, Beers DR, Appel SH, et al. Regulatory T-lymphocytes mediate amyotrophic lateral sclerosis progression and survival. *EMBO Mol Med* 2013;**5**:64–79.
71. Turner MR, Bowser R, Bruijn L, Dupuis L, Ludolph A, Fischbeck KH, et al. Mechanisms, models and biomarkers in amyotrophic lateral sclerosis. *Amyotroph Lateral Scler* 2013;**14**:19–32.
72. Doble A. The pharmacology and mechanism of action of riluzole. *Neurology* 1996;**47**:S233–41.
73. Vucic S, Lin CS, Cheah BC, Murray J, Menon P, Kiernan MC, et al. Riluzole exerts central and peripheral modulating effects in amyotrophic lateral sclerosis. *Brain* 2013;**136**:1361–70.
74. Dorst J, Dupuis L, Petri S, Kollewe K, Abdulla S, Ludolph AC, et al. Percutaneous endoscopic gastrostomy in amyotrophic lateral sclerosis: a prospective observational study. *J Neurol* 2015;**262**:849–58.

75. Korner S, Hendricks M, Kollewe K, Zapf A, Dengler R, Petri S, et al. Weight loss, dysphagia and supplement intake in patients with amyotrophic lateral sclerosis (ALS): impact on quality of life and therapeutic options. *BMC Neurol* 2013;**13**:84.
76. Bourke SC, Tomlinson M, Williams TL, Bullock RE, Shaw PJ, Gibson GJ. Effects of non-invasive ventilation on survival and quality of life in patients with amyotrophic lateral sclerosis: a randomized controlled trial. *Lancet Neurol* 2006;**5**:140–7.
77. Gruis KL, Lechtzin N. Respiratory therapies for amyotrophic lateral sclerosis: a primer. *Muscle Nerve* 2012;**46**:313–31.

MOLECULAR AND CELLULAR THERAPIES FOR MOTOR NEURON DISEASES

3

Genetics of Amyotrophic Lateral Sclerosis

K.P. Kenna[1], J.E. Landers[1] and N. Ticozzi[2,3]

[1]University of Massachusetts Medical School, Worcester, MA, United States [2]IRCCS Istituto Auxologico Italiano, Milan, Italy [3]'Dino Ferrari' Center – Università degli Studi di Milano, Milan, Italy

Molecular and Cellular Therapies for Motor Neuron Diseases.
DOI: http://dx.doi.org/10.1016/B978-0-12-802257-3.00003-1

43

INTRODUCTION

Studies of disease concordance rates among monozygotic and dizygotic twins suggest that 53–84% of amyotrophic lateral sclerosis (ALS) population risk is genetically determined.[*][1] Ten percent of patients exhibit a readily identifiable family history of the disorder, and the first-degree relatives of ALS patients develop disease at ~10 times the rate seen in the general population.[2,3] The majority of ALS cases are believed to result from the combined effects of multiple genetic and nongenetic risk factors that individually confer only minor to modest increases of risk. Despite this, much of the current understanding of ALS etiopathogenesis comes from the study of very rare Mendelian subtypes, where disease appears primarily, if not exclusively, attributable to single gene defects that segregate with disease in families.

The presence or absence of a prior family history is used to divide ALS into familial (FALS) and sporadic (SALS) forms. Strict criteria for this division, in terms of minimal level of relatedness or consistency of phenotype, are not in place but the majority of observable FALS pedigrees would include only a small number of affected persons and are not usually associated with striking patterns of Mendelian segregation. In keeping with the elevated recurrence rate among relatives, mutations of major effect are believed to make a more significant contribution to FALS than to SALS burden. However, high-impact mutations are also observed within an important proportion of SALS cases. In these instances, the absence of a prior family history might reflect the small patient family size (providing limited opportunity for the observation of affected relatives), variant penetrance,[†] and associated age of disease onset.[4] Alternatively, patients might develop disease due to the presence of de novo mutations, which refers to mutations of patient DNA that are absent from the germline of both parents. Examples of ALS-associated de novo mutations have been reported,[5] but are quite rare so far. In general, variants of minor to modest effect, or variants that exert high-risk effects only in the context of compounding environmental or genetic risk factors, are thought to make a greater contribution to the SALS burden than high-impact variants. Owing to their lower independent risk effects, such variants need not necessarily be very rare among the general population, and best estimates put the proportion of ALS genetic risk explained by common genetic variation as 12%.[6]

---

[*]Refers to the proportion of disease risk that is genetically determined both within and across all individuals of a population, not the proportion of cases for which genetic risk factors are involved.

[†]Proportion of individuals carrying a mutation that present with the associated clinical phenotype.

AMYOTROPHIC LATERAL SCLEROSIS GENES

Since the discovery of the first ALS gene in 1993, over 100 disease loci have been proposed to influence ALS susceptibility or clinical phenotype (http://alsod.iop.kcl.ac.uk). The pathogenic relevance of many of these is well supported (Table 3.1), but whether and how most contribute to the ALS burden is far from certain. The difficulty in establishing clinical relevance is multifactorial, but stems primarily from the relatively low frequency of disease, the heterogeneity of causative factors, and the fact that every human genome contains a considerable number of potentially disease-related genetic variants. These issues can also complicate the interpretation of mutations observed at well-established disease genes, where factors such as gene size can mean the probability of observing entirely incidental patient variants is not negligible.[7] Another issue concerning multiple ALS genes is that of pleiotropy, which refers to the association of one gene with multiple phenotypes. In the case of ALS, reported genes have also been associated with frontotemporal dementia (FTD), motor neuropathy, spastic paraplegia, progressive bulbar palsy, glaucoma, spinal muscular atrophy, spinocerebellar ataxia, oculomotor apraxia, and schizophrenia.[8] In certain cases, even individual mutations can associate with multiple seemingly distinct clinical presentations. A prime example of this is a point mutation within the gene *valosin-containing protein (VCP)* which, even within a single family, associates with variable combinations of ALS, FTD, Paget's disease of bone, and inclusion body myopathy.[9] Contrary to this, certain ALS mutations associate with very specific clinical profiles, such as the P525L mutation of fused in sarcoma (FUS) that is consistently observed in the context of aggressive early onset disease,[5] and much of the clinical heterogeneity seen in the disease may reflect variation in causative as well as modifying factors.

The importance of individual ALS genes varies considerably according the ancestral background. Cumulatively, mutations within genes of major effect are identifiable in ~11% of patients of European ancestry (38–67% of FALS; 5–15% of SALS).[10-12] Several common variants associated with ALS susceptibility have also been identified; however, interpretation of these variants is complicated by factors such as linkage disequilibrium‡ and the net contributions of identified common variant associations is not entirely clear.

‡Refers to the nonrandom association of two or more distinct genetic variants.

TABLE 3.1 ALS genes and loci. Main genetic causes of familial ALS

ALS-type	Onset	Inheritance	Locus	Gene	Protein
ALS1	**Adult**	**AD (AR)**	**21q22.1**	**SOD1**	**Cu/Zn superoxide dismutase**
ALS2	Juvenile	AR	2q33–35	ALS2	Alsin
ALS3	Adult	AD	18q21	Unknown	–
ALS4	Juvenile	AD	9q34	SETX	Senataxin
ALS5	Juvenile	AR	15q15–21	SPG11	Spatacsin
ALS6	**Adult**	**AD (AR)**	**16p11.2**	**FUS**	**Fused in sarcoma**
ALS7	Adult	AD	20p13	Unknown	–
ALS8	Adult	AD	20q13.33	VAPB	VAMP-associated protein B
ALS9	Adult	AD	14q11	ANG	Angiogenin
ALS(FTD)10	**Adult**	**AD**	**1q36**	**TARDBP**	**TAR DNA-binding protein**
ALS11	Adult	AD	6q21	FIG4	PI(3,5)P(2)5-phosphatase

ALS12	Adult	AR (AD)	10p15–p14	*OPTN*	Optineurin
ALS13	Adult	Susceptibility?	12q24.12	*ATXN2*	Ataxin-2
ALS(FTD)14	Adult	AD	9p13.3	*VCP*	Valosin-containing protein
ALS(FTD)15	Adult	AD	Xp11.23–p11.1	*UBQLN2*	Ubiquilin-2
ALS(FTD)16	Adult	AD	9p13.3	*SIGMAR1*	Sigma nonopioid intracellular receptor 1
ALS(FTD)17	Adult	AD	3p11.2	*CHMP2B*	Charged Multivesicular Body Protein 2B
ALS–FTD	**Adult**	**AD**	**9p21.2**	***C9orf72***	**C9ORF72**
ALS18	Adult	AD	17p13.2	*PFN1*	profilin 1
ALS19	Adult	AD	2q34	*ERBB4*	Receptor Tyr-kinase erbB-4
ALS(FTD)20	Adult	AD	12q13.13	*hnRNPA1*	Het. nuclear ribonucleoprotein A1
ALS21	Adult	AD	5q31.2	*MATR3*	Matrin 3
ALS(FTD)22	Adult	AD	2q35	*TUBA4A*	Tubulin alpha-4A

PROTEIN AGGREGATION: SUPEROXIDE DISMUTASE 1

Superoxide dismutase 1 (SOD1) was the first gene identified to be mutated in a Mendelian type of ALS in 1993.[13] The gene spans 9.3 kb on chr21q22.1, is composed of 5 exons, and encodes for the 153-amino-acid-long protein Cu/Zn superoxide dismutase, a cytoplasmic enzyme responsible for the catabolism of superoxide radicals to hydrogen peroxide and molecular oxygen. The protein is constituted of two identical monomers, each one consisting of an eight-stranded beta-barrel, and binding a copper and a zinc ion.[14] SOD1 is ubiquitously expressed, highly conserved, and represents ~1% of all cytoplasmic proteins.

To date, more than 170 different *SOD1* mutations have been reported (http://alsod.iop.kcl.ac.uk), the vast majority of which are missense substitutions equally distributed throughout the gene. Several indels, mainly clustered in the C-terminal region and leading to a premature truncation of the mature protein have also been described. Although the mutational frequency varies among different ethnicities, *SOD1* accounts for ~20% of FALS, and 2–3% of SALS patients.[15] The clinical phenotype is characterized by a considerable interfamilial and intrafamilial variability with regards to the age at onset, site of onset, and disease duration.[16] Conclusive genotype–phenotype correlations are possible only for the most frequent *SOD1* mutations. For instance, A4V is consistently associated with a high-penetrant, early-onset lower motor neuron disease with very rapid course.[16] Conversely, H46R and G93D display a very mild phenotype, with carriers often surviving more than 20 years after disease onset.[17,18] It must be noted, however, that the majority of the *SOD1* variants described so far are private mutations, for which no genotype–phenotype correlation can be drawn. Moreover, the pathogenic role of some of these variants has recently started to be questioned. For instance, Felbecker et al. have described four families in which the E100K and D90A mutations do not segregate with the disease.[19] These findings must be taken into serious consideration with regards to genetic testing and counseling in clinical practice. All *SOD1* mutations are inherited as dominant traits, with the exception of the D90A variant, observed both in recessive pedigrees in Scandinavia, and in dominant pedigrees in the rest of the world.[20,21] While dominant families develop classic ALS, recessive individuals have a milder phenotype characterized by upper motor neuron signs and prolonged survival, indicating the possible existence of a protective genetic factor in linkage with D90A in Scandinavian populations.[22]

The pathogenic effects of *SOD1* mutations are most likely not secondary to the abolition of the physiological functions of the wild type protein. In fact, several mutants retain full catalytic activity, and is there no correlation between residual enzymatic activity and/or protein stability with the disease phenotype.[23] The observation that *SOD1* knockout mice do not develop motor neuron disease,[24] while transgenic animals over-expressing the human mutant *SOD1* gene do,[25] also argues against a loss-of-function hypothesis. Conversely, it is believed that SOD1 mutations result into the acquisition of a novel function that is toxic to motor neurons. Several studies showed that mutant SOD1 is prone to misfolding and forms cytoplasmic aggregates. In turn, aggregates may lead to cell death by sequestering other cytoplasmic proteins essential for neuronal survival, by clogging the ubiquitin/proteasome system, by chaperone depletion, or by disrupting mitochondria, the cytoskeleton, and/or axonal transport.[26,27] Interestingly, there is evidence that posttranslational modifications may induce misfolding and increase aggregation propensities also of wild-type SOD1, thus suggesting a possible pathogenic role also in SALS.[28]

DYSFUNCTION OF mRNA METABOLISM: TAR DNA-BINDING PROTEIN AND FUSED IN SARCOMA

A major step into the understanding of ALS pathogenesis has been the discovery of the protein TAR DNA-binding protein 43 (TDP-43) as the main component of ubiquitinated cytoplasmic inclusions in ALS and in frontotemporal lobe dementia with ubiquitin inclusions (FTLD-U).[29] TDP-43 is a 43-kDa, 414 amino-acid-long multifunctional DNA/RNA binding protein encoded by the *TAR DNA-binding protein* (*TARDBP*) gene, and belonging to the heterogeneous ribonucleoprotein (hnRNP) family. After this breakthrough discovery, mutations in the *TARDBP* gene have been found to be a major cause of Mendelian ALS in several populations of different geographic origins.[30–32] Subsequently, disease-causing mutations were identified also in the FUS gene, which encodes for another RNA binding protein,[33,34] thus suggesting that an alteration of mRNA homeostasis may represent a key event in ALS pathogenesis. Mutations in *TARDBP* and *FUS* account for ~5% of FALS and 1% of SALS cases each.[35]

TDP-43 and FUS share many structural, functional, genetic, and neuropathological similarities. Both proteins have been demonstrated to play a role in several biological processes, including transcriptional regulation, splicing, nucleo-cytoplasmic shuttling, transport and stabilization of mRNAs.[36,37] In the central nervous system, both proteins are involved in mRNA transport toward the dendrites and in regulating synaptic plasticity.[38,39]

Given their biologic role, TDP-43 and FUS contain similar protein domains. TDP-43 contains two highly conserved RNA recognition motifs, flanked by an N-terminal domain and a C-terminal tail. The latter element contains a glycine-rich region that is reputed to mediate protein–protein interactions, mainly with others hnRNPs.[40] FUS is composed of an N-terminal transactivating domain, a central domain that contains a RNA recognition and a zinc finger motif, and a C-terminal region rich in arginine and glycine.[41] Both proteins have nuclear localization (NLS) and nuclear exporting (NES) signals, allowing for nucleo-cytoplasmic shuttling.

The vast majority of *TARDBP* mutations are clustered in the C-terminal glycine-rich region of the protein, while pathogenic *FUS* variants are located within or disrupt the NLS.[42] The observation of cytoplasmic inclusion immunoreactive for TDP-43 or FUS in carriers indicates that mutations decrease their solubility, thus increasing aggregation propensity[43]. Moreover, while in unaffected neurons both proteins localize in the cell nucleus, they are absent from the nuclei of inclusion-bearing cells, suggesting a nucleo-cytoplasmic redistribution[44]. These observations lead to intense speculation on the pathogenic role of TDP-43 and FUS in ALS: toxicity might be caused by the aggregating protein being sequestered away from its normal nuclear function (loss-of-function hypothesis) or, conversely, insoluble aggregates might have a toxic gain-of-function independent of the protein's physiological cellular activities.[30,42] These findings collectively suggest that *TARDBP*- and *FUS*-mediated toxicity in ALS and FTD may occur through common cellular pathways, thus underscoring the need for a comprehensive identification of overlaps and differences between the mRNA target binding profiles of the two.[45,46] Notwithstanding these similarities, it must be noted that TDP-43 positive inclusions are observed not only in mutation carriers, but also in the vast majority of SALS and in ~55% of FTLD-U cases. Conversely, FUS aggregates are observed exclusively in ALS patients harboring mutations and in a minority of nonmutated FTD cases displaying atypical neuropathological phenotypes.[47–49]

Significant differences between the two genes also exist at the phenotypic level. *TARDBP*-mutated patients usually develop a motor neuron disease indistinguishable from classic ALS with regards to age at onset, disease duration and distribution of upper and lower motor neuron signs (http://alsod.iop.kcl.ac.uk). The onset of the disease is often spinal, with a preferential involvement of the upper limbs. Since most *TARDBP* mutations are private, it is difficult to establish clear genotype–phenotype correlation. It has been suggested that A382T, which is the variant most commonly observed, may be associated with a low penetrant, predominantly lower motor neuron disease with an asymmetrical onset in the distal muscles of the limbs, subsequently spreading to proximal muscles, with relative sparing of the bulbar muscles.[47] This variant has also been observed in patients with atypical phenotypes including parkinsonism and/or ataxia.[50,51] Conversely, *FUS* mutations are consistently associated with a severe type of motor neuron disease characterized by onset in the third or fourth decade, very rapid disease course, and predominant lower motor neuron signs.[12] Mutations affecting the NLS, in particular R521C, may result in an uncommon phenotype characterized by a symmetrical proximal spinal onset, with early involvement of the axial muscles and head drop.[52]

DYSFUNCTION OF mRNA AND PROTEIN HOMEOSTASIS: C9ORF72

A major breakthrough in our understanding of ALS pathogenesis came with the identification of an expanded $(G_4C_2)_n$ hexanucleotide repeat in the promoter/first intron of the *C9orf72* gene as the major genetic cause of ALS and FTLD-U.[53,54] The repeat is highly polymorphic in the normal population (2–23 units), but is expanded both in ALS and FTD patients up to 4400 units.[55] In the inbred Finnish population, the mutational frequency of *C9orf72* was reported to be as high as 46% in FALS and 21% in SALS.[54] In other populations of European descent the mutational frequencies ranged were 23–47% in FALS, 4–5% in SALS, 12–29% in FTD, and 6–86% in ALS–FTD patients.[56]

The motor phenotype of *C9orf72*+ patients is often characterized by an earlier age at onset and shorter survival time compared to nonmutated individuals, possibly due to a prominent bulbar involvement in a majority of cases.[57–59] Similarly to *TARDBP*, and unlike *SOD1*, the upper limbs are more frequently affected at onset compared to the lower limbs. Patients with concurrent ALS and FTD or with a family history positive for one of both diseases have a significantly higher risk of harboring $(G_4C_2)_n$ repeats (33–86%), further indicating that the two disorders belong to the same pathogenic continuum.[57]

The cognitive deficit of *C9orf72*+ patients is usually consistent with a diagnosis of behavioral variant of FTD (bvFTD), and characterized by socially inappropriate, impulsive behavior and general deterioration in ability to perform routine daily tasks, or apathy, social isolation, and emotional lability. Patients often display prominent psychiatric features such as visual hallucinations, paranoid behavior, persecutory delusions, aggressive behavior, and/or suicidal thoughts.[57–61] Occasionally, patients carrying the mutation may display concurrent extrapyramidal and/or cerebellar signs, and in rare instances $(G_4C_2)_n$ repeat expansions have been identified in individuals with corticobasal syndrome, progressive supranuclear palsy, cerebellar ataxia, or olivopontocerebellar degeneration, suggesting that *C9orf72* may contribute to the pathogenesis of a broad spectrum of neurodegenerative diseases beyond ALS and FTD.[53,54,59,62]

Although the physiological functions of the C9orf72 protein are currently poorly understood, several possible mechanisms have been proposed to explain the $(G_4C_2)_n$-mediated toxicity. In silico and in vitro models suggest that the protein is involved in regulating endosomal trafficking, and that a reduction of mRNA levels due to the presence of the expansion may lead to impaired autophagy and endocytosis (loss-of-function hypothesis).[63] Increasing evidence, however, indicates that $(G_4C_2)_n$ repeats may cause ALS and FTD through the acquisition of a novel toxic function (gain-of-function hypothesis). C9orf72 is bidirectionally transcribed into long pre-mRNA containing sense $(G_4C_2)_n$ and antisense repeats $(C_4G_2)_n$, with a propensity to form stable DNA and RNA G-quadruplex and RNA/DNA hybrid conformations. These abnormal structures, in turn, have a high affinity for ribonucleoproteins which are sequestered and coprecipitated into nuclear RNA foci in a fashion similar to myotonic dystrophy.[64] Alternatively, sense and antisense repeats may undergo an abnormal repeat-associated non-ATG translation (RAN translation), thus generating C9RAN proteins composed of repeated GA, GP, GR, PA, and PR dipeptides which in turn form insoluble neuronal inclusions.[65-67]

Similar to SALS, the neuropathological pattern of *C9orf72*+ individuals is characterized by the presence of widespread TDP-43-immunoreactive inclusions in the brain and spinal cord. In addition to that, however, TDP-43-negative, p62/C9RAN-positive inclusions are consistently observed in the cerebellar cortex, hippocampus, and cerebellar cortex.[68] Thus, although C9RAN inclusions are highly specific for the nervous tissues, their concentrations appear to be inversely proportional to the burden of TDP-43 inclusions and the extent of regional neurodegeneration.[69] Conversely, RNA foci are more represented in brain areas most affected by neurodegenerative processes, but are also present to a various degree in all peripheral tissues.[70] As such, it is still undetermined whether C9orf72 causes ALS and FTD through RNA-mediated toxicity, formation of protein aggregates or both.

BEYOND C9ORF72: DYSFUNCTION OF PROTEIN DEGRADATION SYSTEMS AND THE AMYOTROPHIC LATERAL SCLEROSIS–FRONTOTEMPORAL DEMENTIA ASSOCIATED GENES

The discovery of *C9orf72* gene as the main cause of ALS and FTD definitively consolidated the hypothesis that the two diseases belong to the same clinicopathological spectrum. Even before that, however, similar neuropathological features and a shared genetic background had been described in both diseases. TDP-43-immunoreactive ubiquitinated inclusions are in fact present in >50% of all FTD cases (FTLD-U),[29] while an *FUS* pathology similar to mutated ALS cases is also observed in atypical FTD, basophilic inclusion body dementia, and neuronal intermediate filament inclusion disease.[47,71,72] Although the role of *TARDBP* and *FUS* mutations in causing FTD is currently disputed, several other genes are consistently associated with both diseases. For instance, mutations in the ALS-causing genes *ubiquilin-2* (*UBQLN2*) and *optineurin* (*OPTN*) may be associated with phenotypes characterized by the concurrence of motor signs associated with executive and behavioral dysfunction.[59,73,74] Conversely, other genes initially identified in FTD pedigrees have subsequently been associated with ALS, with or without dementia. *Charged multivesicular body protein 2B (CHMP2B)* mutations have been described in bvFTD patients showing an insidious change in personality and behavior, memory loss, apathy, aggressiveness, stereotyped behavior, disinhibition, dysgraphia, and dyscalculia, as well as in ALS cases.[75,76] Mutations in the *VCP, p62/sequestosome 1 (SQSTM1)* and *hnRNPA1* genes are associated with a complex disease spectrum including ALS, FTD, inclusion body myopathy, and Paget's disease of the bone.[9,77,78]

From a biological standpoint, it is worth noticing that the majority of genes associated with both ALS and FTD are involved in regulating several pathways responsible for maintaining protein homeostasis within motor neurons, such as protein quality control (*VCP*), activation of ER stress (vesicle-associated membrane protein (*VAPB*)), and targeting of misfolded proteins toward the ubiquitin-proteasome system (*UBQLN2, p62/SQSTM1*) and/or autophagy (*OPTN, CHMP2B*). As such, it can be hypothesized that a dysfunction of neuronal protein degradation systems plays a key role in TDP-43 proteinopathies.[46]

DISRUPTION OF CYTOSKELETON: PFN1 AND TUBA4A

Despite having a cell body only 50 μm in diameter, the axonal projections of motor neurons can exceed 1 m in length.[79] This unique feature makes cytoskeletal integrity and axonal transport two key areas where motor neurons might be selectively vulnerable to toxic insult. In keeping with this, early neuropathological studies revealed the aberrant accumulation of neurofilaments, a class of cytoskeletal protein, to be a hallmark of the disease in patients.[80] Studies of ALS-linked mutant *SOD1* overexpression systems have also shown that adverse effects on axonal transport may at least partially explain the toxicity of certain patient variants.[81] Moreover, multiple genes with direct functions in cytoskeletal structure, dynamics or transport have been determined to act as ALS susceptibility loci.

The first cytoskeletal gene associated with ALS susceptibility was *neurofilament heavy polypeptide (NEFH)*[82] The gene encodes an intermediate filament protein, and tail domain deletions were reported as candidate

ALS mutations in SALS and later FALS. Years later another intermediate filament gene, *peripherin 1 (PRPH1)*, was also associated with ALS,[83] though as with *NEFH*, the clinical relevance of identified variants remains incompletely established. More recently, *tubulin alpha-4a (TUBA4A)*, which encodes the 4A isoform of the microtubule protein alpha tubulin, was discovered to play a causal role in FALS.[47] This discovery was made through a large exome-wide rare variant burden analysis involving 635 FALS cases and almost 10,000 non-ALS controls. Identified *TUBA4A* variants were found to introduce deficits in microtubule repolymerization capacity and adversely effect global microtubule network integrity. At least seven other members of the tubulin gene family have also been implicated in neurodevelopmental or neurodegenerative disease phenotypes. The reasons why *TUBA4A* variation might contribute to ALS as opposed an earlier onset phenotype associated with other tubulin genes are not known, but could at least partially reflect that notable *TUBA4A* expression only commences relatively late in development.

In addition to genes encoding *bona fide* components of the cytoskeleton itself, multiple genes associated with cytoskeletal dynamics and function have also been implicated in ALS. Positional cloning identified a G59S variant of *dynactin1 (DCTN1)*,[84] the gene encoding the largest subcomponent of the dynactin complex, as causal for a slowly progressive lower motor neuron phenotype. The dynactin complex plays an important role in retrograde transport, binding both microtubules and dynein motors and the G59S mutation was found to notably reduce microtubule affinity. Moreover, transgenic overexpression of the G59S mutant in mice precipitated a striking motor phenotype that was not observed upon overexpression of wild-type *DCNT1*.[85] Additional variants of *DCTN1* have subsequently been identified in SALS and FALS cases but the clinical relevance of these remains uncertain. Other cytoskeleton-associated genes associated with ALS-like disorders include *kinesin family member 5A (KIF5A)* associated with hereditary spastic paraplegia, *microtubule associated protein tau (MAPT)* associated with FTD and an ALS–FTD–parkinsonism complex, and *spastin (SPAST)* associated with hereditary spastic paraplegia and potentially ALS. However, perhaps the only gene of this class associated with a more typical ALS presentation is *profilin1 (PFN1)*. This gene encodes a protein with multiple functions, one of which includes facilitating the conversion of monomeric G-actin to filamentous F-actin. ALS-associated mutations were identified through exome-wide segregation analysis of two large families.[59] Five additional ALS families harboring mutation carriers were subsequently identified and mutations have also been reported in independent replication studies. The molecular mechanisms of *PFN1* variant pathogenicity are still under investigation, but initial work has shown that certain mutations reduce actin binding affinity, inhibit neurite outgrowth and disrupt the size and morphology of axon growth cones.[59]

FUTURE PERSPECTIVES AND CONCLUSIONS

Mapping the remainder of the unidentified ALS genetic risk factors is an area of ongoing and active research. Due to the probable relevance of very rare and very lowly penetrant mutations, this task is expected to prove extremely challenging but a deeper understanding of genetic risk factors could have many advantages. It has already been shown that certain ALS mutations associate with distinctive clinical profiles, raising the potential utility of mutation screening in diagnostic and prognostic evaluation as well as in risk prediction and tailoring personalized strategies for patient care. Several ALS variants have also already been used to develop cell and animal models of the disease. Such models have many important limitations but provide a platform to characterize disease-related biological pathways and benchmark the potential efficacy of therapeutic strategies. Despite this, a complete understanding of ALS may require much greater knowledge of the environmental components of the disease. Several studies have reported the potential relevance of specific neurotoxins[86] and physical exercise,[87] but very little has been firmly established. It also remains to be seen how much of what can be garnered from studies of rare familial forms of the disorder can be generalized and the extent to which ALS represents a spectrum of related but distinct clinical disorders.

References

1. Al-Chalabi A, Fang F, Hanby MF, et al. An estimate of amyotrophic lateral sclerosis heritability using twin data. *J Neurol Neurosurg Psychiatry* 2010;**81**:1324–6.
2. Hanby MF, Scott KM, Scotton W, et al. The risk to relatives of patients with sporadic amyotrophic lateral sclerosis. *Brain* 2011;**134**:3454–7.
3. Fang F, Kamel F, Lichtenstein P, et al. Familial aggregation of amyotrophic lateral sclerosis. *Ann Neurol* 2009;**66**:94–9.
4. Al-Chalabi A, Lewis CM. Modelling the effects of penetrance and family size on rates of sporadic and familial disease. *Hum Hered* 2011;**71**:281–8.
5. Conte A, Lattante S, Zollino M, et al. P525L FUS mutation is consistently associated with a severe form of juvenile amyotrophic lateral sclerosis. *Neuromuscul Disord* 2012;**22**:73–5.
6. Fogh I, Ratti A, Gellera C, et al. A genome-wide association meta-analysis identifies a novel locus at 17q11.2 associated with sporadic amyotrophic lateral sclerosis. *Hum Mol Genet* 2014;**23**(8):2220–31.
7. Kenna KP, McLaughlin RL, Hardiman O, Bradley DG. Using reference databases of genetic variation to evaluate the potential pathogenicity of candidate disease variants. *Hum Mutat* 2013;**34**:836–41.
8. Al-Chalabi A, Jones A, Troakes C, King A, Al-Sarraj S, van den Berg LH. The genetics and neuropathology of amyotrophic lateral sclerosis. *Acta Neuropathol* 2012;**124**:339–52.
9. Johnson JO, Mandrioli J, Benatar M, et al. Exome sequencing reveals VCP mutations as a cause of familial ALS. *Neuron* 2010;**68**:857–64.
10. Kenna KP, McLaughlin RL, Byrne S, et al. Delineating the genetic heterogeneity of ALS using targeted high-throughput sequencing. *J Med Genet* 2013;**50**:776–83.

11. Chio A, Calvo A, Mazzini L, et al. Extensive genetics of ALS: a population-based study in Italy. *Neurology* 2012;**79**:1983–9.
12. Lattante S, Conte A, Zollino M, et al. Contribution of major amyotrophic lateral sclerosis genes to the etiology of sporadic disease. *Neurology* 2012;**79**:66–72.
13. Rosen DR, Siddique T, Patterson D, et al. Mutations in Cu/Zn superoxide dismutase gene are associated with familial amyotrophic lateral sclerosis. *Nature* 1993;**362**:59–62.
14. Getzoff ED, Tainer JA, Stempien MM, Bell GI, Hallewell RA. Evolution of CuZn superoxide dismutase and the Greek key beta-barrel structural motif. *Proteins* 1989;**5**:322–36.
15. Andersen PM, Sims KB, Xin WW, et al. Sixteen novel mutations in the Cu/Zn superoxide dismutase gene in amyotrophic lateral sclerosis: a decade of discoveries, defects and disputes. *Amyotroph Lateral Scler Other Motor Neuron Disord* 2003;**4**:62–73.
16. Cudkowicz ME, McKenna-Yasek D, Sapp PE, et al. Epidemiology of mutations in superoxide dismutase in amyotrophic lateral sclerosis. *Ann Neurol* 1997;**41**:210–21.
17. Aoki M, Ogasawara M, Matsubara Y, et al. Mild ALS in Japan associated with novel SOD mutation. *Nat Genet* 1993;**5**:323–4.
18. Luigetti M, Conte A, Madia F, et al. A new single-nucleotide deletion of PMP22 in an HNPP family without recurrent palsies. *Muscle Nerve* 2008;**38**:1060–4.
19. Felbecker A, Camu W, Valdmanis PN, et al. Four familial ALS pedigrees discordant for two SOD1 mutations: are all SOD1 mutations pathogenic? *J Neurol Neurosurg Psychiatry* 2010;**81**:572–7.
20. Andersen PM, Forsgren L, Binzer M, et al. Autosomal recessive adult-onset amyotrophic lateral sclerosis associated with homozygosity for Asp90Ala CuZn-superoxide dismutase mutation. A clinical and genealogical study of 36 patients. *Brain* 1996;**119**(Pt 4):1153–72.
21. Robberecht W, Aguirre T, Van den Bosch L, Tilkin P, Cassiman JJ, Matthijs G. D90A heterozygosity in the SOD1 gene is associated with familial and apparently sporadic amyotrophic lateral sclerosis. *Neurology* 1996;**47**:1336–9.
22. Al-Chalabi A, Andersen PM, Chioza B, et al. Recessive amyotrophic lateral sclerosis families with the D90A SOD1 mutation share a common founder: evidence for a linked protective factor. *Hum Mol Genet* 1998;**7**:2045–50.
23. Radunovic A, Shaw CE, Akman-Demir G, Idrisoglu H, Leigh PN. CuZnSOD-associated amyotrophic lateral sclerosis. *Ann Neurol* 1997;**42**:273–4.
24. Reaume AG, Elliott JL, Hoffman EK, et al. Motor neurons in Cu/Zn superoxide dismutase-deficient mice develop normally but exhibit enhanced cell death after axonal injury. *Nat Genet* 1996;**13**:43–7.
25. Gurney ME, Pu H, Chiu AY, et al. Motor neuron degeneration in mice that express a human Cu,Zn superoxide dismutase mutation. *Science* 1994;**264**:1772–5.
26. Pasinelli P, Brown RH. Molecular biology of amyotrophic lateral sclerosis: insights from genetics. *Nat Rev Neurosci* 2006;**7**:710–23.
27. Boillee S, Vande Velde C, Cleveland DW. ALS: a disease of motor neurons and their nonneuronal neighbors. *Neuron* 2006;**52**:39–59.
28. Bosco DA, Morfini G, Karabacak NM, et al. Wild-type and mutant SOD1 share an aberrant conformation and a common pathogenic pathway in ALS. *Nat Neurosci* 2010;**13**:1396–403.
29. Neumann M, Sampathu DM, Kwong LK, et al. Ubiquitinated TDP-43 in frontotemporal lobar degeneration and amyotrophic lateral sclerosis. *Science* 2006;**314**:130–3.
30. Fournier C, Bedlack B, Hardiman O, et al. ALS Untangled No. 20: the Deanna protocol. *Amyotroph Lateral Scler Frontotemporal Degener* 2013;**14**:319–23.
31. Sreedharan J, Blair IP, Tripathi VB, et al. TDP-43 mutations in familial and sporadic amyotrophic lateral sclerosis. *Science* 2008;**319**:1668–72.
32. Van Deerlin VM, Leverenz JB, Bekris LM, et al. TARDBP mutations in amyotrophic lateral sclerosis with TDP-43 neuropathology: a genetic and histopathological analysis. *Lancet Neurol* 2008;**7**:409–16.

33. Kwiatkowski Jr. TJ, Bosco DA, Leclerc AL, et al. Mutations in the FUS/TLS gene on chromosome 16 cause familial amyotrophic lateral sclerosis. *Science* 2009;**323**:1205–8.
34. Vance C, Rogelj B, Hortobagyi T, et al. Mutations in FUS, an RNA processing protein, cause familial amyotrophic lateral sclerosis type 6. *Science* 2009;**323**:1208–11.
35. van Es MA, Schelhaas HJ, van Vught PW, et al. Angiogenin variants in Parkinson disease and amyotrophic lateral sclerosis. *Ann Neurol* 2011;**70**:964–73.
36. Law WJ, Cann KL, Hicks GG. TLS, EWS and TAF15: a model for transcriptional integration of gene expression. *Brief Funct Genomic Proteomic* 2006;**5**:8–14.
37. Buratti E, Baralle FE. Multiple roles of TDP-43 in gene expression, splicing regulation, and human disease. *Front Biosci* 2008;**13**:867–78.
38. Fujii R, Okabe S, Urushido T, et al. The RNA binding protein TLS is translocated to dendritic spines by mGluR5 activation and regulates spine morphology. *Curr Biol* 2005;**15**:587–93.
39. Wang IF, Wu LS, Chang HY, Shen CK. TDP-43, the signature protein of FTLD-U, is a neuronal activity-responsive factor. *J Neurochem* 2008;**105**:797–806.
40. Buratti E, Brindisi A, Giombi M, Tisminetzky S, Ayala YM, Baralle FE. TDP-43 binds heterogeneous nuclear ribonucleoprotein A/B through its C-terminal tail: an important region for the inhibition of cystic fibrosis transmembrane conductance regulator exon 9 splicing. *J Biol Chem* 2005;**280**:37572–84.
41. Morohoshi F, Ootsuka Y, Arai K, et al. Genomic structure of the human RBP56/hTAFII68 and FUS/TLS genes. *Gene* 1998;**221**:191–8.
42. Lagier-Tourenne C, Polymenidou M, Cleveland DW. TDP-43 and FUS/TLS: emerging roles in RNA processing and neurodegeneration. *Hum Mol Genet* 2010;**19**:R46–64.
43. Blokhuis AM, Groen EJ, Koppers M, van den Berg LH, Pasterkamp RJ. Protein aggregation in amyotrophic lateral sclerosis. *Acta Neuropathol* 2013;**125**:777–94.
44. Lagier-Tourenne C, Cleveland DW. Rethinking ALS: the FUS about TDP-43. *Cell* 2009;**136**:1001–4.
45. Lagier-Tourenne C, Polymenidou M, Hutt KR, et al. Divergent roles of ALS-linked proteins FUS/TLS and TDP-43 intersect in processing long pre-mRNAs. *Nat Neurosci* 2012;**15**:1488–97.
46. Ling SC, Polymenidou M, Cleveland DW. Converging mechanisms in ALS and FTD: disrupted RNA and protein homeostasis. *Neuron* 2013;**79**:416–38.
47. Smith BN, Ticozzi N, Fallini C, et al. Exome-wide rare variant analysis identifies TUBA4A mutations associated with familial ALS. *Neuron* 2014;**84**:324–31.
48. Neumann M, Rademakers R, Roeber S, Baker M, Kretzschmar HA, Mackenzie IR. A new subtype of frontotemporal lobar degeneration with FUS pathology. *Brain* 2009;**132**:2922–31.
49. Seelaar H, Klijnsma KY, de Koning I, et al. Frequency of ubiquitin and FUS-positive, TDP-43-negative frontotemporal lobar degeneration. *J Neurol* 2010;**257**:747–53.
50. Cannas A, Borghero G, Floris GL, et al. The p.A382T TARDBP gene mutation in Sardinian patients affected by Parkinson's disease and other degenerative parkinsonisms. *Neurogenetics* 2013;**14**:161–6.
51. Ticozzi N, Tiloca C, Mencacci NE, et al. Oligoclonal bands in the cerebrospinal fluid of amyotrophic lateral sclerosis patients with disease-associated mutations. *J Neurol* 2013;**260**:85–92.
52. Ticozzi N, Silani V, LeClerc AL, et al. Analysis of FUS gene mutation in familial amyotrophic lateral sclerosis within an Italian cohort. *Neurology* 2009;**73**:1180–5.
53. DeJesus-Hernandez M, Mackenzie IR, Boeve BF, et al. Expanded GGGGCC hexanucleotide repeat in noncoding region of C9ORF72 causes chromosome 9p-linked FTD and ALS. *Neuron* 2011;**72**:245–56.
54. Renton AE, Majounie E, Waite A, et al. A hexanucleotide repeat expansion in C9ORF72 is the cause of chromosome 9p21-linked ALS-FTD. *Neuron* 2011;**72**:257–68.

55. Beck J, Poulter M, Hensman D, et al. Large C9orf72 hexanucleotide repeat expansions are seen in multiple neurodegenerative syndromes and are more frequent than expected in the UK population. *Am J Hum Genet* 2013;**92**:345–53.
56. Majounie E, Abramzon Y, Renton AE, et al. Repeat expansion in C9ORF72 in Alzheimer's disease. *N Engl J Med* 2012;**366**:283–4.
57. Byrne S, Elamin M, Bede P, et al. Cognitive and clinical characteristics of patients with amyotrophic lateral sclerosis carrying a C9orf72 repeat expansion: a population-based cohort study. *Lancet Neurol* 2012;**11**:232–40.
58. Gijselinck I, Van Langenhove T, van der Zee J, et al. A C9orf72 promoter repeat expansion in a Flanders-Belgian cohort with disorders of the frontotemporal lobar degeneration-amyotrophic lateral sclerosis spectrum: a gene identification study. *Lancet Neurol* 2012;**11**:54–65.
59. Wu CH, Fallini C, Ticozzi N, et al. Mutations in the profilin 1 gene cause familial amyotrophic lateral sclerosis. *Nature* 2012;**488**:499–503.
60. Snowden JS, Rollinson S, Thompson JC, et al. Distinct clinical and pathological characteristics of frontotemporal dementia associated with C9ORF72 mutations. *Brain* 2012;**135**:693–708.
61. Simon-Sanchez J, Dopper EG, Cohn-Hokke PE, et al. The clinical and pathological phenotype of C9ORF72 hexanucleotide repeat expansions. *Brain* 2012;**135**:723–35.
62. Lindquist SG, Duno M, Batbayli M, et al. Corticobasal and ataxia syndromes widen the spectrum of C9ORF72 hexanucleotide expansion disease. *Clin Genet* 2013;**83**:279–83.
63. Farg MA, Sundaramoorthy V, Sultana JM, et al. C9ORF72, implicated in amytrophic lateral sclerosis and frontotemporal dementia, regulates endosomal trafficking. *Hum Mol Genet* 2014;**23**:3579–95.
64. Haeusler AR, Donnelly CJ, Periz G, et al. C9orf72 nucleotide repeat structures initiate molecular cascades of disease. *Nature* 2014;**507**:195–200.
65. Ash PE, Bieniek KF, Gendron TF, et al. Unconventional translation of C9ORF72 GGGGCC expansion generates insoluble polypeptides specific to c9FTD/ALS. *Neuron* 2013;**77**:639–46.
66. Mori K, Weng SM, Arzberger T, et al. The C9orf72 GGGGCC repeat is translated into aggregating dipeptide-repeat proteins in FTLD/ALS. *Science* 2013;**339**:1335–8.
67. Sareen D, O'Rourke JG, Meera P, et al. Targeting RNA foci in iPSC-derived motor neurons from ALS patients with a C9ORF72 repeat expansion. *Sci Transl Med* 2013;**5** 208ra149.
68. Al-Sarraj S, King A, Troakes C, et al. p62 positive, TDP-43 negative, neuronal cytoplasmic and intranuclear inclusions in the cerebellum and hippocampus define the pathology of C9orf72-linked FTLD and MND/ALS. *Acta Neuropathol* 2011;**122**:691–702.
69. Mackenzie IR, Arzberger T, Kremmer E, et al. Dipeptide repeat protein pathology in C9ORF72 mutation cases: clinico-pathological correlations. *Acta Neuropathol* 2013;**126**:859–79.
70. Lagier-Tourenne C, Baughn M, Rigo F, et al. Targeted degradation of sense and antisense C9orf72 RNA foci as therapy for ALS and frontotemporal degeneration. *Proc Natl Acad Sci U S A* 2013;**110**:E4530–9.
71. Neumann M, Rademakers R, Roeber S, Baker M, Kretzschmar HA, Mackenzie IR. Frontotemporal lobar degeneration with FUS pathology. *Brain* 2009;**132**(Pt 11):2922–31.
72. Seelaar H, Klijnsma KY, de Koning I Frequency of ubiquitin and FUS-positive, TDP-43-negative frontotemporal lobar degeneration. *J Neurol* 2010;**257**:747–753.
73. Deng HX, Chen W, Hong ST, et al. Mutations in UBQLN2 cause dominant X-linked juvenile and adult-onset alS and ALS/dementia. *Nature* 2011;**477**:211–5.
74. Del Bo R, Tiloca C, Pensato V, et al. Novel optineurin mutations in patients with familial and sporadic amyotrophic lateral sclerosis. *J Neurol Neurosurg Psychiatry* 2011;**82**:1239–43.

75. Skibinski G, Parkinson NJ, Brown JM, et al. Mutations in the endosomal ESCRTIII-complex subunit CHMP2B in frontotemporal dementia. *Nat Genet* 2005;**37**:806–8.
76. Parkinson N, Ince PG, Smith MO, et al. ALS phenotypes with mutations in CHMP2B (charged multivesicular body protein 2B). *Neurology* 2006;**67**:1074–7.
77. Watts GD, Wymer J, Kovach MJ, et al. Inclusion body myopathy associated with Paget disease of bone and frontotemporal dementia is caused by mutant valosin-containing protein. *Nat Genet* 2004;**36**:377–81.
78. Kim HJ, Kim NC, Wang YD, et al. Mutations in prion-like domains in hnRNPA2B1 and hnRNPA1 cause multisystem proteinopathy and ALS. *Nature* 2013;**495**:467–73.
79. Shaw PJ, Eggett CJ. Molecular factors underlying selective vulnerability of motor neurons to neurodegeneration in amyotrophic lateral sclerosis. *J Neurol* 2000;**247**(Suppl. 1): I17–27.
80. Delisle MB, Carpenter S. Neurofibrillary axonal swellings and amyotrophic lateral sclerosis. *J Neurol Sci* 1984;**63**:241–50.
81. Williamson TL, Cleveland DW. Slowing of axonal transport is a very early event in the toxicity of ALS-linked SOD1 mutants to motor neurons. *Nat Neurosci* 1999;**2**:50–6.
82. Al-Chalabi A, Andersen PM, Nilsson P, et al. Deletions of the heavy neurofilament subunit tail in amyotrophic lateral sclerosis. *Hum Mol Genet* 1999;**8**:157–64.
83. Gros-Louis F, Lariviere R, Gowing G, et al. A frameshift deletion in peripherin gene associated with amyotrophic lateral sclerosis. *J Biol Chem* 2004;**279**:45951–6.
84. Puls I, Jonnakuty C, LaMonte BH, et al. Mutant dynactin in motor neuron disease. *Nat Genet* 2003;**33**:455–6.
85. Laird FM, Farah MH, Ackerley S, et al. Motor neuron disease occurring in a mutant dynactin mouse model is characterized by defects in vesicular trafficking. *J Neurosci Off J Soc Neurosci* 2008;**28**:1997–2005.
86. Spencer PS, Nunn PB, Hugon J, et al. Guam amyotrophic lateral sclerosis-parkinsonism-dementia linked to a plant excitant neurotoxin. *Science* 1987;**237**:517–22.
87. Huisman MH, Seelen M, de Jong SW, et al. Lifetime physical activity and the risk of amyotrophic lateral sclerosis. *J Neurol Neurosurg Psychiatry* 2013;**84**:976–81.

Molecular Mechanisms of Amyotrophic Lateral Sclerosis

M. Collins and R. Bowser

St Joseph's Hospital and Medical Center, Phoenix, AZ, United States

INTRODUCTION

Amyotrophic lateral sclerosis (ALS) is recognized as the most common form of adult-onset motor neuron disease. This progressive, fatal neuro-degenerative disorder occurs in approximately two persons per 100,000.[1-5] Since the initial description of the symptoms and associated pathology in 1874, considerable insights into the genetic, molecular, and biochemical mechanisms of ALS have been gained. The pathological hallmark of ALS is the death of pyramidal motor neurons of the corticospinal pathway in the motor cortex and spinal column. This leads to a myriad of clinical symptoms, such as muscle weakness, muscle atrophy, and spasticity. Considerable variability in site of onset, rate of progression, and survival time occurs in ALS patients and underscores the overall heterogeneity of the disease, which can result from multiple etiologies. This suggests that ALS may represent a collection of disorders that produce similar pathological and clinical phenotypes. The purpose of this chapter is to discuss various mechanisms of ALS, understand how these mechanisms contribute to selective motor neuron vulnerability/degeneration, discuss mechanistic insights gained from ALS model systems, and develop a systematic view of how these mechanisms converge to produce the disease phenotype.

Broadly, ALS can be separated into two categories based on etiology. The vast majority of cases are classified as sporadic ALS (SALS) and are of unknown cause. Approximately 5–10% of all ALS cases are the result of inherited genetic abnormalities and are thus classified as familial ALS (FALS). Although FALS cases are a small fraction of the overall ALS population and are further stratified on the basis of the underlying genetic mutation, considerable insight into disease mechanisms have been gained by studying these rare forms of ALS. As we will emphasize throughout this review, FALS and FALS model systems have proven highly valuable for the study of both FALS and SALS. These monogenic disease-causing variants provide an approach to understand how a triggering event (in this case a genetic abnormality) can produce a cascade of molecular events that ultimately lead to motor neuron degeneration. Our approach is thus to discuss individual mechanisms associated with ALS, combining evidence from research on SALS, FALS, and ALS model systems. This overview will serve as groundwork for building a systematic framework that examines how the interplay of various mechanisms lead to ALS and opportunities these mechanisms present for therapeutic intervention.

RNA METABOLISM

Evidence for RNA Binding Proteins in the Disease Process

Since the initial discovery of the TAR DNA-binding protein (TDP-43) as a major component of neuronal cytoplasmic inclusions in ALS[6] and subsequent identification of genetic alterations in the *TARDBP* gene that cause familial forms of ALS and frontotemporal dementia (FTD),[7] the number of RNA/DNA-binding proteins associated with ALS has expanded considerably (Table 4.1). TDP-43-positive neuronal inclusions are a pathologic hallmark of both ALS and FTD. Mutations in *TARDBP* account for approximately 4% of familial cases and a small number of apparently SALS cases.[18]

Shortly after the discovery of *TARDBP* mutations that cause ALS, missense mutations in the *fused in sarcoma (FUS)* gene were identified as the cause of chromosome-16p-linked FALS.[8,9] Mutations in *FUS* also account for ~4% of FALS cases. The observed protein amino acid domain homology between TDP-43 and FUS, with both proteins containing multiple RNA binding motifs, suggested that RNA metabolism may play an important role in ALS. Other key structural elements include the presence of glycine-rich domains and prion-like domains that contribute to their pathological aggregation and impaired function in ALS. This latter element was used to predict other RNA binding proteins associated with ALS. This led to the discovery of disease-causing mutations in *TAF15*, *hnRNPA1*, and *hnRNPA2B1*.[12,16] Key functional properties linking these RNA binding proteins include association with stress granules (SGs) and nucleocytoplasmic translocation during cellular stress.

Genetic studies have identified other RNA binding proteins linked to familial or sporadic forms of ALS. These include disease-causing mutations in *SETX*, *ANG*, and *ELP3*,[10,11,15] repeat expansion of *ATXN2* associated with increased risk of ALS,[13] and others such as *RBM45* that are linked to ALS due to pathologic inclusions of the protein that occur in patients.[17]

TABLE 4.1   RNA/DNA-Binding Proteins Linked to Amyotrophic Lateral Sclerosis

Gene	Locus	Protein	References
TARDBP	1p36.22	TDP-43	7
FUS	16p11.2	Fus	8,9
SETX	9q34.13	Senataxin	10
ANG	14q11.1	Angiogenin	11
TAF15	17q11.1–q11.2	TAF15	12
ATXN2	12q23–q24.1	Ataxin-2	13
EWSR1	22q12.2	EWSR1	14
ELP3	8p21.1	ELP3	15
hnRNPA1	12q13.1	hnRNPA1	16
hnRNPA2B1	7p15.2	hnRNPA2B1	16
RBM45	2p31.2	RBM45	17

ALTERED RNA SPLICING, TRANSPORT, AND TRANSLATION

While most of the genetic alterations of RNA binding proteins impact their subcellular distribution and accumulation into SGs and/or pathologic inclusions (see below), we will focus here on specific effects of disease-causing mutations in TDP-43 and FUS on RNA metabolism (Fig. 4.1). RNA binding proteins have diverse roles in the cell and function within many nuclear substructures and the cytoplasm. At present, the vast majority of evidence for impaired RNA processing in ALS has come from studies of TDP-43 and FUS. However, many other RNA binding proteins linked to ALS interact with TDP-43 and/or FUS, and therefore likely impact RNA metabolism. Since both TDP-43 and FUS bind RNA/DNA (Fig. 4.1), determining the specific binding sequences and effects on gene expression were crucial to understanding how these proteins contribute to cell death in ALS. Using CLIP-SEQ, TDP-43 was shown to bind over 6000 RNA targets in the brain, approximately 30% of the transcriptome.[19–22] TDP-43 binding to long introns (>100 kb) is required for the normal maturation and splice site selection of immature mRNA species.[20,22] TDP-43 binding to the 3′-UTR of mRNAs may impact on stability or transport, whereas binding to long noncoding RNAs (ncRNAs) may influence their regulatory roles. Splicing of many RNA targets of TDP-43 is altered in ALS spinal cord tissue.[23]

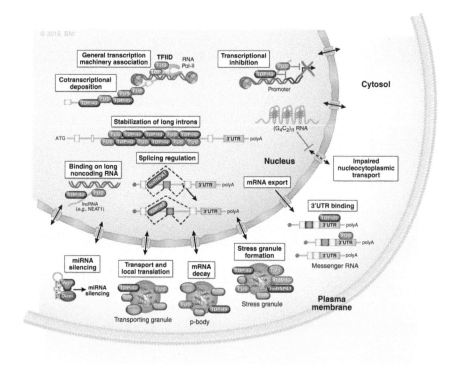

FIGURE 4.1   RNA processing functions of amyotrophic lateral sclerosis (ALS)-associated RNA binding proteins (RBPs). A schematic diagram of the RNA-related functions of ALS-linked RBPs is shown. These RBPs affect the transcription, processing, export, maturation, transport, and turnover of RNA molecules. In addition to their effects on normal transcription/translation, ALS RBPs participate in stress responses via stress granules and modulate miRNAs. The c9ORF72-linked hexanucleotide repeat expansion (G_4C_2) may impair many of these processes via expansion-linked nucleocytoplasmic transport defects.

Likewise, FUS binds over 5500 RNA targets in the brain,[19] and its binding pattern to long introns suggests that FUS remains bound to pre-mRNAs until splicing is complete. Loss of FUS function results in changes of the splicing pattern or abundance in over 1000 RNAs.[19] All three members of the FET gene family (FUS, EWSR1, and TAF-15) are implicated in ALS, suggesting that the functions of these proteins on global RNA splicing and metabolism are particularly important for motor neurons. In addition, both TDP-43 and FUS bind to long ncRNAs (lncRNA) and influence their function and subcellular localization (Fig. 4.1). Both proteins, e.g., bind NEAT1,[24] a lncRNA core component of nuclear paraspeckles, which function in cell stress responses and in the nuclear retention of hyperedited RNAs.[25] FUS directly regulates NEAT1 levels and decreasing FUS levels leads to reduced numbers of paraspeckles.[26] At the same time, FUS-positive inclusions contain other paraspeckle proteins, suggesting that pathological changes in FUS levels or function impair normal paraspeckle formation/function, thereby altering cellular homeostatic responses and increasing motor neuron vulnerability to degeneration.

Finally, both TDP-43 and FUS exhibit neuron-specific functions that further implicate them in neurodegenerative diseases. Both localize to dendrites in response to neuronal activity. Disease-causing mutations that result in functional impairments and protein mis-localization therefore likely impact on synaptic structure and function via loss of localized translation of specific mRNAs.[27,28] Given the remarkable functional diversity of RNA binding proteins such as TDP-43 and FUS, it is unsurprising that mutations that affect their structure and function confer numerous aberrant changes in the regulation of gene expression. Motor neurons, in particular, seem acutely vulnerable to alterations in the levels of ALS-linked RNA binding proteins by either loss of function, gain of toxic function, or both. While much insight into the sequence targets, processing functions, and subcellular associations of ALS-associated RNA binding proteins have been gained in recent years, determining which of these properties contribute directly to motor neuron vulnerability/degeneration remains an unanswered question with considerable therapeutic implications.

ADENOSINE DEAMINASE ACTING ON RNA 2

Further evidence for motor neuron-specific defects in RNA processing have come from studies of RNA editing of the GluA2 subunit of the l-α-amino-3-hydroxy-5-methyl-4-isoxazolepropionic acid (AMPA) receptor. Adenosine deaminase acting on RNA 2 (ADAR2)-mediated conversion of adenosine to inosine (A-to-I editing) of the GluA2 pre-mRNA results in replacement of glutamine with arginine in the translated protein. This editing normally occurs in all motor neurons and results in expression of Ca^{2+}-impermeable AMPA receptors. ADAR2 levels, and consequently A-to-I editing, are dramatically reduced in SALS motor neurons. The resultant enhancement of AMPA Ca^{2+} permeability leads to increased motor neuron vulnerability and the development of TDP-43 pathology.[29]

MICRORNAS

Over the past decade, microRNAs (miRNAs) have been identified as significantly impacting overall gene expression by modulating the stability and/or translational repression of target mRNAs.[30] miRNAs are short noncoding RNAs (~22 nucleotides) and over 1000 have been identified in humans, constituting a large class of regulators of gene expression.

Approximately 60% of all mammalian mRNAs are predicted targets of miRNAs.[31] miRNA binding to mRNA targets results in reduced protein expression from the specific bound mRNA. A single miRNA may have many mRNA transcript targets, therefore impacting on the expression of a large number of genes. Various types of cellular stress impact on the levels of miRNAs and therefore regulate how a cell responds to stress.[32] As we further describe below, stress responses during ALS represent an important pathogenic mechanism and therefore miRNAs likely contribute to the overall ability of a cell to properly respond to acute or chronic stress conditions that exist during ALS.

Recent studies have examined miRNA changes in ALS patients and model systems, including the G93A SOD1 transgenic mouse model, circulating monocytes, skeletal muscle, and lumbar spinal cord tissue from ALS patients, and serum from FALS patients and premanifest carriers of genetic mutations.[31–36] Altered levels of miRNAs may have significant impact on gene expression during ALS (Fig. 4.2). Perhaps due to differences in models, cell/tissue types, and analytical methods used in these studies, a common set of miRNA alterations was not observed. However, two studies detected increased levels of miR-146a and miR-155 in microglial cells.[33,34] In addition, increases in miR-146a were detected in two studies using either peripheral monocytes or spinal cord tissue from ALS patients.[33,35] The results support further studies on miRNA changes in ALS using standardized approaches and model systems, ideally in large collaborative research efforts. Pathway analysis of genes modulated by the miRNAs altered in ALS will provide important information regarding mechanistic pathways regulated by miRNAs during ALS and possibly very early changes in these pathways. In addition, miRNAs may correlate with the rate of ALS disease progression, as miR-206 levels have been associated with the rate of disease progression in the G93A SOD1 mouse model.[36] Finally, both TDP-43 and FUS (and likely other RNA binding proteins implicated in ALS) impact miRNA biogenesis,[37,38] linking ALS disease-causing mutations to miRNA-regulated gene expression.

PROTEIN AGGREGATION AND TOXICITY

The abnormal aggregation of proteins into inclusion bodies in motor neurons is a well-known pathological feature of both SALS[39] and FALS.[40] Early characterizations of inclusions defined their filamentous, skein-like morphology, along with their eosinophilic core (and, hence, proteinaceous composition). Subsequent work revealed a variety of characteristic inclusion types in ALS motor neurons, including larger skein-like inclusions reactive for ubiquitin, smaller filamentous inclusions containing neurofilament proteins, dense spheroids with a Lewy body-like appearance

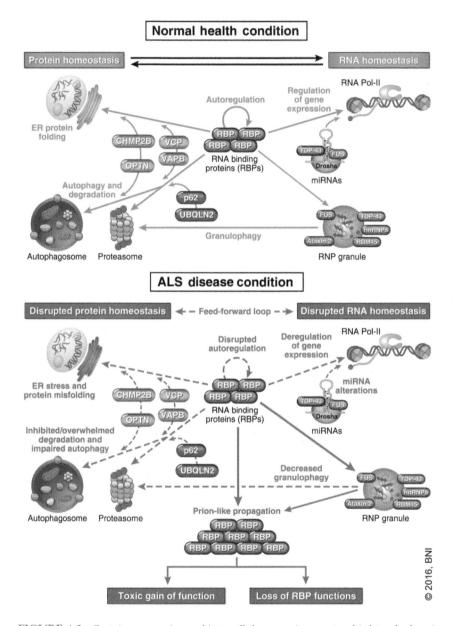

FIGURE 4.2   Protein aggregation and intracellular stress in amyotrophic lateral sclerosis (ALS). A schematic depicting protein/RNA homeostasis in normal healthy and ALS disease conditions are shown. Aggregation of RNA binding proteins (RBPs) in ALS impairs the normal gene expression regulatory effects of these proteins while also impairing protein clearance/degradation pathways. Familial ALS-linked mutations occur in many genes associated with ER protein folding, processing, and sorting, as well as autophagy and proteasome function. Prion-like aggregation of RBPs can also induce toxic gain of function and may contribute to disease propagation.

(compact inclusions), and Bunina bodies, small granular inclusions of lysosomal origin.[41] Since the identification of inclusions as a pathological hallmark of ALS, considerable effort has been devoted to determining the protein constituents and neurotoxic mechanisms.

SUPEROXIDE DISMUTASE 1

The discovery that point mutations in the *superoxide dismutase 1* (*SOD1*) gene can produce familial forms of ALS[42] (approximately 20% of all FALS cases[43]) led to the identification of mutant, misfolded SOD1 protein within inclusions. Subsequent work has shown that SOD1 pathology can also occur in SALS cases.[44,45] SOD1 catalyzes the conversion of the superoxide anion into hydrogen peroxide and molecular oxygen via the cyclical reduction and oxidation of copper.[46,47] Despite altered enzymatic activity in some FALS-linked SOD1 mutant proteins, existing evidence suggests that loss of SOD1 antioxidant function does not contribute to SOD1-linked FALS. Notably, disease progression and severity are not correlated with mutant SOD1 enzymatic activity in human patients[48,49] and mice lacking the SOD1 gene develop normally and do not develop motor deficits,[50] in contrast to mutant-SOD1-expressing mice that develop progressive motor abnormalities and SOD1 aggregates.[51,52] FALS-linked mutant SOD1 aggregates into insoluble amyloid-like fibrils in vitro,[53–55] in transgenic mutant-SOD1-expressing mice,[52,56] and in FALS patients.[45,57] The mutant SOD1 aggregation is driven by mutation-induced misfolding, which can result from the protein sequence itself, reduced capacity to bind metal ions, or both.[45–55]

While human ALS patients and transgenic G93A SOD1 mice develop insoluble, ubiquitinated SOD1 aggregates in motor neurons, current models implicate soluble, misfolded SOD1 as the toxic species, similar to models proposed in other aggregating proteins, such as amyloid beta. Soluble, misfolded SOD1 can form oligomeric pore structures and exerts many deleterious effects within the cell.[53,58] The protein is capable of inducing endoplasmic reticulum (ER) stress, which overwhelms the cell's capacity to provide normal clearance of cytoplasmic proteins and can ultimately lead to apoptotic cell death.[59–61] Mutant SOD1 also aberrantly accumulates in the mitochondrial intermembrane space, impairing normal mitochondrial function.[60–63] Where the mutation reduces metal ion binding, mutant SOD1 also disrupts calcium homeostasis, enhancing susceptibility to cellular stress.[64] More recent work suggests that cells also secrete misfolded SOD1, which can then seed the aggregation SOD1 (mutant or wild-type native) in adjacent cells via a prion-like process,[44,65] providing a possible mechanism for spread of disease. Therefore, mutant SOD1 may induce intracellular mechanisms of cell death and propagate spread of disease via cell–cell communication pathways.

TAR DNA-BINDING PROTEIN-43

Subsequent work demonstrated that many ubiquitinated, skein-like inclusions in SALS motor neurons were nonreactive to anti-SOD1 antibodies. The identity of the constituent protein(s) of these inclusions remained elusive until 2006, when TDP-43 was identified as the main component of cytoplasmic inclusions in ALS.[6] Subsequent studies identified mutations in the *TARDBP* gene that induce familial forms of either ALS or frontotemporal lobar dementia (FTLD),[7,66–68] each with extensive TDP-43 inclusion pathology. These disorders show a number of common TDP-43 modifications, the most prevalent of which is the incorporation of the protein into insoluble, ubiquitinated filamentous inclusions in neurons.[69–72] The protein is also frequently depleted from the cell nucleus, hyperphosphorylated, and cleaved into C-terminal fragments (CTFs), though this latter phenomenon occurs in a central nervous system (CNS) region-specific pattern.[69–72]

TDP-43 contains two RNA recognition motifs (RRMs), a nuclear localization sequence (NLS), a nuclear export signal, a glycine-rich C-terminal domain, and a prion-like domain, the latter two of which are aggregation prone and the location of the majority of ALS/FTLD-causing *TARDBP* mutations.[73,74] As described above, TDP-43 has many roles in the regulation of gene expression. The importance of this regulatory capacity is underscored by the observation that *TARDBP* null mice are embryonic lethal.[75,76] TDP-43 also undergoes altered nucleocytoplasmic transport in response to cellular stress.[77,78] The resultant association of the protein with cytoplasmic SGs, stress-induced protein–RNA complexes containing stalled translational components (Fig. 4.2), has emerged as a plausible mechanism for TDP-43 aggregate formation.[79]

Unlike SOD1, TDP-43 toxicity is mediated by both loss of function and toxic gain of function mechanisms.[71,73] Loss of the normal gene expression regulatory functions of TDP-43 may occur through hypofunction induced by mutation or posttranslational modification (e.g., phosphorylation or C-terminal cleavage). Similarly, trapping of the protein in cytoplasmic aggregates is expected to contribute to loss of function by depleting the protein from the nucleus and impairing both normal nuclear and cytoplasmic TDP-43 functions. The importance of maintaining homeostatic TDP-43 levels to cell viability is highlighted by studies demonstrating TDP-43's ability to self-regulate its expression in cells via an auto-regulatory feedback loop[20] and the observation of neuronal and behavioral abnormalities in a variety of model systems over- or underexpressing TDP-43 (wild-type or mutant), including mice,[20,75–81] Drosophila,[82–84] and zebrafish.[85,86]

In addition, misfolded, aberrantly modified TDP-43 displays toxic properties unrelated to its normal functions that are primarily related to the cytoplasmic accumulation and processing of the protein. Increased cytoplasmic accumulation is an early event in human TDP-43 proteinopathies and is thought to precede inclusion formation.[87] When recapitulated in cell culture and animal models, the phenomenon induces TDP-43 aggregation/misfolding, proteasomal and autophagic activation, and is toxic to cultured neurons.[88–90] Cellular stress together with aberrant cytoplasmic localization of TDP-43 promotes cleavage of the protein into CTFs by the activation of caspases.[91] The precise function(s) of the fragments remains elusive; however, CTFs are hyperphosphorylated and toxic to cultured cells.[92] Likewise, the effects of TDP-43 phosphorylation are not entirely clear, though these modifications are strongly associated with misfolded, insoluble forms of the protein.[93] Misfolded, cytoplasmic TDP-43 also appears capable of seeding further misfolding of endogenous TDP-43 in a prion-like manner, as observed for SOD1.[94]

FUSED IN SARCOMA

FUS/translocated in sarcoma is a related RNA binding protein also mutated in rare instances of FALS (approximately 4% of all FALS cases) and found in skein-like inclusions in FALS and SALS motor neurons.[8,9,95,96] Structurally, the protein contains an RRM, an NLS, a nuclear export sequence, two arginine/glycine-rich regions, a zinc finger domain, a prion-like domain, and a glycine-rich domain. These latter two features confer a propensity for aggregation to the protein.[97] Like TDP-43, it has numerous roles in the regulation of gene expression.[95,96] FUS binding to DNA regulates transcription/splicing and the protein is recruited to sites of DNA damage, where it contributes to break repair.[95] FUS also regulates transcription by direct binding to DNA. FUS also binds long intronic regions of pre-mRNAs[98] and contributes to splicing and the protein also regulates the subcellular localization of RNA molecules.[95] FUS, like TDP-43, is capable of nucleocytoplasmic shuttling by virtue of its structure and, when localized to the cytoplasm, associates with SGs.[99] As it does for TDP-43, this association provides a plausible mechanism for FUS inclusion formation during ALS.

Both loss of function and toxic gain of function mechanisms are thought to contribute to FUS-mediated neurodegeneration.[95] The trapping of FUS in inclusions is expected to impair the normal functions of FUS and neurons appear particularly sensitive to mutations that alter FUS levels or confine the protein to the cytoplasm. Like TDP-43, several ALS-causing *FUS* mutations are found in the glycine-rich region of the protein, while, in contrast to TDP-43, numerous additional disease-causing mutations have been found in the NLS.[95,96] This underscores the vital role of intranuclear FUS functions to cell viability and the toxic properties of excessive cytoplasmic FUS. Cytoplasmic-confined FUS forms inclusions and these are positive for ubiquitin and p62.[100–103] Intriguingly, these same inclusions are negative for TDP-43, implying separate mechanisms of inclusion formation, despite the association of both proteins with SGs. FUS inclusions are nonamyloid, though the protein can form fibrils in vitro.[97] Insoluble FUS can be found in human ALS tissue and FUS inclusions are frequently ubiquitinated.[103] Collectively, these processes suggest that FUS induced neurodegeneration includes altered protein distribution to the cytoplasm, misfolding, and inclusion formation in both FUS-linked FALS and some SALS patients.

OTHER AGGREGATING PROTEINS

In recent years, gene mutations in a number of other aggregation prone proteins have been found in FALS. These include mutations in the UBQLN2, OPTN, SQSTM1, VCP, HNRNPA1, HNRNPA2B1, and PFN1 genes.[41] Despite diverse functions of each protein, aggregates of the mutant protein are consistently found in FALS cases. Moreover, recent work has shown that non-ATG translation of the FALS-associated C9ORF72 repeat expansion leads to the expression of dipeptide repeat proteins (DPRs) that are aggregated in FALS tissue and model systems. These findings underscore the preponderant role of protein aggregation in motor neuron degeneration, especially when considered together with the observation that similar aggregates of these proteins are frequently found in SALS cases.[41] Moreover, some FALS genes encode proteins associated with the ubiquitin proteasome system, demonstrating the indispensable role of protein quality control and degradation in maintaining neuronal viability. While the protein composition of aggregates in SALS and FALS varies considerably, proteasomal stress and impairment of normal cellular trafficking of biomolecules are expected to occur widely in the various forms the disorder takes and, thus, these processes represent promising therapeutic targets for future drug development and clinical trials.

C9ORF72

The most common genetic cause of ALS and FTLD is a GGGGCC repeat expansion in the C9ORF72 gene.[104,105] The C9ORF72 repeat expansion is believed to underlie approximately 40% of FALS and 6–7% of SALS cases.[106] Patients with C9ORF72 repeat expansions that develop disease display variable amounts of TDP-43 pathology and abundant TDP-43 negative, p62-positive neuronal and glial inclusions in multiple regions of the CNS, including the cerebellum.[107] At present, the function(s) of the c9ORF72 protein is not fully characterized; however, the protein's high degree of homology with the DENN protein, a GDP/GTP exchange factor that acts on Rab-GTPases,[108] and reported ability to regulate endosomal trafficking[109] suggest that the protein has multiple functions.

Proposed neurotoxic mechanisms of the repeat expansion include haploinsufficiency and a toxic gain of function. Studies of the former have produced conflicting results: conditional knockout of the c9ORF72 gene does not result in the development of motor neuron degeneration or an overt motor/behavioral phenotype in mice;[110] however, knockdown of the zebrafish ortholog (zc9ORF72) led to motor deficits and axonal pathology.[111] Differences in model system, genetic ablation strategy, and repeat expansion size in these studies could account for these discrepancies and underscore the complexity of C9-mediated neurodegeneration. Experimental evidence for a repeat expansion induced toxic gain of function have implicated both the accumulation of RNA foci within the nucleus and cytoplasmic translation of dipeptide proteins via repeat-associated non-AUG translation (RAN) translation products and resultant accumulation of DPR cytoplasmic inclusions.[112] The presence of sense and antisense foci is a common observation in both patient tissues/cell lines and C9 model systems,[113] and C9 transcripts are capable of forming G-quadruplex structures that may aberrantly influence gene expression by sequestering RNA binding proteins.[114] As noted above, c9ORF72 FALS patients exhibit DPR cytoplasmic inclusions resulting from non-ATG (RAN) translation of the repeat. DPRs are neurotoxic in a variety of ALS model systems[115] and coculture of astrocytes taken from C9 patients with wild-type motor neurons results in extensive neuronal death,[116] suggesting intracellular and noncell autonomous mechanisms of neurodegeneration in C9-linked ALS. Studies in ALS patients, however, have noted the presence of DPR-containing inclusions in regions of the CNS that do not degenerate in ALS,[117] suggesting that DPR inclusions are not directly neurotoxic.

Recent studies examining mechanisms of C9-mediated toxicity have consistently identified impairments in nucleocytoplasmic transport of proteins.[118,119] It remains unclear whether all proteins that enter/exit the nucleus are equally impacted by C9 repeat expansion induced transport defects. Therapeutic approaches that enhance nucleocytoplasmic transport or remedy defective transport-induced pathologies may thus be an effective treatment for C9-linked ALS. However, nucleocytoplasmic transport deficits may be present from birth, whereas ALS or FTLD typically occurs in late to middle age, raising the question of whether additional factors cause disease and C9 repeat expansion induced nucleocytoplasmic transport defects merely make the cell more susceptible to further insult. Antisense oligonucleotides (ASOs) and small molecules that target production of C9 RAN proteins have also been identified as therapeutics with clinical potential.[120,121] ASOs have been shown to ameliorate or prevent the development of C9-associated pathology and pathophysiology in some model systems. However, since C9 repeat expansion BAC transgenic mice harbor both nuclear RNA foci and C9 RAN proteins throughout life yet exhibit no phenotype, the role of both RNA foci and C9 RAN protein inclusions in ALS neurodegeneration remains uncertain. Moreover, while loss of C9ORF72 function does not induce motor neuron degeneration or impair CNS function, the necessity of the c9ORF72 protein for the proper functioning of other organ systems/cell types has not been evaluated. Thus, while knockdown-based approaches (such as ASOs) have shown early promise in model systems, much further research is needed to determine their safety/efficacy.

AXONOPATHY AND AXONAL TRANSPORT DEFECTS

Synaptic activity at the level of the axon is energetically demanding and requires a diverse array of biochemical reactions, cellular organelles, and molecules for function. Anterograde axonal transport facilitates these processes by transporting proteins, mRNAs, organelles, vesicles, and other signaling molecules to the axon terminal.[122] At the same time, the axon requires a means to retrogradely transport misfolded proteins, damaged organelles, vesicles, and signaling molecules from the axon to the cell soma. These functions are accomplished by axonal transport using protein molecular motor complexes in conjunction with the axonal cytoskeleton. Because motor neuron axons may be up to 1 m in length, they are particularly vulnerable to impairments in axonal transport and defects in cytoskeletal structure.

The pathological aggregation of cytoskeletal intermediate filament proteins is a common pathological observation in SALS and FALS. Spheroidal axonal aggregates, Lewy body-like inclusions, and hyaline conglomerate inclusions are frequently positive for peripherin and the neurofilament proteins in FALS and SALS cases.[123–125] These proteins are critical components of the neuronal cytoskeleton and modulate intracellular transport. Peripherin forms cytoskeleton-associated homopolymers that can also interact with other cytoskeletal proteins.[126] These cytoskeletal intermediate filament proteins are largely localized to the axon where they provide structural strength and regulate axonal diameter. The neurofilament cytoskeletal assembly is highly dynamic and the rate of transport of neurofilament proteins is inversely related to their phosphorylation state.[127] The abnormal accumulation and aggregation of these intermediate filament proteins occur in both the cell body and the axon of ALS motor neurons.[123] In the case of neurofilaments, the aggregated proteins are generally hyperphosphorylated, and hence, immobile, suggesting impaired cytoskeletal dynamics. Genetic evidence for motor neuron vulnerability to impaired cytoskeletal structure comes from the identification of rare mutations in the peripherin or neurofilament heavy chain genes in ALS patients,[128–131] and the observation that peripherin-overexpressing mice develop intermediate filament protein inclusions and progressive motor neuron loss.[132]

The remarkable length of motor neuron axons and the necessity of axonal transport for synaptic function implicates axonal transport as a susceptibility factor in motor neuron degeneration. A dying-back mechanism was initially suggested by the identification of decreased retrograde transport in motor neuron axons from ALS patients.[133] Impairments in axonal transport were further implicated as a susceptibility factor for motor neuron degeneration when mice overexpressing the human neurofilament heavy protein were found to develop selective, progressive motor neuron degeneration.[134] Mice overexpressing dynamitin showed deficits in retrograde axonal transport related to the disassembly of the dynactin complex and motor neuron degeneration.[135] Thus, impairment of axonal transport in either direction is sufficient to cause motor neuron degeneration.

Transgenic mice expressing G93A SOD1 also demonstrate that impaired axonal transport contribute to motor neuron degeneration. Impairments in anterograde axonal transport is one of the earliest pathologic events that occurs prior to symptom onset.[136] Slowed anterograde transport in these mice, particularly of tubulin and neurofilament proteins, leads to axonal swellings containing these proteins.[137] Other studies have suggested that impairments in fast axonal transport leads to depletion of axonal mitochondria, due to intact retrograde transport.[138] However, others have implicated reduced retrograde transport in transgenic SOD1 mice by virtue of the fact that mutant SOD1, but not wild-type protein, interacts directly with the dynein complex.[139] Intriguingly, expression of mutant SOD1 alters not simply the rate of retrograde transport, but the cargo as well. The dynein-associated fraction of axoplasm from mutant SOD1 transgenic mice shows a significant reduction of neurotrophic factors (including nerve growth factor and brain-derived neurotrophic factor) and significant increases in stress-associated signaling proteins (including caspase-8 and pJNK).[140] Despite apparent discrepancies, these studies all indicate that impaired axonal transport is an early, presymptomatic event in transgenic SOD1 mice.

Recent work has shown that TDP-43 also associates with mRNAs in the cytoplasm and is transported via slow axonal transport. FALS-linked mutant TDP-43 misfolds and reduces the net movement of mRNA granules and shifts the balance of axonal transport in the retrograde direction.[141] Thus, impaired axonal transport and cytoskeletal abnormalities at the level of the axon are susceptibility factors for motor neuron degeneration in various forms of ALS. The loss of normal cytoskeletal structure and trafficking of cargoes impairs neuronal function, while protein aggregates and enhanced stress signaling are associated with neuronal toxicity. Defects in axonal transport are consistently an early event observed in ALS model systems, suggesting it represents an early point for therapeutic intervention. Altered nucleocytoplasmic transport in patient with the C9 repeat expansion may disrupt axonal transport and lead to cognitive deficits in FTD due to axonopathy in cortical brain regions.

CELLULAR STRESS

Intracellular stress occurring in motor neurons is widely recognized as a molecular mechanism leading to ALS pathology and motor neuron death. Studies in human ALS tissues and model systems have documented the occurrence of multiple forms of intracellular stress occurring within motor neurons and the ways these stressors may contribute to cell death. Though it is likely that multiple cellular stresses contribute to cell death in concert, understanding the effects of a single cellular stressor is valuable in defining the time course of events that lead to cell death and in identifying potential therapeutic targets.

STRESS GRANULES

Cytoplasmic SGs are protein–RNA complexes that form in response to a variety of cellular stressors. The onset of cellular stress causes a rapid cessation of mRNA translation. The stalled translational complex rapidly disassembles and the components coalesce into SGs. The purpose of SGs appears to be two-fold: (1) to prioritize translation of mRNAs that favor the cellular response to stress, and (2) to remove other mRNAs and RNA binding proteins from the harmful cellular environment and stop their translation.[142,143] SGs are dynamic, transient structures that dissipate following the removal of the stressor and a constant turnover of protein and mRNA occurs during the life of the SG. Multiple lines of evidence now suggest SG formation and duration are vital determinants of cell fate decisions following stressor onset or in disease.[142–145]

The notion that this adaptive response has a role in ALS pathogenesis has roots in two discoveries: (1) that TDP-43 is a primary component of ubiquitinated inclusions in ALS/FTLD patients,[6] and (2) that TDP-43 translocates from the nucleus and associates with SGs during conditions of cellular stress.[78] Subsequent identification of FUS as a component of inclusions,[103] FUS mutations leading to FALS,[96] and its association with SGs[101] furthered the notion that SGs represent potential sites of cytoplasmic inclusion formation. Analysis of the domain structure of TDP-43 and FUS (and other SG-associated proteins) provided evidence of the structural elements mediating the association of these proteins with SGs, as well as the mechanisms by which these proteins may be irreversibly converted from a transient SG assembly into insoluble aggregates. TDP-43 incorporation into SGs requires RRM1 and the glycine-rich C-terminal component of the protein,[100] which is aggregation prone.[146] For FUS, the protein's zinc finger RNA binding domain is required and the degree of association is mediated by the protein's RRM and glycine-rich domain.[100] The necessity of intact RNA binding for incorporation of both proteins suggests that their association with SGs is part of the normal cellular response to stress and that loss of this function adversely affects motor neuron viability.

Though the aggregation of proteins into fibrils can serve beneficial functions in yeast in certain environments, the persistent association of TDP-43, FUS, and other RNA binding proteins in SGs in ALS is thought to result in the generation of toxic oligomers and fibrils and the loss of DNA/RNA binding functions by misfolding and removal of these proteins from their normal cellular milieu. Moreover, emerging evidence suggests that a host of RNA binding proteins may participate in this process. Analysis of prion-like domain harboring proteins in the human proteome shows a striking enrichment of DNA/RNA binding proteins, with as many as 20% of proteins with prion-like, low complexity domains having RNA binding function.[147] The identification of mutations in the genes encoding these proteins (e.g., *TARDP*, *HNRNPA1*, and others[148]) in FALS cases and the association of these proteins with SGs provides additional evidence in support of this concept. Thus, SGs contribute to ALS pathogenesis by both toxic loss of normal function and toxic gain of function and may thus be a point of therapeutic intervention.

AUTOPHAGY AND PROTEASOMAL STRESS

Autophagy is a degradative pathway by which intracellular material, including proteins and organelles, are enveloped and degraded via fusion with lysosomes. The function of autophagy and the observation of numerous types of proteinaceous inclusions in ALS suggests a role for autophagy in the disease process. Many studies consistently find autophagy defects in human ALS tissue and ALS model systems. While many signaling molecules mediate various steps in autophagy, among the more well-characterized are phosphatidylinositol 3-kinase signaling during autophagosome formation and p62/SQSTM1 tagging of proteins/structures targeted for degradation.[149,150] Autophagy is also crucial for the clearance of misfolded or aggregated proteins and is activated by the unfolded protein response (UPR). The protein targets of autophagy appear to be relatively long-lived proteins, which includes RNA binding proteins, in contrast with the ubiquitin/proteasome system, which is relatively specific for short-lived proteins.[151]

Autophagic clearance of proteins and organelles is particularly important for neuronal homeostasis and survival, as neurons have considerable metabolic needs and are acutely sensitive to the presence of misfolded or aggregated proteins. ALS patients show accumulation of p62-positive inclusions in motor neurons and glia,[41] suggesting insufficient clearance of misfolded and aggregated proteins. The identification of mutations in genes encoding autophagy-associated proteins as causative factors in subsets of FALS population strengthened the proposed role of autophagy in motor neuron homeostasis (Fig. 4.2). Mutations in *OPTN*, the gene encoding optineurin, a membrane trafficking protein and autophagy receptor for damaged mitochondria,[152] cause rare forms of FALS.[153] *OPTN*-linked FALS cases harbor optineurin-positive inclusions, which also contain TDP-43, p62, and ubiquitin. Optineurin-positive inclusions are relatively rare in SALS cases.[154]

Mutations in the *valosin containing protein (VCP)* gene, which encodes valosin-containing protein, are also found in a subset of FALS patients.[155] VCP protein is essential for autophagy and loss of VCP function (either via expression of a nonfunctional mutant protein or gene silencing) leads to the accumulation of unfused autophagosomes and neurodegeneration.[156,157] Intriguingly, *OPTN* and/or *VCP* mutations also lead to the accumulation of cytoplasmic TDP-43 inclusions, suggesting that autophagic hypofunction is sufficient to promote TDP-43 nuclear translocation, aggregation, and ultimately, cell death. Therefore, SALS and FALS are united by the common motor neuron susceptibility factor of impaired or overwhelmed autophagic function and TDP-43 aggregation. VCP protein is not incorporated into cytoplasmic inclusions in ALS patients or model systems, further suggesting that loss of autophagic function is the primary pathogenic mechanism of *VCP* mutations.[157,158] Finally, FALS-causing mutations in genes for the autophagy receptor protein p62[159] (*SQSTM1*), UPS and autophagy-associated protein ubiquilin 1[160] (*UBQLN1*), and the endosomal trafficking component charged multivesicular protein 2B[161] (*CHMP2B*) provide further evidence that impaired protein/organelle clearance and vesicular trafficking contribute to motor neuron death in ALS patients and model systems.

Activation of autophagy via rapamycin reduces TDP-43 aggregation and ameliorates motor deficits in both transgenic mice expressing TDP-43[162] and Drosophila overexpressing the TDP-43 ortholog dTDP,[163] suggesting that autophagic balance is a key determinant of motor neuron health and survival in ALS model systems. These results suggest that autophagy activation is a potential ALS therapeutic target. However, administration of rapamycin to SOD1-G93A mice accelerated motor neuron degeneration, suggesting that some caution is necessary in targeting autophagy as a therapy and that autophagic balance is a key determinant of motor neuron health.[164]

ENDOPLASMIC RETICULUM STRESS

The ER is an integral component in the synthesis, folding, modification, and transport of proteins through the secretory pathway. Perturbation of normal ER function and, particularly, the accumulation of un- or misfolded proteins causes ER stress. ER stress signals that the normal capacity of the ER to modify and fold proteins has been saturated, which triggers the UPR. Key components of the UPR include a general decrease in the rate of protein synthesis, an increase in the expression of ER-associated chaperone proteins, and activation of the endoplasmic reticulum-associated (ERAD) protein degradation pathway and ER-induced autophagy.[165,166] ERAD initiates translocation of misfolded protein into the cytoplasm for polyubiquitin tagging and proteasomal degradation.

ER function and stress are monitored by three ER membrane proteins with signaling functions: IRE1, PERK, and ATF6. The function of the latter two proteins is regulated by the ER chaperone BiP, which is normally bound to the luminal domains of PERK and ATF6.[165,166] The accumulation of misfolded proteins releases BiP and allows cytosolic translocation of active PERK and ATF6. PERK inhibits translation (reducing the folding load on the ER) by phosphorylating its target, eIF2α. ATF6 is proteolytically processed to a transcriptional activating form and translocates to the nucleus where it increases the transcription of ER-folding-associated genes. IRE1 is activated by the direct binding of misfolded proteins and also undergoes cytosolic translocation.[165,166] Once in the cytosol, its RNAse domain leads to removal of an intron in the XBP1 transcript, leading to production of the associated protein, which is a transcriptional activator of ER-fold-associated genes. The ER is also a major reservoir for intracellular Ca^{2+} and contains numerous calcium channels and sensors. When UPR and ERAD are insufficient to restore ER homeostasis, the above ER proteins and Ca^{2+} signaling can induce apoptosis.[166,167]

Accumulating evidence suggests that impaired UPR and excessive ER stress are contributing factors to motor neuron vulnerability in ALS. ER abnormalities are common in SALS motor neurons, with irregularities in ER structure, chromatolysis of ER membrane, and ribosomal detachment from ER membrane evident.[168] Spinal cord motor neurons of SALS patients show several signs of ER stress, including enhanced PERK-induced eIF2α phosphorylation and increased expression of ER chaperone proteins.[169] Further evidence that ER homeostasis and UPR are components of selective motor neuron vulnerability in ALS comes from longitudinal studies of vulnerable and resistant motor neurons in G93A SOD1 transgenic mice. Vulnerable motor neurons were consistently more susceptible to the onset of ER stress and showed upregulation of UPR markers compared to resistant motor neurons, despite comparable accumulation of ubiquitinated cytosolic proteins.[170] Cytoplasmic SOD1 inclusions in G93A SOD1 mice are also positive for ER-resident chaperone proteins, suggesting that misfolded SOD1 accumulates in the ER, activates the UPR, and may ultimately overwhelm its capacity, resulting in apoptotic cell death.[171] This, together with evidence that the ER stress-reducing drug salubrinal preserves neuromuscular function and increases survival in SOD1 transgenic mice, argues that the role of ER stress and UPR in disease is complex. Maintenance of ER homeostasis is dependent on multiple sensors and signaling pathways and the relative balance of these, together with other processes like proteasomal and autophagic protein degradation, likely determines whether UPR activation is beneficial or harmful to motor neurons in ALS.

INFLAMMATION AND GLIAL FUNCTION

A host of noncell autonomous mechanisms contribute to neurodegeneration in ALS.[172] Among the most well studied are those involving glial cells (astrocytes, microglia, and oligodendrocytes), and more recently T-cells and other peripheral cell types that modulate immunity and inflammation occurring in ALS (Fig. 4.3).

Neuroinflammation within the CNS is a common pathologic feature in many neurodegenerative diseases, including ALS (reviewed in[173]). Both animal models of ALS and post mortem tissue from ALS patients exhibit cellular signs of inflammation, namely activation of astrocytes and microglia within affected regions of the brain and spinal cord. Inflammation, therefore, has been a frequent therapeutic target for ALS. Many anti-inflammatory drugs have shown promise in the G93A SOD1 mouse model of disease, but all have failed human clinical trials. These include celecoxib, ceftriaxone, thalidomide, and minocycline.[174–176] Reasons may include inadequate powering of the clinical trial, suboptimal dosage, or administration to ALS patients after disease onset and therefore after closing of a therapeutic window. It is also possible that these therapies failed to reach the appropriate target, as pharmacodynamic biomarkers that demonstrate target engagement have only been recently introduced into ALS clinical trials. However, not all neuroinflammation during ALS may be detrimental, as early inflammatory responses can be neuroprotective. Transfer of wild-type microglia into a transgenic SOD1 mouse model of ALS slows disease progression,[177] suggesting that supportive glial responses that release neurotrophic factors and decrease neuronal

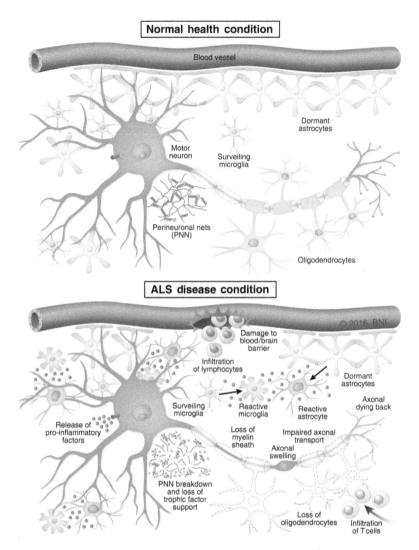

FIGURE 4.3 Noncell autonomous mechanisms of amyotrophic lateral sclerosis (ALS). Top: under healthy conditions in the parenchyma, astrocytes, and microglia surveil the local environment and aid in homeostatic maintenance. Lymphocytes and T cells are typically absent from the local environment and perineuronal nets provide structural and trophic support to motor neurons. Bottom: during ALS, activation of microglia and astrocytes by secretion of inflammatory factors from neuronal and nonneuronal cells exacerbates the degenerative motor neuron phenotype and continued gliosis. Oligodendrocyte hypofunction and degeneration, impaired axonal transport, and perineuronal net breakdown also contribute to motor neuron death. Lastly, infiltration of circulating T cells and lymphocytes may occur via breakdown of the blood–brain barrier or choroid plexus.

stress may be beneficial. Below is an overview of cell types that contribute to the genesis and modulation of inflammation during ALS.

ASTROCYTES

Astrocytes are the most abundant cell type in the CNS and perform a variety of important functions, including regulating levels of extracellular neurotransmitters, modulating metabolic or ionic homeostasis, providing structural and trophic support to neurons, having direct contact with the blood vasculature, and participating in immune responses.[178] Astrocytes also aid in maintaining neuronal viability by secreting growth and trophic factors, the secretion of which may be increased in response to neuronal injury or stress.[179–181] Astrocytes contact many cell types in addition to neurons, including other astrocytes and cells of the vascular system and regulate the functions of these cells as well (Fig. 4.3). Collectively, then, astrocytes act as an extensively connected support network that adapts its activities to the needs of adjacent neurons.

Several lines of evidence implicate astrocytes in the degeneration of motor neurons in ALS.[173,182,183] Dying motor neurons in the spinal cord and motor cortex of ALS patients are frequently surrounded by reactive astrocytes and astrocytes exhibit pathological hallmarks of ALS, including intracellular inclusion bodies. Astrocytes in ALS patients also show elevated expression of inflammatory markers, including nitric oxide synthase (NOS), tumor necrosis factor-α (TNFα), and nuclear factor-kappa-β, suggestive of an astrocytic response to motor neuron degeneration.[184–187] One of the main findings of studies in G93A SOD1 transgenic mice is that astrocytes secrete numerous factors that predispose motor neurons toward apoptotic cell death.[182] Astrocytic production of nitric oxide via increased expression of NOS leads to uptake of the molecule by motor neurons, where it impairs mitochondrial function and can lead to cell death.[184,188] Likewise, transgenic mutant SOD1 mice show enhanced expression of receptors for p75 and TNFα.[189–191] Astrocytic secretion of NGF can activate p75 receptors and trigger apoptotic cell death.[192,193] Astrocytes in ALS patients show markedly reduced levels of EAAT2 due to mRNA splicing abnormalities,[194,195] thus contributing to glutamate excitotoxicity.

An important question for resolving the role of astrocytes in ALS is whether reactive astrogliosis is a cause of degeneration, a consequence, or both. Conflicting reports from mutant SOD1 transgenic mouse lines have emerged, with reactive astrocytosis evident several weeks prior to symptom onset in G37R SOD1 transgenic mice,[196] but following motor neuron loss by several weeks in the more extensively studied G93A SOD1 mouse model.[197] Astrocytes that harbor SOD1 mutations are directly toxic to motor neurons both in vitro and in the G93A SOD1 mouse model of disease.[198] However, astrocytes expressing mutant TDP-43 are not toxic to cultured motor neurons, indicating that different genetic mutations induce different astrocyte-mediated mechanisms of motor neuron death. The contribution of astrocytes to neurodegeneration in ALS can therefore be viewed as two-fold. First, the development of pathology within astrocytes (e.g., proteinaceous inclusions) is expected to impair their normal function of supporting neuronal health. Motor neurons, with their considerable energy demands, are likely to be particularly susceptible to impairment of the astrocytic network that maintains a homeostatic environment in the CNS. Second, activation of astrocytes by the secretion of chemical factors and misfolded proteins from dying motor neurons results in profound activation of astrocytes with accompanying reactive astrogliosis. The creation of numerous iPSC lines from SALS, FALS, and healthy control subjects offers tremendous promise toward delineating the contribution of astrocytes and other glia to motor neuron degeneration and identifying potential therapeutics targeting astrocytes or other glia.[199]

MICROGLIA

Microglia are resident macrophage cells of the CNS and a key component of the CNS immune system. Their branching processes are used as mobile probes to evaluate the local cellular environment and microglia express receptors for numerous types of molecules (e.g., complement proteins, cytokines, ions, plasma proteins, and neurotrophic factors) to aid this process.[200,201] Microglia exhibit considerable phenotypic heterogeneity in relation to such environmental signals and the density of microglia varies considerably throughout the CNS, with microglial numbers being particularly high in myelinated regions.[202,203]

Microglial activation results in phenotypic and morphological changes and occurs in response to signaling molecules. Activation generally occurs in response to a pathological environmental change, such as infection, disease, or inflammation and is associated with a thickening and retraction of processes and the assumption of a more rounded morphology.[200,201] Activated microglia also exhibit different "active" phenotypes based on the stimulating molecule(s). The "M1" activation is commonly neurotoxic and is accompanied by microglial release of reactive oxygen species, nitric oxide, and inflammatory cytokines. In contrast, neuroprotective microglial activation ('M2') is associated with greater phagocytic activity (e.g., for removing protein aggregates), release of trophic factors, and the uptake of potentially neurotoxic substances (e.g., glutamate via GLT-1). Generally, the balance of pro- versus antiinflammatory factors will determine the nature and extent of the response (M1 or M2).[200–202]

Microglia are implicated in the progression of ALS from studies in both ALS patients and ALS transgenic mouse models. Activated microglia are consistently detected in proximity to degenerating motor neurons in ALS tissue samples and these microglia are frequently positive for inflammatory markers.[204–207] Microglial activation is greater with increasing disease severity, consistent with a microglial role in disease progression. An important question is whether microglial activation is a consequence of the neurodegenerative process or a primary cause of neurodegeneration, and whether microglial responses can be modulated to favor neuroprotection. Answers to these questions have largely come, as they have for astrocytes, from studies of mutant SOD1 transgenic mice.

Microglial activation is recapitulated in mutant SOD1 transgenic mice.[207] Cultured microglia from G93A SOD1 transgenic mice have enhanced inflammatory marker expression.[208] This phenotype can also be conferred by microglial uptake of secreted, misfolded SOD1,[209,210] suggesting that the mutant SOD1 misfolded conformation is sufficient to activate and shift microglia to a pro-inflammatory phenotype. Similar results obtained for mutant, misfolded TDP-43 and TDP-43 CTFs,[211] further the notion that extracellular misfolded proteins contribute to microglial activation.

Given the signal-dependent variability in microglial activation and response (M1 vs M2), determining whether microglia can be induced to assume a benign or neuroprotective phenotype in ALS is a question with important therapeutic implications. In vitro studies have shown that treatment of microglia with interleukin (IL)-4 following lipopolysaccharide activation reduces microglial-associated motor neuron toxicity.[212] Extracellular mutant SOD1 activation of microglia can be attenuated by blocking signaling through CD14 and Toll-like receptors. In vitro coculture systems fail to fully capture the complexity of the CNS and associated complex signaling crosstalk between cell types, however. Consequently, as with astrocytes, therapeutic targeting of microglia responses in the ALS population will likely require modulation of multiple signaling pathways.

OLIGODENDROCYTES

Oligodendrocytes are a highly specialized glia found exclusively in the CNS. They ensheath axons in myelin, a hydrophobic lipid mixture with protein components, thus insulating them and markedly enhancing the speed and energetic efficiency of action potential propagation by saltatory conduction. Because of the length of axons and the energy requirements of conducting action potentials, oligodendrocytes also provide metabolic support to neurons directly at the level of myelin. Oligodendrocytes express the lactate transporter monocarboxylate transporter 1 (MCT1) and removal of the protein by ASO leads axon damage and neurodegeneration.[213] The axons of cortical and spinal motor neurons are among the longest in the CNS and demyelination, reduced metabolic support, and loss of normal oligodendrocyte function have all been hypothesized as contributors to motor neuron selective vulnerability/degeneration in ALS.

In studies of human ALS spinal cord tissue, oligodendrocytes exhibit inclusion pathology and increased levels of oligodendrocyte (NG2+) precursors were detected in the gray matter of ALS patients.[214,215] G93A SOD1 transgenic mice exhibit dysfunction and degeneration of oligodendrocytes prior to loss of motor neurons.[215] MCT1 levels are also reduced in the spinal cords of ALS patients and mutant SOD1 transgenic mice.[213] In addition to compromising oligodendrocyte metabolic support of motor neurons, reduced MCT1 levels are also expected to contribute to myelination abnormalities, as lactate is an essential energy source for the production of myelin. Selective ablation of mutant SOD1 expression from oligodendrocytes delayed disease onset and prolonged survival in transgenic mice.[214] These results suggest that oligodendrocytes are a key contributor to motor neuron viability and that expression of ALS mutant proteins impairs their normal function and maturation to an extent sufficient to produce neurodegeneration.

T CELLS

Recent evidence strongly implicates CNS-infiltrating immune cells via breakdown of the blood–brain barrier or via the choroid plexus contribute to ALS and other neurodegenerative diseases (Fig. 4.3). Infiltrating T lymphocytes have been shown to play important roles in regulating progression and neuroprotection in mouse models of ALS.[216] Fox3p positive regulatory T-cells (Tregs) infiltration in the CNS plays an important role in modulating the rate of disease progression, working together with M2 microglia to suppress inflammation.[217] Passive transfer of Tregs into G93A SOD1 mice lacking functional T lymphocytes prolonged survival and decreased Tregs in the blood of ALS patients correlated to more rapid disease progression. In addition, early reduced FoxP3 levels could be used to identify rapidly progressing ALS patients.[218]

Nevertheless, in most chronic neurodegenerative diseases, the recruitment of these cells into the CNS seems to be insufficient or delayed, resulting in a pro-inflammatory response that contributes to the exacerbation of pathology.[219] Transport of these lymphocytes into the CNS appears to occur at the level of the choroid plexus.[220,221] Upon entry, Tregs and other infiltrating leukocytes could be transported via the CSF to infiltrate within the brain and spinal cord parenchyma. In addition, the CSF has recently been proposed to be a mechanism for spread of disease.[222] Therefore, both Tregs and the infiltration of other leukocytes into the CNS via the choroid plexus are therapeutic targets for ALS and offer new insight into the mechanisms of disease.

PERINEURONAL NETS AND EXTRACELLULAR MATRIX DYSFUNCTION

The extracellular matrix of the CNS is highly specialized and of unique composition, consisting of chondroitin sulfate proteoglycans, fibronectin, hyaluronan, and several extracellular-matrix-associated proteins, including tenascin-R.[223] The protein components of perineuronal nets (PNNs) are synthesized by both neuronal and glial cells, with the type of neuron (e.g., glutamatergic, GABAergic), determining the specific protein and glycan composition of the PNN. PNNs are highly abundant in motor regions of the brain and spinal cord and surround the soma, axons, and dendrites of motor neurons.[223,224] PNN development is dependent on neuronal activity and PNNs are vital to the establishment of proper synaptic connectivity during development.[225] In adulthood, PNNs are known to be repellant to adjacent dendrites and axons, suggesting that PNNs help maintain established synaptic networks and a homeostatic extracellular environment. Consistent with this latter point, PNNs entrap growth factors and aid in maintaining local ionic homeostasis.[224,226]

Evidence of PNN abnormalities in ALS come from studies using ALS tissue samples and ALS model systems. Recent work in human subjects indicates that levels of numerous extracellular matrix proteins are decreased in the CSF of ALS patients.[227] Decreased CSF levels for one of these PNN proteins, tenascin-R, were correlated with a loss of tenascin-R immunoreactivity around motor neurons in ALS patient spinal cord.[227] A similar loss of tenascin-R immunoreactivity around motor neurons is observed in mutant SOD1 transgenic rats,[228] suggesting that PNN abnormalities and degradation are seen in both the human disease and an animal model. One cause of PNN abnormalities in ALS is the elevation in matrix metalloproteinases (MMPs) observed in ALS brain and spinal cord tissue[229] and in transgenic mouse models of ALS.[230] MMPs are zinc-dependent endopeptidases capable of degrading extracellular matrix and PNN components. They can be activated by cytokines, reactive oxygen species, and other inflammatory factors secreted from activated microglia or reactive astrocytes.[231] MMP-mediated PNN breakdown contributes to disease progression in SOD1 mice and administration of an MMP inhibitor extended survival in these animals.[230] Despite this promising data, PNN alterations in ALS and their contribution to selective motor neuron vulnerability remain a relatively understudied topic in the field of ALS research (Fig. 4.3). Future studies addressing the role of PNN dysfunction in ALS will likely provide mechanistic insights and identify new therapeutic targets.

CONCLUSIONS

We have provided an overview for many of the mechanisms proposed to contribute to motor neuron degeneration in various forms of ALS (Fig. 4.4). Delineating which mechanisms are primary, causative factors in the development of the different forms of disease in human patients remains one of the great challenges for the field. Emerging evidence suggests significant genetic contributions to the development of disease in apparently sporadic patients, with a current heritability estimate for SALS of approximately 20%.[232] Variability in multiple genomic regions is associated with increased risk of disease and approximately 25% of SALS patients exhibit polymorphisms in known FALS-linked genes.[233] Thus, it is likely that a combination of genetic and nongenetic factors interact to produce a cumulative individual risk level that, if exceeded, leads to disease. While incomplete, our understanding of nongenetic factors that

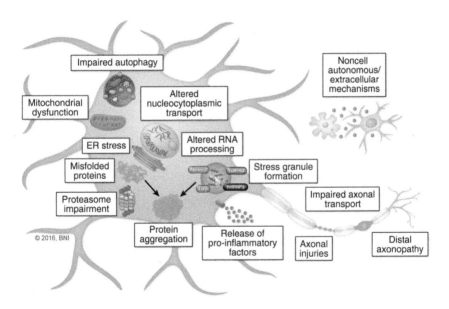

FIGURE 4.4 Systematic overview of amyotrophic lateral sclerosis (ALS)-linked molecular mechanisms within motor neurons. Numerous intracellular mechanisms contribute to motor neuron death. Each mechanism represents a point of motor neuron susceptibility, a concept supported by familial ALS-linked mutations in genes associated with each mechanism. Each represent a therapeutic opportunity, though drug combinations that target multiple mechanisms may be necessary to impede disease progression.

contribute to sporadic forms of disease continues to grow. The unique morphology and metabolic requirements of motor neurons contribute to the cell-type specific degeneration seen in ALS. As we have described, motor neurons are acutely vulnerable to defects in RNA processing, protein folding/clearance, and axonal transport. The accumulation of such defects, individually or concurrently, likely leads to further exacerbation of ongoing pathological processes and, potentially, spread of disease to neighboring cells via secretion of toxic factors or misfolded proteins with prion-like properties. This accumulative model of disease is consistent with the middle to late age of onset observed in a majority of ALS patients. The identification of genetic loci that increase cumulative risk of disease will provide mechanistic insight into the initiation of SALS. Interplay between various mechanisms, genetic variants and/or mutations, and environmental factors likely contribute to the onset, progression rate, and selective neurodegeneration observed in ALS. The complexity of ALS makes drug development challenging and continued biomarker development is necessary for stratification of the patient population for clinical trials, demonstrating target engagement, and to monitor efficacy and patient disease progression. Due to the heterogeneity of the disease, a combination of drug treatments targeting multiple pathways, and possibly cell replacement strategies, will likely be necessary to effectively impact disease course within patients. Our ever-increasing understanding of ALS disease mechanisms is vital to designing and implementing these therapeutic approaches.

References

1. Logroscino G, Traynor BJ, Hardiman O, et al. Incidence of amyotrophic lateral sclerosis in Europe. *J Neurol Neurosurg Psychiatry* 2010;**81**(4):385–90.
2. Uenal H, Rosenbohm A, Kufeldt J, et al. Incidence and geographical variation of amyotrophic lateral sclerosis (ALS) in Southern Germany--completeness of the ALS registry Swabia. *PLoS One* 2014;**9**(4):e93932.
3. Vazquez MC, Ketzoian C, Legnani C, et al. Incidence and prevalence of amyotrophic lateral sclerosis in Uruguay: a population-based study. *Neuroepidemiology* 2008;**30**(2):105–11.
4. Logroscino G, Beghi E, Zoccolella S, et al. Incidence of amyotrophic lateral sclerosis in southern Italy: a population based study. *J Neurol Neurosurg Psychiatry* 2005;**76**(8):1094–8.
5. Wolfson C, Kilborn S, Oskoui M, Genge A. Incidence and prevalence of amyotrophic lateral sclerosis in Canada: a systematic review of the literature. *Neuroepidemiology* 2009;**33**(2):79–88.
6. Neumann M, Sampathu DM, Kwong LK, et al. Ubiquitinated TDP-43 in frontotemporal lobar degeneration and amyotrophic lateral sclerosis. *Science* 2006;**314**(5796):130–3.
7. Sreedharan J, Blair IP, Tripathi V, et al. TDP-43 mutations in familial and sporadic amyotrophic lateral sclerosis. *Science* 2008;**319**(5870):1668–72.
8. Kwiatkowski Jr. TJ, Bosco DA, Leclerc AL, et al. Mutations in the FUS/TLS gene on chromosome 16 cause familial amyotrophic lateral sclerosis. *Science* 2009;**323**(5918):1205–8.
9. Vance C, Rogelj B, Hortobagyi T, et al. Mutations in FUS, an RNA processing protein, cause familial amyotrophic lateral sclerosis type 6. *Science* 2009;**323**(5918):1208–11.

10. Chance PF, Rabin BA, Ryan SG, et al. Linkage of the gene for an autosomal dominant form of juvenile amyotrophic lateral sclerosis to chromosome 9q34. *Am J Hum Genet* 1998;**62**:633–40.
11. Greenway MJ, Alexander MD, Ennis S, et al. A novel candidate region for ALS on chromosome 14q11.2. *Neurology* 2004;**63**:1936–8.
12. Couthouis J, Hart MP, Shorter J, et al. A yeast functional screen predicts new candidate ALS disease genes. *Proc Natl Acad Sci* 2011;**108**(52):20881–90.
13. Elden AC, Kim H-J, Hart MP, et al. Ataxin-2 intermediate-length polyglutamine expansions are associated with increased risk for ALS. *Nature* 2010;**466**(7310):1069–75.
14. Couthouis J, Hart MP, Erion R, et al. Evaluating the role of the FUS/TLS-related gene EWSR1 in amyotrophic lateral sclerosis. *Hum Mol Genet* 2012;**21**(13):2899–911.
15. Simpson CL, Lemmens R, Miskiewicz K, et al. Variants of the elongator protein 3 (ELP3) gene are associated with motor neuron degeneration. *Hum Mol Genet* 2009;**18**(3):472–81.
16. Kim HJ, Kim NC, Wang YD, et al. Prion-like domain mutations in hnRNPs cause multisystem proteinopathy and ALS. *Nature* 2013;**495**:467–73.
17. Collins M, Riascos D, Kovalik T, et al. The RNA-binding motif 45 (RBM45) protein accumulates in inclusion bodies in amyotrophic lateral sclerosis (ALS) and frontotemporal lobar degeneration with TDP-43 inclusions (FTLD-TDP) patients. *Acta Neuropathol* 2012;**124**(5):717–32.
18. Chio A, Calvo A, Mazzini L, et al. Extensive genetics of ALS: a population-based study in Italy. *Neurology* 2012;**79**(19):1983–9.
19. Lagier-Tourenne C, Polymenidou M, Hutt KR, et al. Divergent roles of ALS-linked proteins FUS/TLS and TDP-43 intersect in processing long pre-mRNAs. *Nat Neurosci* 2012;**15**(11):1488–97.
20. Polymenidou M, Lagier-Tourenne C, Hutt KR, et al. Long pre-mRNA depletion and RNA missplicing contribute to neuronal vulnerability from loss of TDP-43. *Nat Neurosci* 2011;**14**(4):459–68.
21. Sephton CF, Cenik C, Kucukural A, et al. Identification of neuronal RNA targets of TDP-43-containing ribonucleoprotein complexes. *J Biol Chem* 2011;**286**(2):1204–15.
22. Tollervey JR, Curk T, Rogelj B, et al. Characterizing the RNA targets and position-dependent splicing regulation by TDP-43. *Nat Neurosci* 2011;**14**(4):452–8.
23. Xiao S, Sanelli T, Dib S, et al. RNA targets of TDP-43 identified by UV-CLIP are deregulated in ALS. *Mol Cell Neurosci* 2011;**47**(3):167–80.
24. Nishimoto Y, Nakagawa S, Hirose T, et al. The long non-coding RNA nuclear-enriched abundant transcript 1_2 induces paraspeckle formation in the motor neuron during the early phase of amyotrophic lateral sclerosis. *Mol Brain* 2013;**6**(31)http://dx.doi.org/10.1186/1756-6606-1186-1131
25. Naganuma T, Hirose T. Paraspeckle formation during the biogenesis of long non-coding RNAs. *RNA Biol* 2013;**10**(3):456–61.
26. Shelkovnikova TA, Robinson HK, Troakes C, Ninkina N, Buchman VL. Compromised paraspeckle formation as a pathogenic factor in FUSopathies. *Hum Mol Genet* 2014;**23**(9):2298–312.
27. Wang I-F, Wu L-S, Chang H-Y, Shen CKJ. TDP-43, the signature protein of FTLD-U, is a neuronal activity-responsive factor. *J Neurochem* 2008;**105**:797–806.
28. Fujii R, Okabe S, Urushido T, et al. The RNA binding protein TLS is translocated to dendritic spines by mGluR5 activation and regulates spine morphology. *Curr Biol* 2005;**15**:587–93.
29. Yamashita T, Kwak S. The molecular link between inefficient GluA2 Q/R site-RNA editing and TDP-43 pathology in motor neurons of sporadic amyotrophic lateral sclerosis patients. *Brain Res* 2014;**1584**:28–38.
30. Fabian MR, Sonenberg N, Filipowicz W. Regulation of mRNA translation and stability by microRNAs. *Annu Rev Biochem* 2010;**79**:351–79.
31. Friedman RC, Farh KK, Burge CB, Bartel DP. Most mammalian mRNAs are conserved targets of microRNAs. *Genome Res* 2009;**19**:92–105.

32. Leung AKL, Sharp PA. MicroRNA functions in stress response. *Mol Cell* 2010;**40**:205–15.
33. Butovsky O, Siddiqui S, Gabriely G, et al. Modulating inflammatory monocytes with a unique microRNA gene signature ameliorates murine ALS. *J Clin Invest* 2012;**122**:3063–87.
34. Parisi C, Arisi I, D'Ambrosi N, et al. Dysregulated microRNAs in amyotrophic lateral sclerosis microglia modulate genes linked to neuroinflammation. *Cell Death Dis* 2013;**4**(e959)http://dx.doi.org/10.1038/cddis.2013.1491
35. Campos-Melo D, Droppelmann CA, He Z, Volkening K, Strong MJ. Altered microRNA expression profile in amyotrophic lateral sclerosis: a role in the regulation of NFL mRNA levels. *Mol Brain* 2013;**6**(26)http://dx.doi.org/10.1186/1756-6606-1186-1126
36. Williams AH, Valdez G, Moresi V, et al. MicroRNA-206 delays ALS progression and promotes regeneration of neuromuscular synapses in mice. *Science* 2009;**326**:1549–54.
37. Morlando M, Dini Modigliani S, Torrelli G, et al. FUS stimulates microRNA biogenesis by facilitating co-transcriptional Drosha recruitment. *EMBO J* 2012;**31**(24):4502–10.
38. Kawahara Y, Mieda-Sato A. TDP-43 promotes microRNA biogenesis as a component of the Drosha and Dicer compelexes. *Proc Natl Acad Sci USA* 2012;**109**(9):3347–52.
39. Sun CN, Araoz C, Lucas G, Morgan PN, White HJ. Amyotrophic lateral sclerosis. Inclusion bodies in a case of the classic sporadic form. *Ann Clin Lab Sci* 1975;**5**(1):38–44.
40. Takahashi K, Nakamura H, Okada E. Hereditary amyotrophic lateral sclerosis. Histochemical and electron microscopic study of hyaline inclusions in motor neurons. *Arch Neurol* 1972;**27**(4):292–9.
41. Blokhuis AM, Groen EJ, Koppers M, van den Berg LH, Pasterkamp RJ. Protein aggregation in amyotrophic lateral sclerosis. *Acta Neuropathol* 2013;**125**(6):777–94.
42. Rosen DR, Siddique T, Patterson D, et al. Mutations in Cu/Zn superoxide dismutase gene are associated with familial amyotrophic lateral sclerosis. *Nature* 1993;**362**(6415):59–62.
43. de Belleroche J, Orrell R, King A. Familial amyotrophic lateral sclerosis/motor neurone disease (FALS): a review of current developments. *J Med Genet* 1995;**32**(11):841–7.
44. Rotunno MS, Bosco DA. An emerging role for misfolded wild-type SOD1 in sporadic ALS pathogenesis. *Front Cell Neurosci* 2013;**7**:253.
45. Chattopadhyay M, Valentine JS. Aggregation of copper-zinc superoxide dismutase in familial and sporadic ALS. *Antioxid Redox Signal* 2009;**11**(7):1603–14.
46. Fridovich I. Superoxide dismutases. *Annu Rev Biochem* 1975;**44**:147–59.
47. Rotilio G, Calabrese L, Bossa F, Barra D, Agro AF, Mondovi B. Properties of the apoprotein and role of copper and zinc in protein conformation and enzyme activity of bovine superoxide dismutase. *Biochemistry* 1972;**11**(11):2182–7.
48. Andersen PM, Nilsson P, Keranen ML, et al. Phenotypic heterogeneity in motor neuron disease patients with CuZn-superoxide dismutase mutations in Scandinavia. *Brain* 1997;**120**(Pt 10):1723–37.
49. Radunovic A, Delves HT, Robberecht W, et al. Copper and zinc levels in familial amyotrophic lateral sclerosis patients with CuZnSOD gene mutations. *Ann Neurol* 1997;**42**(1):130–1.
50. Reaume AG, Elliott JL, Hoffman EK, et al. Motor neurons in Cu/Zn superoxide dismutase-deficient mice develop normally but exhibit enhanced cell death after axonal injury. *Nat Genet* 1996;**13**(1):43–7.
51. Bruijn LI, Houseweart MK, Kato S, et al. Aggregation and motor neuron toxicity of an ALS-linked SOD1 mutant independent from wild-type SOD1. *Science* 1998;**281**(5384):1851–4.
52. Wang J, Xu G, Gonzales V, et al. Fibrillar inclusions and motor neuron degeneration in transgenic mice expressing superoxide dismutase 1 with a disrupted copper-binding site. *Neurobiol Dis* 2002;**10**(2):128–38.
53. Elam JS, Taylor AB, Strange R, et al. Amyloid-like filaments and water-filled nanotubes formed by SOD1 mutant proteins linked to familial ALS. *Nat Struct Biol* 2003;**10**(6):461–7.

54. Furukawa Y, Kaneko K, Yamanaka K, O'Halloran TV, Nukina N. Complete loss of post-translational modifications triggers fibrillar aggregation of SOD1 in the familial form of amyotrophic lateral sclerosis. *J Biol Chem* 2008;**283**(35):24167–76.

55. Banci L, Bertini I, Durazo A, et al. Metal-free superoxide dismutase forms soluble oligomers under physiological conditions: a possible general mechanism for familial ALS. *Proc Natl Acad Sci USA* 2007;**104**(27):11263–7.

56. Sasaki S, Warita H, Murakami T, et al. Ultrastructural study of aggregates in the spinal cord of transgenic mice with a G93A mutant SOD1 gene. *Acta Neuropathol* 2005;**109**(3):247–55.

57. Kato S, Shimoda M, Watanabe Y, Nakashima K, Takahashi K, Ohama E. Familial amyotrophic lateral sclerosis with a two base pair deletion in superoxide dismutase 1: gene multisystem degeneration with intracytoplasmic hyaline inclusions in astrocytes. *J Neuropathol Exp Neurol* 1996;**55**(10):1089–101.

58. Allen MJ, Lacroix JJ, Ramachandran S, et al. Mutant SOD1 forms ion channel: implications for ALS pathophysiology. *Neurobiol Dis* 2012;**45**(3):831–8.

59. Rabizadeh S, Gralla EB, Borchelt DR, et al. Mutations associated with amyotrophic lateral sclerosis convert superoxide dismutase from an antiapoptotic gene to a proapoptotic gene: studies in yeast and neural cells. *Proc Natl Acad Sci USA* 1995;**92**(7):3024–8.

60. Vehvilainen P, Koistinaho J, Gundars G. Mechanisms of mutant SOD1 induced mitochondrial toxicity in amyotrophic lateral sclerosis. *Front Cell Neurosci* 2014;**8**:126.

61. Tan W, Naniche N, Bogush A, Pedrini S, Trotti D, Pasinelli P. Small peptides against the mutant SOD1/Bcl-2 toxic mitochondrial complex restore mitochondrial function and cell viability in mutant SOD1-mediated ALS. *J Neurosci* 2013;**33**(28):11588–98.

62. Vijayvergiya C, Beal MF, Buck J, Manfredi G. Mutant superoxide dismutase 1 forms aggregates in the brain mitochondrial matrix of amyotrophic lateral sclerosis mice. *J Neurosci* 2005;**25**(10):2463–70.

63. Carri MT, Cozzolino M. SOD1 and mitochondria in ALS: a dangerous liaison. *J Bioenerg Biomembr* 2011;**43**(6):593–9.

64. Mondola P, Santillo M, Seru R, et al. Cu,Zn superoxide dismutase increases intracellular calcium levels via a phospholipase C-protein kinase C pathway in SK-N-BE neuroblastoma cells. *Biochem Biophys Res Commun* 2004;**324**(2):887–92.

65. Grad LI, Cashman NR. Prion-like activity of Cu/Zn superoxide dismutase: implications for amyotrophic lateral sclerosis. *Prion* 2014;**8**(1):33–41.

66. Kabashi E, Valdmanis PN, Dion P, et al. TARDBP mutations in individuals with sporadic and familial amyotrophic lateral sclerosis. *Nat Genet* 2008;**40**(5):572–4.

67. Gitcho MA, Baloh RH, Chakraverty S, et al. TDP-43 A315T mutation in familial motor neuron disease. *Ann Neurol* 2008;**63**(4):535–8.

68. Borroni B, Bonvicini C, Alberici A, et al. Mutation within TARDBP leads to frontotemporal dementia without motor neuron disease. *Hum Mutat* 2009;**30**(11):E974–83.

69. Kwong LK, Neumann M, Sampathu DM, Lee VM, Trojanowski JQ. TDP-43 proteinopathy: the neuropathology underlying major forms of sporadic and familial frontotemporal lobar degeneration and motor neuron disease. *Acta Neuropathol* 2007;**114**(1):63–70.

70. Baralle M, Buratti E, Baralle FE. The role of TDP-43 in the pathogenesis of ALS and FTLD. *Biochem Soc Trans* 2013;**41**(6):1536–40.

71. Janssens J, Van Broeckhoven C. Pathological mechanisms underlying TDP-43 driven neurodegeneration in FTLD-ALS spectrum disorders. *Hum Mol Genet* 2013;**22**(R1):R77–87.

72. Arai T, Hasegawa M, Nonoka T, et al. Phosphorylated and cleaved TDP-43 in ALS, FTLD and other neurodegenerative disorders and in cellular models of TDP-43 proteinopathy. *Neuropathology* 2010;**30**(2):170–81.

73. Buratti E, Baralle FE. Multiple roles of TDP-43 in gene expression, splicing regulation, and human disease. *Front Biosci J Virtual Library* 2008;**13**:867–78.

74. Lagier-Tourenne C, Polymenidou M, Cleveland DW. TDP-43 and FUS/TLS: emerging roles in RNA processing and neurodegeneration. *Hum Mol Genet* 2010;**19**(R1):R46–64.
75. Wu LS, Cheng WC, Hou SC, Yan YT, Jiang ST, Shen CK. TDP-43, a neuro-pathosignature factor, is essential for early mouse embryogenesis. *Genesis* 2010;**48**(1):56–62.
76. Sephton CF, Good SK, Atkin S, et al. TDP-43 is a developmentally regulated protein essential for early embryonic development. *J Biol Chem* 2010;**285**(9):6826–34.
77. Colombrita C, Zennaro E, Fallini C, et al. TDP-43 is recruited to stress granules in conditions of oxidative insult. *J Neurochem* 2009;**111**(4):1051–61.
78. Liu-Yesucevitz L, Bilgutay A, Zhang YJ, et al. Tar DNA binding protein-43 (TDP-43) associates with stress granules: analysis of cultured cells and pathological brain tissue. *PLoS One* 2010;**5**(10):e13250.
79. Dewey CM, Cenik B, Sephton CF, Johnson BA, Herz J, Yu G. TDP-43 aggregation in neurodegeneration: are stress granules the key? *Brain Res* 2012;**1462**:16–25.
80. Wegorzewska I, Bell S, Cairns NJ, Miller TM, Baloh RH. TDP-43 mutant transgenic mice develop features of ALS and frontotemporal lobar degeneration. *Proc Natl Acad Sci USA* 2009;**106**(44):18809–14.
81. Wils H, Kleinberger G, Janssens J, et al. TDP-43 transgenic mice develop spastic paralysis and neuronal inclusions characteristic of ALS and frontotemporal lobar degeneration. *Proc Natl Acad Sci USA* 2010;**107**(8):3858–63.
82. Feiguin F, Godena VK, Romano G, D'Ambrogio A, Klima R, Baralle FE. Depletion of TDP-43 affects Drosophila motoneurons terminal synapsis and locomotive behavior. *FEBS Lett* 2009;**583**(10):1586–92.
83. Li Y, Ray P, Rao EJ, et al. A Drosophila model for TDP-43 proteinopathy. *Proc Natl Acad Sci USA* 2010;**107**(7):3169–74.
84. Lin MJ, Cheng CW, Shen CK. Neuronal function and dysfunction of Drosophila dTDP. *PloS One* 2011;**6**(6):e20371.
85. Kabashi E, Lin L, Tradewell ML, et al. Gain and loss of function of ALS-related mutations of TARDBP (TDP-43) cause motor deficits in vivo. *Hum Mol Genet* 2010;**19**(4):671–83.
86. Schmid B, Hruscha A, Hogl S, et al. Loss of ALS-associated TDP-43 in zebrafish causes muscle degeneration, vascular dysfunction, and reduced motor neuron axon outgrowth. *Proc Natl Acad Sci USA* 2013;**110**(13):4986–91.
87. Giordana MT, Piccinini M, Grifoni S, et al. TDP-43 redistribution is an early event in sporadic amyotrophic lateral sclerosis. *Brain Pathol* 2010;**20**(2):351–60.
88. Winton MJ, Igaz LM, Wong MM, Kwong LK, Trojanowski JQ, Lee VM. Disturbance of nuclear and cytoplasmic TAR DNA-binding protein (TDP-43) induces disease-like redistribution, sequestration, and aggregate formation. *J Biol Chem* 2008;**283**(19):13302–9.
89. Scotter EL, Vance C, Nishimura AL, et al. Differential roles of the ubiquitin proteasome system and autophagy in the clearance of soluble and aggregated TDP-43 species. *J Cell Sci* 2014;**127**(Pt 6):1263–78.
90. Picher-Martel V, Dutta K, Phaneuf D, Sobue G, Julien JP. Ubiquilin-2 drives NF-kappaB activity and cytosolic TDP-43 aggregation in neuronal cells. *Mol Brain* 2015;**8**(1):71.
91. Suzuki H, Lee K, Matsuoka M. TDP-43-induced death is associated with altered regulation of BIM and Bcl-xL and attenuated by caspase-mediated TDP-43 cleavage. *J Biol Chem* 2011;**286**(15):13171–83.
92. Zhang YJ, Xu YF, Cook C, et al. Aberrant cleavage of TDP-43 enhances aggregation and cellular toxicity. *Proc Natl Acad Sci USA* 2009;**106**(18):7607–12.
93. Tsuji H, Arai T, Kametani F, et al. Molecular analysis and biochemical classification of TDP-43 proteinopathy. *Brain J Neurol* 2012;**135**(Pt 11):3380–91.
94. Nonaka T, Masuda-Suzukake M, Arai T, et al. Prion-like properties of pathological TDP-43 aggregates from diseased brains. *Cell Rep* 2013;**4**(1):124–34.
95. Sama RR, Ward CL, Bosco DA. Functions of FUS/TLS from DNA repair to stress response: implications for ALS. *ASN Neuro* 2014;**6**(4)

96. Ling SC, Polymenidou M, Cleveland DW. Converging mechanisms in ALS and FTD: disrupted RNA and protein homeostasis. *Neuron* 2013;**79**(3):416–38.
97. Sun Z, Diaz Z, Fang X, et al. Molecular determinants and genetic modifiers of aggregation and toxicity for the ALS disease protein FUS/TLS. *PLoS Biol* 2011;**9**(4):e1000614.
98. Rogelj B, Easton LE, Bogu GK, et al. Widespread binding of FUS along nascent RNA regulates alternative splicing in the brain. *Sci Rep* 2012;**2**:603.
99. Andersson MK, Stahlberg A, Arvidsson Y, et al. The multifunctional FUS, EWS and TAF15 proto-oncoproteins show cell type-specific expression patterns and involvement in cell spreading and stress response. *BMC Cell Biol* 2008;**9**:37.
100. Bentmann E, Neumann M, Tahirovic S, Rodde R, Dormann D, Haass C. Requirements for stress granule recruitment of fused in sarcoma (FUS) and TAR DNA-binding protein of 43 kDa (TDP-43). *J Biol Chem* 2012;**287**(27):23079–94.
101. Bosco DA, Lemay N, Ko HK, et al. Mutant FUS proteins that cause amyotrophic lateral sclerosis incorporate into stress granules. *Hum Mol Genet* 2010;**19**(21):4160–75.
102. Gal J, Zhang J, Kwinter DM, et al. Nuclear localization sequence of FUS and induction of stress granules by ALS mutants. *Neurobiol Aging* 2011;**32**(12) 2323.e2327-e2340.
103. Deng HX, Zhai H, Bigio EH, et al. FUS-immunoreactive inclusions are a common feature in sporadic and non-SOD1 familial amyotrophic lateral sclerosis. *Ann Neurol* 2010;**67**(6):739–48.
104. DeJesus-Hernandez M, Mackenzie IR, Boeve BF, et al. Expanded GGGGCC hexanucleotide repeat in noncoding region of C9ORF72 causes chromosome 9p-linked FTD and ALS. *Neuron* 2011;**72**(2):245–56.
105. Renton AE, Majounie E, Waite A, et al. A hexanucleotide repeat expansion in C9ORF72 is the cause of chromosome 9p21-linked ALS-FTD. *Neuron* 2011;**72**(2):257–68.
106. Rademakers R. C9orf72 repeat expansion in patients with ALS and FTLD. *Lancet Neurol* 2012;**11**:297–8.
107. Al-Sarraj S, King A, Troakes C, et al. p62 positive, TDP-43 negative, neuronal cytoplasmic and intranuclear inclusions in the cerebellum and hippocampus define the pathology of C9orf72-linked FTLD and MND/ALS. *Acta Neuropathol* 2011;**122**:691–702.
108. Levine TP, Daniels RD, Gatta AT, Wong LH, Hayes MJ. The product of C9orf72, a gene strongly implicated in neurodegeneration, is structurally related to DENN Rab-GEFs. *Bioinformatics* 2013;**29**(4):499–503.
109. Farg MA, Sundaramoorthy V, Sultana JM, et al. C9ORF72, implicated in amytrophic lateral sclerosis and frontotemporal dementia, regulates endosomal trafficking. *Hum Mol Genet* 2014;**23**(13):3579–95.
110. Koppers M, Blokhuis AM, Westeneng H-J, et al. C9orf72 ablation in mice does not cause motor neuron degeneration or motor deficits. *Ann Neurol* 2015;**78**(3):426–38.
111. Ciura S, Lattante S, Le Ber J, et al. Loss of function of C9orf72 causes motor deficits in a zebrafish model of amyotrophic lateral sclerosis. *Ann Neurol* 2013;**74**(2):180–7.
112. Rohrer JD, Isaacs AM, Mizielinska S, et al. C9orf72 expansions in frontotemporal dementia and amyotrophic lateral sclerosis. *Lancet Neurol* 2015;**14**(3):291–301.
113. Vatovec S, Kovanda A, Rogelj B. Unconventional features of C9ORF72 expanded repeat in amyotrophic lateral sclerosis and frontotemporal lobar degeneration. *Neurobiol Aging* 2014;**35**(10):2421.e1–2421.e12.
114. Simone R, Fratta P, Neidle S, Parkinson GN, Isaacs AM. G-quadruplexes: emerging roles in neurodegenerative diseases and the non-coding transcriptome. *FEBS Lett* 2015;**589**(14):1653–68.
115. Wen X, Tan W, Westergard T, et al. Antisense proline-arginine RAN dipeptides linked to C9ORF72-ALS/FTD form toxic nuclear aggregates that initiate in vitro and in vivo neuronal death. *Neuron* 2014;**84**(6):1213–25.
116. Meyer K, Ferraiuolo L, Miranda CJ, et al. Direct conversion of patient fibroblasts demonstrates non-cell autonomous toxicity of astrocytes to motor neurons in familial and sporadic ALS. *Proc Natl Acad Sci USA.* 2014;**111**(2):829–32.

117. Mackenzie I, Arzberger T, Kremmer E, et al. Dipeptide repeat protein pathology in C9ORF72 mutation cases: clinico-pathological correlations. *Acta Neuropathol* 2013;**126**(6):859–79. 2013/12/01.
118. Freibaum BD, Lu Y, Lopez-Gonzalez R, et al. GGGGCC repeat expansion in C9orf72 compromises nucleocytoplasmic transport. *Nature* 2015;**525**(7567):129–33.
119. Zhang K, Donnelly CJ, Haeusler AR, et al. The C9orf72 repeat expansion disrupts nucleocytoplasmic transport. *Nature* 2015;**525**(7567):56–61.
120. Su Z, Zhang Y, Gendron TF, et al. Discovery of a biomarker and lead small molecules to target r(GGGGCC)-associated defects in c9FTD/ALS. *Neuron* 2014;**83**(5):1043–50.
121. Zu T, Liu Y, Banez-Coronel M, et al. RAN proteins and RNA foci from antisense transcripts in C9ORF72 ALS and frontotemporal dementia. *Proc Natl Acad Sci USA* 2013;**110**(51):E4968–77.
122. Millecamps S, Julien JP. Axonal transport deficits and neurodegenerative diseases. *Nat Rev Neurosci* 2013;**14**(3):161–76.
123. Xiao S, McLean J, Robertson J. Neuronal intermediate filaments and ALS: a new look at an old question. *Biochim Biophys Acta* 2006;**1762**(11–12):1001–12.
124. Kondo A, Iwaki T, Tateishi J, Kirimoto K, Morimoto T, Oomura I. Accumulation of neurofilaments in a sporadic case of amyotrophic lateral sclerosis. *Jpn J Psychiatry Neurol* 1986;**40**(4):677–84.
125. He CZ, Hays AP. Expression of peripherin in ubiquinated inclusions of amyotrophic lateral sclerosis. *J Neurol Sci* 2004;**217**(1):47–54.
126. Lariviere RC, Julien JP. Functions of intermediate filaments in neuronal development and disease. *J Neurobiol* 2004;**58**(1):131–48.
127. Chou YH, Goldman RD. Intermediate filaments on the move. *J Cell Biol* 2000;**150**(3):F101–6.
128. Leung CL, He CZ, Kaufmann P, et al. A pathogenic peripherin gene mutation in a patient with amyotrophic lateral sclerosis. *Brain Pathol* 2004;**14**(3):290–6.
129. Gros-Louis F, Lariviere R, Gowing G, et al. A frameshift deletion in peripherin gene associated with amyotrophic lateral sclerosis. *J Biol Chem* 2004;**279**(44):45951–6.
130. Figlewicz DA, Krizus A, Martinoli MG, et al. Variants of the heavy neurofilament subunit are associated with the development of amyotrophic lateral sclerosis. *Hum Mol Genet* 1994;**3**(10):1757–61.
131. Tomkins J, Usher P, Slade JY, et al. Novel insertion in the KSP region of the neurofilament heavy gene in amyotrophic lateral sclerosis (ALS). *Neuroreport* 1998;**9**(17):3967–70.
132. Beaulieu JM, Nguyen MD, Julien JP. Late onset of motor neurons in mice overexpressing wild-type peripherin. *J Cell Biol* 1999;**147**(3):531–44.
133. Breuer AC, Lynn MP, Atkinson MB, et al. Fast axonal transport in amyotrophic lateral sclerosis: an intra-axonal organelle traffic analysis. *Neurology* 1987;**37**(5):738–48.
134. Julien JP, Cote F, Collard JF. Mice overexpressing the human neurofilament heavy gene as a model of ALS. *Neurobiol Aging* 1995;**16**(3):487–90. discussion 490-482.
135. LaMonte BH, Wallace KE, Holloway BA, et al. Disruption of dynein/dynactin inhibits axonal transport in motor neurons causing late-onset progressive degeneration. *Neuron* 2002;**34**(5):715–27.
136. Warita H, Itoyama Y, Abe K. Selective impairment of fast anterograde axonal transport in the peripheral nerves of asymptomatic transgenic mice with a G93A mutant SOD1 gene. *Brain Res* 1999;**819**(1–2):120–31.
137. Williamson TL, Cleveland DW. Slowing of axonal transport is a very early event in the toxicity of ALS-linked SOD1 mutants to motor neurons. *Nat Neurosci* 1999;**2**(1):50–6.
138. De Vos KJ, Chapman AL, Tennant ME, et al. Familial amyotrophic lateral sclerosis-linked SOD1 mutants perturb fast axonal transport to reduce axonal mitochondria content. *Hum Mol Genet* 2007;**16**(22):2720–8.

139. Zhang F, Strom AL, Fukada K, Lee S, Hayward LJ, Zhu H. Interaction between familial amyotrophic lateral sclerosis (ALS)-linked SOD1 mutants and the dynein complex. *J Biol Chem* 2007;**282**(22):16691–9.
140. Perlson E, Jeong GB, Ross JL, et al. A switch in retrograde signaling from survival to stress in rapid-onset neurodegeneration. *J Neurosci* 2009;**29**(31):9903–17.
141. Alami NH, Smith RB, Carrasco MA, et al. Axonal transport of TDP-43 mRNA granules is impaired by ALS-causing mutations. *Neuron* 2014;**81**(3):536–43.
142. Li YR, King OD, Shorter J, Gitler AD. Stress granules as crucibles of ALS pathogenesis. *J Cell Biol* 2013;**201**(3):361–72.
143. Anderson P, Kedersha N. Stress granules: the Tao of RNA triage. *Trends Biochem Sci* 2008;**33**(3):141–50.
144. Moeller BJ, Cao Y, Li CY, Dewhirst MW. Radiation activates HIF-1 to regulate vascular radiosensitivity in tumors: role of reoxygenation, free radicals, and stress granules. *Cancer Cell* 2004;**5**(5):429–41.
145. Esclatine A, Taddeo B, Roizman B. Herpes simplex virus 1 induces cytoplasmic accumulation of TIA-1/TIAR and both synthesis and cytoplasmic accumulation of tristetraprolin, two cellular proteins that bind and destabilize AU-rich RNAs. *J Virol* 2004;**78**(16):8582–92.
146. Johnson BS, Snead D, Lee JJ, McCaffery JM, Shorter J, Gitler AD. TDP-43 is intrinsically aggregation-prone, and amyotrophic lateral sclerosis-linked mutations accelerate aggregation and increase toxicity. *J Biol Chem* 2009;**284**(30):20329–39.
147. King OD, Gitler AD, Shorter J. The tip of the iceberg: RNA-binding proteins with prion-like domains in neurodegenerative disease. *Brain Res* 2012;**1462**:61–80.
148. Kim HJ, Kim NC, Wang YD, et al. Mutations in prion-like domains in hnRNPA2B1 and hnRNPA1 cause multisystem proteinopathy and ALS. *Nature* 2013;**495**(7442):467–73.
149. Bjorkoy G, Lamark T, Brech A, et al. p62/SQSTM1 forms protein aggregates degraded by autophagy and has a protective effect on huntingtin-induced cell death. *J Cell Biol* 2005;**171**(4):603–14.
150. Zavodszky E, Vicinanza M, Rubinsztein DC. Biology and trafficking of ATG9 and ATG16L1, two proteins that regulate autophagosome formation. *FEBS Lett* 2013;**587**(13):1988–96.
151. Schreiber A, Peter M. Substrate recognition in selective autophagy and the ubiquitin-proteasome system. *Biochim Biophys Acta* 2014;**1843**(1):163–81.
152. Wong YC, Holzbaur EL. Optineurin is an autophagy receptor for damaged mitochondria in parkin-mediated mitophagy that is disrupted by an ALS-linked mutation. *Proc Natl Acad Sci USA* 2014;**111**(42):E4439–48.
153. Maruyama H, Morino H, Ito H, et al. Mutations of optineurin in amyotrophic lateral sclerosis. *Nature* 2010;**465**(7295):223–6.
154. Hortobagyi T, Troakes C, Nishimura AL, et al. Optineurin inclusions occur in a minority of TDP-43 positive ALS and FTLD-TDP cases and are rarely observed in other neurodegenerative disorders. *Acta Neuropathol* 2011;**121**(4):519–27.
155. Johnson JO, Mandrioli J, Benatar M, et al. Exome sequencing reveals VCP mutations as a cause of familial ALS. *Neuron* 2010;**68**(5):857–64.
156. Ju JS, Fuentealba RA, Miller SE, et al. Valosin-containing protein (VCP) is required for autophagy and is disrupted in VCP disease. *J Cell Biol* 2009;**187**(6):875–88.
157. Rodriguez-Ortiz CJ, Hoshino H, Cheng D, et al. Neuronal-specific overexpression of a mutant valosin-containing protein associated with IBMPFD promotes aberrant ubiquitin and TDP-43 accumulation and cognitive dysfunction in transgenic mice. *Am J Pathol* 2013;**183**(2):504–15.
158. Ayaki T, Ito H, Fukushima H, et al. Immunoreactivity of valosin-containing protein in sporadic amyotrophic lateral sclerosis and in a case of its novel mutant. *Acta Neuropathol Commun* 2014;**2**:172.

159. Fecto F, Yan J, Vemula SP, et al. SQSTM1 mutations in familial and sporadic amyotrophic lateral sclerosis. *Arch Neurol* 2011;**68**(11):1440–6.
160. Deng HX, Chen W, Hong ST, et al. Mutations in UBQLN2 cause dominant X-linked juvenile and adult-onset alS and ALS/dementia. *Nature* 2011;**477**(7363):211–5.
161. Cox LE, Ferraiuolo L, Goodall EF, et al. Mutations in CHMP2B in lower motor neuron predominant amyotrophic lateral sclerosis (ALS). *PloS One* 2010;**5**(3):e9872.
162. Wang IF, Guo BS, Liu YC, et al. Autophagy activators rescue and alleviate pathogenesis of a mouse model with proteinopathies of the TAR DNA-binding protein 43. *Proc Natl Acad Sci USA* 2012;**109**(37):15024–9.
163. Cheng CW, Lin MJ, Shen CK. Rapamycin alleviates pathogenesis of a new Drosophila model of ALS-TDP. *J Neurogenet* 2015;**29**(2–3):59–68.
164. Zhang X, Li L, Chen S, et al. Rapamycin treatment augments motor neuron degeneration in SOD1(G93A) mouse model of amyotrophic lateral sclerosis. *Autophagy* 2011;**7**(4):412–25.
165. Gardner BM, Pincus D, Gotthardt K, Gallagher CM, Walter P. Endoplasmic reticulum stress sensing in the unfolded protein response. *Cold Spring Harb Perspect Biol* 2013;**5**(3):a013169.
166. Sano R, Reed JC. ER stress-induced cell death mechanisms. *Biochim Biophys Acta* 2013;**1833**(12):3460–70.
167. Hoyer-Hansen M, Jaattela M. Connecting endoplasmic reticulum stress to autophagy by unfolded protein response and calcium. *Cell Death Differ* 2007;**14**(9):1576–82.
168. Kanekura K, Suzuki H, Aiso S, Matsuoka M. ER stress and unfolded protein response in amyotrophic lateral sclerosis. *Mol Neurobiol* 2009;**39**(2):81–9.
169. Ilieva EV, Ayala V, Jove M, et al. Oxidative and endoplasmic reticulum stress interplay in sporadic amyotrophic lateral sclerosis. *Brain J Neurol* 2007;**130**(Pt 12):3111–23.
170. Saxena S, Cabuy E, Caroni P. A role for motoneuron subtype-selective ER stress in disease manifestations of FALS mice. *Nat Neurosci* 2009;**12**(5):627–36.
171. Wate R, Ito H, Zhang JH, Ohnishi S, Nakano S, Kusaka H. Expression of an endoplasmic reticulum-resident chaperone, glucose-regulated stress protein 78, in the spinal cord of a mouse model of amyotrophic lateral sclerosis. *Acta Neuropathol* 2005;**110**(6):557–62.
172. Illieva H, Polymenidou M, Cleveland DW. Non-cell autonomous toxicity in neurodegenerative disorders: ALS and beyond. *J Cell Biol* 2009;**187**:761–72.
173. Philips T, Robberecht W. Neuroinflammation in amyotrophic lateral sclerosis: role of glial activation in motor neuron disease. *Lancet Neurol* 2011;**10**(3):253–63.
174. Berry JD, Shefner JM, Conwit R, et al. Design and initial results of a multi-phase randomized trial of ceftriaxone in amyotrophic lateral sclerosis. *PloS One* 2013;**8**(4):e61177.
175. Cudkowicz M, Shefner J, Schoenfeld DA, et al. Trial of celecoxib in amyotrophic lateral sclerosis. *Ann Neurol* 2006;**60**(1):22–31.
176. Gordon PH, Moore DH, Miller RG, et al. Efficacy of minocycline in patients with amyotrophic lateral sclerosis: a phase III randomized trial. *Lancet Neurol* 2007;**6**(12):1045–53.
177. Beers DR, Henkel JS, Xiao Q, et al. Wild-type microglia extend survival in PU.1 knockout mice with familial amyotrophic lateral sclerosis. *Proc Natl Acad Sci USA* 2006;**103**(43):16021–6.
178. Sofroniew MV, Vinters HV. Astrocytes: biology and pathology. *Acta Neuropathol* 2010;**119**:7–35.
179. Weber B, Barros LF. The astrocyte: powerhouse and recycling center. *Cold Spring Harb Perspect Biol* 2015;**7**(12)
180. Bayraktar OA, Fuentealba LC, Alvarez-Buylla A, Rowitch DH. Astrocyte development and heterogeneity. *Cold Spring Harb Perspect Biol* 2015;**7**(1):a020362.
181. Parpura V, Heneka MT, Montana V, et al. Glial cells in (patho)physiology. *J Neurochem* 2012;**121**(1):4–27.
182. Vargas MR, Johnson JA. Astrogliosis in amyotrophic lateral sclerosis: role and therapeutic potential of astrocytes. *Neurotherapeutics* 2010;**7**(4):471–81.

183. Valori CF, Brambilla L, Martorana F, Rossi D. The multifaceted role of glial cells in amyotrophic lateral sclerosis. *Cell Mol Life Sci* 2014;**71**(2):287–97.
184. Almer G, Vukosavic S, Romero N, Przedborski S. Inducible nitric oxide synthase up-regulation in a transgenic mouse model of familial amyotrophic lateral sclerosis. *J Neurochem* 1999;**72**(6):2415–25.
185. Poloni M, Facchetti D, Mai R, et al. Circulating levels of tumour necrosis factor-alpha and its soluble receptors are increased in the blood of patients with amyotrophic lateral sclerosis. *Neurosci Lett* 2000;**287**(3):211–4.
186. Veglianese P, Lo Coco D, Bao Cutrona M, et al. Activation of the p38MAPK cascade is associated with upregulation of TNF alpha receptors in the spinal motor neurons of mouse models of familial ALS. *Mol Cell Neurosci* 2006;**31**(2):218–31.
187. Frakes AE, Ferraiuolo L, Haidet-Phillips AM, et al. Microglia induce motor neuron death via the classical NF-kappaB pathway in amyotrophic lateral sclerosis. *Neuron* 2014;**81**(5):1009–23.
188. Sasaki S, Warita H, Abe K, Iwata M. Inducible nitric oxide synthase (iNOS) and nitro-tyrosine immunoreactivity in the spinal cords of transgenic mice with a G93A mutant SOD1 gene. *J Neuropathol Exp Neurol* 2001;**60**(9):839–46.
189. Hensley K, Floyd RA, Gordon B, et al. Temporal patterns of cytokine and apoptosis-related gene expression in spinal cords of the G93A-SOD1 mouse model of amyotrophic lateral sclerosis. *J Neurochem* 2002;**82**(2):365–74.
190. Seeburger JL, Tarras S, Natter H, Springer JE. Spinal cord motoneurons express p75NGFR and p145trkB mRNA in amyotrophic lateral sclerosis. *Brain Res* 1993;**621**(1):111–5.
191. Lowry KS, Murray SS, McLean CA, et al. A potential role for the p75 low-affinity neurotrophin receptor in spinal motor neuron degeneration in murine and human amyotrophic lateral sclerosis. *Amyotroph Lateral Scler Other Motor Neuron Disord* 2001;**2**(3):127–34.
192. Frade JM, Rodriguez-Tebar A, Barde YA. Induction of cell death by endogenous nerve growth factor through its p75 receptor. *Nature* 1996;**383**(6596):166–8.
193. Pehar M, Cassina P, Vargas MR, et al. Astrocytic production of nerve growth factor in motor neuron apoptosis: implications for amyotrophic lateral sclerosis. *J Neurochem* 2004;**89**(2):464–73.
194. Lin CL, Bristol LA, Jin L, et al. Aberrant RNA processing in a neurodegenerative disease: the cause for absent EAAT2, a glutamate transporter, in amyotrophic lateral sclerosis. *Neuron* 1998;**20**(3):589–602.
195. Meyer T, Fromm A, Munch C, et al. The RNA of the glutamate transporter EAAT2 is variably spliced in amyotrophic lateral sclerosis and normal individuals. *J Neurol Sci* 1999;**170**(1):45–50.
196. Wong PC, Pardo CA, Borchelt DR, et al. An adverse property of a familial ALS-linked SOD1 mutation causes motor neuron disease characterized by vacuolar degeneration of mitochondria. *Neuron* 1995;**14**(6):1105–16.
197. Hall ED, Oostveen JA, Gurney ME. Relationship of microglial and astrocytic activation to disease onset and progression in a transgenic model of familial ALS. *Glia* 1998;**23**(3):249–56.
198. Phatnani HP, Guarnieri P, Friedman BA, et al. Intricate interplay between astrocytes and motor neurons in ALS. *Proc Natl Acad Sci* 2013;**110**(8):E756–65.
199. Ferraiuolo L. The non-cell-autonomous component of ALS: new in vitro models and future challenges. *Biochem Soc Trans* 2014;**42**(5):1270–4.
200. Wirenfeldt M, Babcock AA, Vinters HV. Microglia - insights into immune system structure, function, and reactivity in the central nervous system. *Histol Histopathol* 2011;**26**(4):519–30.
201. Hanisch UK, Kettenmann H. Microglia: active sensor and versatile effector cells in the normal and pathologic brain. *Nat Neurosci* 2007;**10**(11):1387–94.
202. Wu Y, Dissing-Olesen L, MacVicar BA, Stevens B. Microglia: dynamic mediators of synapse development and plasticity. *Trends Immunol* 2015;**36**(10):605–13.

203. Olah M, Biber K, Vinet J, Boddeke HW. Microglia phenotype diversity. *CNS Neurol Disord Drug Targets* 2011;**10**(1):108–18.
204. Engelhardt JI, Appel SH. IgG reactivity in the spinal cord and motor cortex in amyotrophic lateral sclerosis. *Arch Neurol* 1990;**47**(11):1210–6.
205. Maihofner C, Probst-Cousin S, Bergmann M, Neuhuber W, Neundorfer B, Heuss D. Expression and localization of cyclooxygenase-1 and -2 in human sporadic amyotrophic lateral sclerosis. *Eur J Neurosci* 2003;**18**(6):1527–34.
206. Henkel JS, Engelhardt JI, Siklos L, et al. Presence of dendritic cells, MCP-1, and activated microglia/macrophages in amyotrophic lateral sclerosis spinal cord tissue. *Ann Neurol* 2004;**55**(2):221–35.
207. Henkel JS, Beers DR, Zhao W, Appel SH. Microglia in ALS: the good, the bad, and the resting. *J Neuroimmune Pharmacol* 2009;**4**(4):389–98.
208. Weydt P, Yuen EC, Ransom BR, Moller T. Increased cytotoxic potential of microglia from ALS-transgenic mice. *Glia* 2004;**48**(2):179–82.
209. Urushitani M, Sik A, Sakurai T, Nukina N, Takahashi R, Julien JP. Chromogranin-mediated secretion of mutant superoxide dismutase proteins linked to amyotrophic lateral sclerosis. *Nat Neurosci* 2006;**9**(1):108–18.
210. Roberts K, Zeineddine R, Corcoran L, Li W, Campbell IL, Yerbury JJ. Extracellular aggregated Cu/Zn superoxide dismutase activates microglia to give a cytotoxic phenotype. *Glia* 2013;**61**(3):409–19.
211. Zhao W, Beers DR, Bell S, et al. TDP-43 activates microglia through NF-kappaB and NLRP3 inflammasome. *Exp Neurol* 2015;**273**:24–35.
212. Zhao W, Xie W, Xiao Q, Beers DR, Appel SH. Protective effects of an anti-inflammatory cytokine, interleukin-4, on motoneuron toxicity induced by activated microglia. *J Neurochem* 2006;**99**(4):1176–87.
213. Lee Y, Morrison BM, Li Y, et al. Oligodendroglia metabolically support axons and contribute to neurodegeneration. *Nature* 2012;**487**(7408):443–8.
214. Kang SH, Li Y, Fukaya M, et al. Degeneration and impaired regeneration of gray matter oligodendrocytes in amyotrophic lateral sclerosis. *Nat Neurosci* 2013;**16**(5):571–9.
215. Philips T, Bento-Abreu A, Nonneman A, et al. Oligodendrocyte dysfunction in the pathogenesis of amyotrophic lateral sclerosis. *Brain J Neurol* 2013;**136**(Pt 2):471–82.
216. Chiu IM, Chen A, Zheng Y, et al. T lymphocytes potentiate endogenous neuroprotective inflammation in a mouse model of ALS. *Proc Natl Acad Sci USA* 2008;**105**(46):17913–8.
217. Beers DR, Henkel JS, Zhao W, et al. Endogenous regulatory T lymphocytes ameliorate amyotrophic lateral sclerosis in mice and correlate with disease progression in patients with amyotrophic lateral sclerosis. *Brain J Neurol* 2011;**134**(5):1293–314.
218. Henkel JS, Beers DR, Wen S, et al. Regulatory T-lymphocytes mediate amyotrophic lateral sclerosis progression and survival. *EMBO Mol Med* 2013;**5**(1):64–79.
219. Schwartz M, Baruch K. The resolution of neuroinflammation in neurodegeneration: leukocyte recruitment via the choroid plexus. *EMBO J* 2014;**33**(1):7–22.
220. Baruch K, Schwartz M. CNS-specific T cells shape brain function via the choroid plexus. *Brain Behav Immun* 2013;**34**:11–16.
221. Kunis G, Baruch K, Rosenzweig N, et al. IFN-γ-dependent activation of the brain's choroid plexus for CNS immune surveillance and repair. *Brain* 2013;**136**:3427–40.
222. Smith R, Myers K, Ravits J, Bowser R. Amyotrophic lateral sclerosis: is the spinal fluid pathway involved in seeding and spread? *Med Hypotheses* 2015;**85**(5):576–83.
223. Kwok JC, Dick G, Wang D, Fawcett JW. Extracellular matrix and perineuronal nets in CNS repair. *Dev Neurobiol* 2011;**71**(11):1073–89.
224. Karetko M, Skangiel-Kramska J. Diverse functions of perineuronal nets. *Acta Neurobiol Exp (Wars)* 2009;**69**(4):564–77.
225. Reimers S, Hartlage-Rubsamen M, Bruckner G, Rossner S. Formation of perineuronal nets in organotypic mouse brain slice cultures is independent of neuronal glutamatergic activity. *Eur J Neurosci* 2007;**25**(9):2640–8.

226. Deepa SS, Umehara Y, Higashiyama S, Itoh N, Sugahara K. Specific molecular inter-actions of oversulfated chondroitin sulfate E with various heparin-binding growth factors. Implications as a physiological binding partner in the brain and other tissues. *J Biol Chem* 2002;**277**(46):43707–16.

227. Collins MA, An J, Hood BL, Conrads TP, Bowser RP. Label-Free LC-MS/MS proteomic analysis of cerebrospinal fluid identifies protein/pathway alterations and candidate biomarkers for amyotrophic lateral sclerosis. *J Proteome Res* 2015;**14**(11):4486–501.

228. Forostyak S, Homola A, Turnovcova K, Svitil P, Jendelova P, Sykova E. Intrathecal delivery of mesenchymal stromal cells protects the structure of altered perineuronal nets in SOD1 rats and amends the course of ALS. *Stem Cells* 2014;**32**(12):3163–72.

229. Lim GP, Backstrom JR, Cullen MJ, Miller CA, Atkinson RD, Tokes ZA. Matrix metal-loproteinases in the neocortex and spinal cord of amyotrophic lateral sclerosis patients. *J Neurochem* 1996;**67**(1):251–9.

230. Lorenzl S, Narr S, Angele B, et al. The matrix metalloproteinases inhibitor Ro 28-2653 [correction of Ro 26-2853] extends survival in transgenic ALS mice. *Exp Neurol* 2006;**200**(1):166–71.

231. Brkic M, Balusu S, Libert C, Vandenbroucke RE. Friends or foes: matrix metallopro-teinases and their multifaceted roles in neurodegenerative diseases. *Mediators Inflamm* 2015;**2015**:620581.

232. Keller MF, Ferrucci L, Singleton AB, et al. Genome-wide analysis of the heritability of amyotrophic lateral sclerosis. *JAMA Neurol* 2014

233. Cady J, Allred P, Bali T, et al. Amyotrophic lateral sclerosis onset is influenced by the burden of rare variants in known amyotrophic lateral sclerosis genes. *Ann Neurol* 2015;**77**(1):100–13.

An Introduction to the Natural History, Genetic Mapping, and Clinical Spectrum of Spinal Muscular Atrophy

A. McDonough, L. Urquia and N. Boulis

Emory University, Atlanta, GA, United States

OUTLINE

INTRODUCTION: EPIDEMIOLOGY

The neurodegenerative disease spinal muscular atrophy (SMA) is the leading genetic cause of infant mortality worldwide.[1] Incidence and prevalence estimates vary widely across populations sampled; studies have estimated the incidence of SMA at 7.8 in 100,000 births (Northeast Italy), 9.80 in 100,000 births (West-Thuringen, Germany), and 3.53 per 100,000 births (Cuba).[2-4] The study in Cuba further analyzed incidence by self-reported race, and noted a stark contrast between the incidence of SMA in whites (8 per 100,000 births) and blacks (0.89 per 100,000 births).[4] A summary of incidence studies is provided in Table 5.1.

Overall prevalence of SMA is estimated at 1 per 10,000 people.[5] Twice as many boys as girls are affected, and female cases tend to be less severe, with female incidence of SMA decreasing with age.[6] Approximately 60% of cases are classified as Type I (inability to sit, onset <6 months, early respiratory failure), 27% of cases are Type II (onset 6–18 months, ability to sit but not stand), 12% are Type III (onset >18 months, ability to stand), and 1% are Type IV (adult-onset, proximal limb weakness but no major muscular impairment).[7]

SMA is caused by mutation or deletion of the survival motor neuron (SMN) gene. Since the autosomal recessive inheritance pattern of the condition was discovered, many studies have taken advantage of large sets of screening data to examine rates of carrier frequency across populations. Screening tests identify carriers as individuals with one functional copy of the SMN1 gene. Carrier frequencies vary substantially based on race. A 2012 study of 72,453 specimens at the Genzyme Genetics Molecular Diagnostic Laboratory in Westborough, MA, USA, found the following carrier frequencies across six racial categories: Caucasian, 1 in 47; Ashkenazi Jew, 1 in 67; Asian, 1 in 59; African American, 1 in 72; Hispanic, 1 in 68. The overall carrier frequency was calculated to be 1 in 54.[8] A population-based cohort study in Taiwan found a carrier frequency of 1 in 48 based on screening of 107,611 pregnant women.[9]

HISTORICAL CONTEXT OF SPINAL MUSCULAR ATROPHY: REVIEW OF CASE STUDIES AND HISTORY OF CATEGORIZATION

The first documentation of SMA appeared in the literature in the 1890s. As many additional cases were described over the next century, the disease was gradually organized into subtypes based on severity. By the middle of the 20th century, three subtypes were established, based on onset and clinical progression criteria.[9-11] In 1995, the discovery of the

TABLE 5.1 Incidence of Spinal Muscular Atrophy

Authors	Year	Location	Years	Method	Subtypes	Reported annual incidence	Incidence per 100,000 live births
Pearn	1978	Northeast England	1956–72	Census and hospital registry data	Type II, III	1/24,100	4.15
Spiegler et al.	1990	Warsaw	1976–85	Hospital and genetic center survey	Type I, II, III	1/19,474	5.14
Burd et al.	1991	North Dakota	1980–87	Review of death certificates	Type I	1/6,720	14.88
Mostacciuolo et al.	1992	Veneto, Italy	1960–83	Hospital survey	Type I, II, III	7.8/100,000	7.80
Thieme et al.	1993	West-Thüringen, Germany	1974–87	Hospital survey	Type I	1/10,202	9.80
Zalvidar et al.	2005	Cuba	1996–2002	National database review	Type I	3.53/100,000	3.53

genetic underpinnings of SMA revolutionized our understanding of this disease and offered the promise of new therapeutics.

The First Case Studies

Written documentation of the natural history of SMA dates to 1891, when Guido Werdnig, an Austrian neurologist at the University of Graz, wrote case studies on 3 year old Wilhelm Bauer and his 1 year old brother.[12] Wilhelm developed weakness in his proximal limbs at about 1 year of age, and over the subsequent 2 years progressively lost muscle tone, voluntary movements, and the ability to swallow and hold up his head. Just before Wilhelm's 3rd birthday, Werdnig noted that the boy exhibited weakened muscles, large fat deposits, flexed legs, tremor in hands and arms, and very limited, labored movement.

Less than a month after this clinical examination, Wilhelm developed rales and exhibited retraction of intercostal muscles. Two days later, he developed dyspnea and fever; several days after the onset of dyspnea, he passed away. His younger brother lived until 6 years of age and died of similar respiratory complications. Werdnig completed a microscopic examination of the older brother's spinal cord cross sections and gastrocnemius muscle. Within the second and third cervical nerves, he noted several abnormal findings: regions of the lateral tracts lacking in myelinated fibers, degenerated anterior funiculi, indistinct processes in the ganglion cells of the anterior horn of the spinal cord, and many empty cell-beds. Posterior horns and roots were observed to be normal (Fig. 5.1). These findings were consistent across many of the cervical, thoracic, and lumbar sections.

Upon examination of the gastrocnemius muscle, he noted groups of muscle fibers separated from one another by large masses of fatty tissue. These fibers included single fibers with disrupted cross-striations and degraded contractile tissue that resembled flattened tubes with nuclei (Fig. 5.1).

Based on the clinical history and autopsy, Werdnig concluded that Wilhelm, along with his younger brother who similarly lost muscle function in his legs yet maintained normal sensation, had an infantile, familial muscular atrophy. This illness, Werdnig posited, resembled a neurological illness due to its swift course of atrophy. Thus, it was distinct from muscular dystrophy, or a slow degradation of muscle mass due to a lack of functional protein in the muscle.

Over the next decade, German neurologist Johann Hoffmann published five papers in accord with Werdnig's conclusions.[13–16] He documented a total of seven cases across four families, all with onset around 1 year of age. Age of death varied between 14 months and 5 years.[13] Hoffmann's

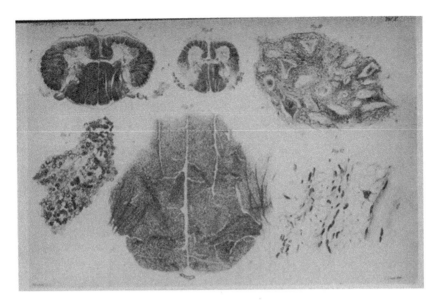

FIGURE 5.1 Histological analysis of Werdnig's first case study. Cross section of seventh cervical level showing honeycombed anterior horn (I); cross section of second lumbar level (II); anterior horn at L2–3 level, demonstrating empty cell beds (b, c) and some shriveled ganglion cells (III); posterior columns at sixth cervical nerve level (IV); posterior columns at sixth cervical level, higher magnification (V); longitudinal section of gastrocnemius muscle displaying fiber atrophy (VI).[12]

autopsies echoed two of Werdnig's key observations: atrophy of muscle in the extremities (with proximal onset) and damaged cells in the anterior horn of the spinal cord.[13] Hoffmann agreed with the majority of Werdnig's evidence, disputing only his finding that affected children exhibited hand tremor.[17] Through these case studies, Hoffmann reinforced evidence for a familial, early-onset motor neuron disease, which he called *spinale muskelatrophie*, or "spinal muscular atrophy."[14]

The Emergence of Subtypes

Over the next century, additional case studies helped solidify the medical definition of SMA and divide the illness into broad severity categories (severe, intermediate, mild).

Following Werdnig and Hoffmann's influential findings, two papers published at the turn of the 20th century shed light on the most severe form of infantile SMA. In 1899, Sylvestre presented "Paralysie flasque de quatre membres et des muscles du tronc (sauf le diaphragme) chez un nouveau-né" (translation: Flaccid paralysis of the four limbs and intercostal muscles (except the diaphragm) in a newborn) at the Pediatric Society of Paris. The case, which chronicles a 2 month old with spinal muscle atrophy since birth, was later published in the *Bulletins de la Société de Pédiatrie de Paris*. Two other children in the infant's family exhibited similar symptoms and passed away before 6 months of age.[18]

Four years later, British neurologist Charles Edward Beevor published a comparative study of one case of severe infantile SMA and one case of spinal cord hemorrhage at birth.[19] Two infants were admitted to the National Hospital, Queen Square, in London, with similar patterns of paralysis in the legs and torso. Beevor included detailed notes on the family history, case history, and postmortem examination of both infants with accompanying illustrations by Dr. F Batten.

Within the comprehensive description of Case 1, the child with SMA, several details stand out that link the case to previous documentation of SMA. Of the 5 week old infant's seven siblings, three had exhibited similar paralysis and died before 8 months of age, demonstrating familial inheritance. In terms of clinical history, Beevor's description of muscle degradation paralleled that of Sylvestre: flaccid limbs, paralysis of limb and intercostal muscles, and normal diaphragm function. Finally, the postmortem examination revealed atrophy of anterior horn cells at the cervical, thoracic, and lumbar levels that echoed Werdnig and Hoffmann's findings a decade earlier. A comparison of Case 1 and normal tissue revealed that Case 1's anterior horn cells at the level of the second lumbar were significantly smaller, measuring less than 0.03 mm, while those of the normal tissue measured between 0.04 and 0.05 mm. All peripheral nerves appeared normal in Case 1 with the exception of degenerate nerves in the brachial plexuses. Additional spinal cord degeneration was found in lumbar to cervical posterior columns and in the 5th right cervical root. Staining of right bicep muscle tissue showed extremely atrophied muscle fibers (0.008 mm in diameter compared to a normal diameter of 0.02 mm).

Although Beevor's case paralleled that of Werdnig and Hoffmann in many of its clinical details, it was distinct in its age of onset and clinical progression. The infant's mother reported that she felt no movement during pregnancy, suggesting that the paralysis developed in utero. The newborn was also fully paralyzed at birth. The infant passed away at 8 weeks, a considerably more rapid progression than that of Werdnig's and Hoffmann's cases. Thus, Beevor established comprehensive evidence for the existence of a congenital form of SMA. In our modern classification system, this case is representative of type 0 or type 1 SMA.

The third major subtype of SMA, a milder variety with age of onset in childhood and adolescence, emerged in the literature in the 1950s. Two research groups initially described this variant of SMA as a condition simulating muscular dystrophy. Swedish neurologists Wohlfart, Eliason, and Fex published a study of three families with hereditary muscle weakness and atrophy caused by nerve damage, and concluded that it could possibly be a "benign variant" of severe SMA.[20] One year later, Kugelberg and Welander presented 12 cases of progressive limb weakness originally misdiagnosed as muscular dystrophy but later found to be neurogenic through muscle biopsy and electromyographic data.[21] The study found a pattern of onset of muscle weakness in thigh and pelvic muscles and loss of knee reflexes as the first neurological positive finding. Muscle weakness gradually progressed in these cases over 20–40 years. Kugelberg and Welander also argued that this disease was distinct from Werdnig–Hoffmann syndrome due to its delayed onset and milder symptoms, a theory that genetic analysis has since overturned.[17] Today, type III SMA is interchangeably referred to as Kugelberg–Welander disease, or juvenile SMA.

Modern Clinical Presentation and Diagnostic Criteria

The first classification system was created by Byers and Banker in the 1960s and included Group I, Group II, and Group III designations, in order from most to least severe muscle degradation and earliest to latest onset, based on 52 cases of muscular atrophy at Boston Children's Hospital.[17] Group I cases were characterized by onset before 2 months of age, widespread muscle weakness, and early death. Group II cases were defined by an onset between 2 and 12 months of age, initially localized muscle weakness, and longer survival. Finally, Group III included cases with onset after 1 year of age and survival measured in years as opposed to months.[17]

In the early 1990s, soon after the discovery of gene linkage to chromosome 5q, the International SMA Collaboration convened to establish specific, standard diagnostic criteria for SMA to assist clinicians in recognition and diagnosis of disease.[11] Inclusion criteria were broken up into two categories: weakness, further characterized as symmetrical, proximal > distal, legs >arms, and trunk involved; and denervation, confirmed by electromyographic diagnosis, muscle biopsy, and clinical fasciculations. Exclusions included central nervous system dysfunction, arthrogryposis, involvement of other neurologic system of other organs, sensory loss, eye muscle weakness, and marked facial weakness.[11]

In 1998, the 59th International Workshop of the European Neuromuscular Center revised the above criteria to include electrophysiological, histopathological, and genetic inclusion and exclusion criteria. Notably, when present alongside clinical symptoms, a mutation in or absence of the SMN gene is diagnostic. Additional modified inclusion criteria include abnormal spontaneous electromyographic activity, increased mean duration and amplitude of motor unit action potentials, and atrophic and type I hypertrophic fibers.[10] The revised criteria also accounted for observed exceptions to the 1990 criteria; for instance, severe congenital cases often exhibit some sensory loss and arthrogryposis.

These workshops distilled the volume of SMA studies into the three broad categories of severe, intermediate, and mild (I–III) that echoed Byers and Banker's prior organization (Table 5.2).

Patients with type I SMA develop symptoms by 6 months of age and face a 38.1% survival probability at 2 years of life.[22] These SMA patients account for approximately 50% of cases, making it the most common form of the disease. Clinical characteristics of type I patients include generalized hypotonia, impaired bulbar function, and respiratory insufficiency. These children are defined by their inability to sit unsupported.[20,21] An increase in survival of type I patients has been shown with the use of noninvasive, assisted ventilation.[23] This type of SMA is also referred to as Werdnig–Hoffmann disease, despite the fact that the clinical criteria do not match Werdnig or Hoffmann's case studies.

Type II SMA is categorized today by an onset of proximal limb weakness at 6–18 months of age with progressive weakness, scoliosis, joint contractures, and substantial respiratory morbidity developing into childhood. These children achieve the ability to sit independently but never stand or walk. It is usually not fatal (70% of those affected reach 25 years of age) but necessitates wheelchair assistance. It is also referred to as Dubowitz disease, after the English neurologist Victor Dubowitz, who cataloged many additional intermediate SMA cases in 1960.[24]

Patients with type III SMA develop onset of proximal weakness at >1 year of age and experience milder symptoms and a normal lifespan. Clinical characteristics include heel cord tightness and contractures, tremor, lumbar lordosis, and an increased risk for fractures and scoliosis. Though weakness continues to progress throughout childhood and adolescence, type III SMA patients will achieve independent ambulation.[25]

Additional Diagnostic Categories and Challenges

It is worth noting that several different categorizations of SMA subtypes persist in the literature. Certain clinicians advocate for three solely functional categories: nonsitters, sitters, and walkers.[26] Others utilize a four-category system, with SMA subtypes I, II, III, and IV. These are assigned on the basis of clinical factors such as age of onset, motor milestones, and prognosis. Types I through IV are, respectively, infantile (Werdnig–Hoffmann), intermediate, mild, and adult-onset. Type IV SMA, or adult-onset SMA, is characterized by mild proximal limb weakness and an onset after 18 years of age. Onset of limb weakness usually occurs after 30 years of age and occurs gradually, with proximal lower limbs affected first. Quadriceps, ileo-psoas, glutei, triceps brachialis, and deltoid are

TABLE 5.2   Characterization of Spinal Muscular Atrophy Phenotypes

Nomenclature	Age of onset	Natural lifespan	Maximum milestone of development	Clinical manifestations (age of development)
TYPE 0				
Arthrogryposis multiplex congenita type	Prenatal	<6 months	Unable to breath unsupported	Congenital hypotonia, weakness, proximal joint contractures, respiratory failure
TYPE I				
Severe infantile SMA, Werdnig–Hoffmann disease	Birth–6 months	38.1% survival at 2 years of life	Typically breathes unsupported, unable to sit unsupported	Generalized hypotonia, impaired bylbar function, respiratory insufficiency
TYPE II				
Intermediate SMA, Dubowitz-type	6–18 months	70% survival to adulthood	Sits unsupported, unable to ambulate independently	Progressive weakness, scoliosis, joint contractures, respiratory morbidity
TYPE III				
Mild SMA, Kugelberg–Welander disease	>18 months	Normal	Achieves independent ambulation	Heel cord tightness and contractures, tremor and lumbar lordosis, increased risk for fractures and scoliosis
TYPE IV				
Adult-type	Adulthood	Normal	Achieves independent ambulation	Proximal leg weakness

the muscles that most commonly show signs of atrophy.[27,28] Additional symptoms vary widely but may include hand tremors and tongue fasciculations.[28] There are no major motor impairments and the natural lifespan is normal. SMN1 mutations are implicated in some cases of Type IV SMA, but in many cases there are no clear genetic causes.[27]

Type I can be further divided into IA (congenital onset), IB (0–3 months of age onset), and IC (3–6 months of age onset). IB and IC are referred to as "nonsitters" due to a lack of ability to sit unsupported. In a five-subtype system, such as that proposed by MacLeod in 1999, antenatal onset is distinctly referred to as Type 0, while early onset (0–6 months) is named Type 1.[24,29] Type 0, known also as arthrogryposis multiplex congenital type, is at the most severe end of the SMA spectrum with age of onset in utero and death within weeks of birth. The average lifespan of these infants is less than 6 months. Patients exhibit congenital hypotonia, weakness, proximal joint contractures, and respiratory failure. These patients require respiratory support measures to maintain life.

A growing body of evidence suggests the substantial variability within diagnostic categories and calls into question the reliability of certain diagnostic criteria. For instance, a 2002 study by Borkowska and colleagues followed 349 SMA Type 1 patients and observed that 36, or approximately 10%, survived until their 5th birthdays.[23] Classic diagnostic criteria maintain that Type I patients will die before their 2nd birthdays. This finding therefore disputes the notion that data on age of death should be used as the basis for creation of subtypes.

Victor Dubrowitz proposed a new numerical system of classification that further breaks down types I, II, and III into intervals of 0.1.[25] Therefore, a child with early onset and severe paralysis would not necessarily fall under the same umbrella type as a child with early onset who is barely unable to sit without aid. This system could account for the variability of survival, motor, and respiratory function within broad types.

Our current knowledge of SMN protein function supports Dubrowitz's notion of an SMA severity spectrum. The SMN protein is likely multifunctional, and its effects are thus likely dependent on specific protein levels. Select inhibitions of function may correlate to specific degrees of severity of disease progression. Our growing understanding of SMN protein function will help to further elucidate the complex variation of SMA presentation in the future. A more precise classification of the many forms of this illness will assist in accurate diagnosis and tailored treatment for each individual afflicted.

Our modern understanding of SMN protein and SMA subtypes also suggests that the disease may be neurodevelopmental in addition to neurodegenerative. SMN protein levels are particularly high in motor neurons during prenatal nervous system development in healthy individuals, suggesting a critical role for SMN in normal motor function. Furthermore,

SMA type 0 and type I patients exhibit pathological features consistent with incomplete formation of motor neuron pathways: immature motor neurons and myofibers with ongoing apoptosis. In a mouse model of SMA, neuromuscular junction synapses were impaired due to decreased vesicle density and conservation of the fetal achetylcholine receptor subunit.[30] It is posited that Type II and III SMA, associated with higher SMN2 copy number and thus higher levels of SMN protein, are likely more neurodegenerative in nature: motor neurons and muscle tissue develop normally during the gestational period but are vulnerable to degradation over time.[31]

OUR MODERN UNDERSTANDING OF THE DISEASE: MAPPING EFFORTS AND IDENTIFICATION OF THE SURVIVAL MOTOR NEURON GENE

The collection of case studies completed by the end of the 20th century led to predictions that mutations at a single chromosomal region cause differences between types of SMA. In 1990, investigators studied families diagnosed with Types I, II, and III SMA and found that clinically heterogeneous forms of SMA map to a single 10 cM locus of chromosome 5q13.[32,33] Later narrowed to a 1 to 2 cM region using recombinant mapping, this candidate region was flanked by multicopy repeats of microsatellites. Some multicopy microsatellites were shown to be in linkage disequilibrium with the disease locus.[6] Characterization of these sequences in unrelated SMA patients showed large-scale chromosomal rearrangements, including gene duplications, gene conversions, and de novo deletions, encompassing the disease locus. Results from these early studies indicate that rearrangement events in this unstable region are statistically associated with the severe form of SMA.[34–37]

The SMN gene was identified as the SMA-determining gene in 1995 by Lesebvre and colleagues who began by identifying a large inverted duplication of a 500 kb element within the candidate region in normal chromosomes.[38] These elements were termed E^{Tel} and E^{Cen} for their location in the telomeric and centromeric regions, respectively. A comparison of DNA from both control and SMA patients revealed genomic rearrangements in the E^{Tel} of SMA patients. A human fetal cDNA library was probed with genomic DNA from the E^{Tel} candidate region, revealing a 20 kb gene encoding a novel protein of 294 amino acids. The mRNA transcript resulting in this protein was found to have eight exons. In a study of 229 SMA patients with ranging disease severity, this telomeric gene (SMN1, SMN^T) was either absent or truncated due to interruption or deleterious mutation in 98.6% of patients. Of these 229 patients, 213 patients (93%) lacked the SMN1 exons 7 and 8 on both mutant chromosomes.[38]

Evidence for a highly homologous gene in the E^{Cen} of 95% of controls was provided using Southern blot and pulsed-field gel electrophoresis analyses. The centromeric gene (SMN2, SMN^C) differs from SMN1 by only five nucleotides, none of which affect the predicted amino acid sequence. A 1.7 kb mRNA transcript correlating to the SMN2 gene was identified using northern blot analysis in both control and SMA patients. However, when compared to eight controls, the amount of SMN2-specific mRNA was markedly less in four of seven SMA type I patients.[38] Sequence analysis of this mRNA product revealed that most SMN2 gene transcripts undergo alternative splicing of exon 7, resulting in a truncated protein. This protein is nonfunctional and thought to be rapidly degraded.[39–41] Controls lacking SMN2 on both chromosomes expressed only full-length mRNA while controls carrying both SMN2 and SMN1 genes expressed both full-length and truncated mRNA. This suggested that the exon 7-deleted truncated mRNA transcripts were specific to the SMN2 gene.[38]

Molecular Pathogenesis of the Survival Motor Neuron 2 Gene

Implications for SMA disease pathogenesis surfaced in the late 90s with important associations between differential SMN2 expression and clinical phenotype. Study of genomic variation, gene conversion events, and transcription patterns in SMA patients suggested that relative copy number of SMN2 could influence disease severity. Analysis of the SMN gene carried by parents of SMA patients indicated that the number of SMN2 copies is variable and higher in parents of type II and III patients compared to parents of type I patients.[42–48] The majority of parents of type II and III patients (66.5% and 75%, respectively) were found to carry three or more copies of the SMN2 gene, compared to the majority of parents of type I patients who carried one to two SMN2 copies only.[42,43] Survival probability has been found to vary dramatically as a result of SMN2 copy number. A 2015 study by Qu et al. found that, among a cohort of 232 SMA patients, survival probabilities at 5 years were 100%, 91.7%, and 5.1% depending on whether the individual had 4, 3, or 2 SMN2 copies, respectively.[49]

Further insight into the role of SMN2 gene copy number was gained when Campbell et al. found that a large number of type II and III SMA chromosomes contain a converted allele, rather than a deleted allele. The loss of SMN1 by a gene conversion event is more frequently the cause of type II and III cases, whereas the majority of type I cases result from large-scale deletions encompassing the SMN1 gene. In conversion of SMN1 to SMN2, there is a gain of a copy of SMN2 in type II and III patients.[42] These studies and findings suggest that increased copy number of SMN2 is associated with a milder disease presentation.

SMN2 fails to provide complete protection from SMA because of splicing patterns first described by Lorson et al. A single-base-pair difference between SMN1 and SMN2 disrupts an exonic splicing enhancer in SMN2. Although this base change is translationally silent, it significantly increases the frequency of exon 7 skipping, greatly reducing the efficiency of exon 7 inclusion into final transcripts from the SMN2 gene.[39,50] Densitometry analysis of SMA type I, SMA type II, SMA type III, and control patient RNA samples revealed a significant correlation between the ratio of exon 7-deleted transcript products and clinical phenotype. SMA type I and II patients produced 27% and 21% more truncated transcript, respectively, compared to control patients. SMA type III patients produced 54% less truncated transcript than controls. However, each SMA subtype produced approximately 40–45% as much full-length transcript as compared to control cells, and there was no significant difference in the amount of full-length transcript produced among the three SMA subtypes.[51] It can be implied from this data that SMN2 produces a dramatically reduced amount of full-length mRNA and abundant levels of truncated protein from aberrantly spliced mRNAs.

The Survival Motor Neuron Protein

The SMN protein is expressed in the cytoplasm and nucleus of all eukaryotic cells, with widespread but uneven distribution in the central nervous system.[52,53] Particularly high levels of protein were found to localize to the lower motor neurons of the spinal cord and brainstem.[52] Within the nucleus, SMN protein is found in 0.1 to 1-µm structures named gems.[54] Gems, from "Gemini of the coiled bodies," appear to interact directly with coiled bodies, which are subnuclear structures thought to be significant in mRNA metabolism. The SMA phenotype was found to correlate with the number of gems as detected by SMN antibodies, where cells from affected patients had significantly fewer gems than normal carriers or controls. Furthermore, type II patients had significantly more gems compared to type I patients despite having equivalent copy numbers of SMN1 and SMN2.[53]

The physiological role of SMN protein is complex, and the reason that motor neurons are particularly vulnerable to reduced SMN protein levels is not well understood. However, research into the structure and function of SMN protein has proposed a molecular basis to support its pathogenetic role in motor neuron degeneration and death that characterizes SMA disease phenotypes. Qualitative and/or quantitative abnormalities of the SMN protein may play a role in motor neuron function.

TREATMENT AND STANDARD OF CARE

At present, there is no effective medical treatment available for SMA patients. However, recent advances in medical technology have expanded diagnostic options and improved the standard of care for patients with SMA.

The discovery and enhanced understanding of the SMA-determining gene, SMN1, has led to the development of diagnostic strategies for early disease detection. The fetal genotype can be screened for homozygous deletion of SMN1 using chorionic villi sampling and amniocentesis.[55,56] For a less invasive prenatal diagnosis, fetal cells can be isolated from a small amount of peripheral maternal blood and analyzed with polymerase chain reaction (PCR).[57] Neonates and infants who present with SMA-like clinical symptoms should be tested for homozygous deletion of the SMN1 gene using one of several PCR detection strategies.[58–60] This mutation analysis has a sensitivity of approximately 95% in affected homozygotes; however, it cannot detect carriers with heterozygous SMN1 deletions. Potential carriers should be screened using gene dosage analysis. This technique does have an important limitation. Five percent of the population has three copies of SMN1, likely due to a tandem duplication at this locus. This finding has an important implication for dosage analysis. A small fraction of people that have two copies of SMN1 will, in fact, be carriers for SMA because they have one allele with two copies of SMN1 and another where SMN1 is absent.[61]

Despite the limitations of the currently available carrier tests, the American College of Medical Genetics recommends broad-based population carrier screening for parents with or without a family history of SMA. This recommendation was made contingent on the availability of educational materials and genetic counseling to provide parents with information about the implications of false-negative results.[61] Additionally, SMN1 detection via allele-specific PCR assay of embryos grown in culture offers parents the choice of in-vitro fertilization using unaffected embryos to eliminate the possibility of an SMA-affected child.[62,63]

Early intervention and care for patients with SMA is critical to maximize quantity and quality of life. The clinical management of SMA is made difficult by the complexity of medical problems associated with the diagnosis, requiring pulmonary, gastrointestinal/nutrition, orthopedics/ rehabilitation, and palliative care assessments and interventions. Outside of interventions provided by medical professionals, family education and counseling should be a priority to caring for these patients. The physician should provide anticipatory guidance and formulate a plan of multidisciplinary intervention with the family.[64]

Due to a lack of therapies that target the underlying disease, the current standard of care centers on pulmonary monitoring and interventions to reduce the morbidity and mortality in patients. For infants who cannot sit unassisted, respiratory assessment should include evaluation of cough effectiveness, observation of breathing, and monitoring of gas exchange. Additional recommendations for Type II patients include the assessment of forced vital capacity and lung volume tests, assessment of sleep-disordered breathing, and pulse oximetry monitoring. Supportive care is indicated in patients with daytime hypercapnia and sleep-disordered breathing. This care typically includes nocturnal noninvasive ventilation and airway clearance. Techniques for airway clearance include oral suctioning, mechanically assisted cough (mechanical insufflation-exsufflation), daily assisted cough, and secretion mobilization techniques (including chest physiotherapy and postural drainage). An increase in survival of type I patients has been shown with the use of noninvasive, assisted ventilation.[65] For severe respiratory insufficiency or failure, a tracheotomy may be considered by the physician and deserves careful consideration from the family.[66] Lastly, due to the debilitating nature and short life expectancy of the disease, current care standards focus a significant amount of time and resources on delivering palliative care options to severely affected patients.

Advances in pulmonary monitoring have improved prognosis for individuals living with SMA. A retrospective study of 143 type I SMA patients by Oskoui and colleagues found that SMA patients born between 1995 and 2006 exhibited a significantly longer life span and 70% reduction in risk of death compared to those born between 1980 and 1994. Three clinical treatments significantly predicted reduction of risk of death: mechanical insufflation-exsufflation device, gastrostomy tube feeding, and ventilation for more than 16 hours a day.[67]

CONCLUSION

Despite many recent advances in research that aim to target the underlying genetic mechanism of SMA, many questions about the onset and progression of SMA remain. The missing link right now exists between disruption of protein production and the biological attributes of SMA – degradation of α-motor neurons and corresponding muscle atrophy. Why do loss-of-function mutations in SMN1 lead to damage of spinal cord motor neurons but not surrounding tissues and other organ systems? How do different SMN protein levels lead to the observed spectrum of disease severity? By studying the exact mechanism of gene expression and protein functions, researchers hope to identify new therapeutic targets in order to prevent disease progression. An increasing number of prospective patient studies are also being initiated around the world to assess disease natural history with greater precision. Although there are still many unanswered questions about the mechanism of SMA, recent advances provide hope that one day it will be a curable disease.

References

1. Pearn J. Classification of spinal muscular atrophies. *Lancet* 1980;**1**(8174):919–22.
2. Mostacciuolo ML, Danieli GA, Trevisan C, et al. Epidemiology of spinal muscular atrophies in a sample of the Italian population. *Neuroepidemiology* 1992;**11**:34–8.
3. Thieme A, Mitulla B, Schulze F, et al. Epidemiological data on Werdnig-Hoffmann disease in Germany (West-Thuringen). *Hum Genet* 1993;**91**:295–7.
4. Zaldivar T, Montejo Y, Acevedo AM, Guerra R, et al. Evidence of reduced frequency of spinal muscular atrophy type I in the Cuban population. *Neurology* 2005;**65**(4):636–8.
5. Monani UR, De Vivo DC. Neurodegeneration in spinal muscular atrophy: disease phenotype, animal models, therapies, and beyond: spinal muscular atrophy: clinical perspectives. *Future Neurol* 2014;**9**(1):49–65.
6. Hausmanowa-Petrusewicz I, Zaremba J, Borkowska J. Chronic proximal spinal muscular atrophy of childhood and adolescence: sex influence. *J Med Genet* 1984;**21**(6):447–50.
7. Darris BT. Spinal muscular atrophies. *Pediatr Clin N Am* 2015;**62**:743–66.
8. Sugarman EA, Nagan N, Zhu H, et al. Pan-ethnic carrier screening and prenatal diagnosis for spinal muscular atrophy: clinical laboratory analysis of > 72,400 specimens. *Eur J Hum Genet* 2012;**20**:27–32.
9. Su YN, Hung CC, Lin SY, Chen FY, et al. Carrier screening for spinal muscular atrophy (SMA) in 107,611 pregnant women during the period 2005–2009: a prospective population-based cohort study. *PLoS One* 2011;**6**(2):e17067.
10. Zerres K, Davies KE. 59th ENMC International Workshop: Spinal Muscular Atrophies: recent progress and revised diagnostic criteria. 17-19 April 1998. Soest-duinen, The Netherlands. *Neuromuscul Disord* 1999;**9**:272–8.
11. Munsat TL. Workshop report: international SMA collaboration. *Neuromuscul Disord* 1991;**1**:81.
12. Werdnig G. Two early infantile hereditary cases of progressive muscular atrophy simulating dystrophy, but on a neural basis [article in German]. *Arch Psychiat Neurol* 1891;**22**:437–81.
13. Iannaccone S, Caneris O. Johann Hoffmann *The founders of child neurology*. San Francisco, CA: Norman Publishing; 1990.278.83
14. Hoffmann J. Ueber chronische spinale Muskelatrophie im Kindesalter, auf famil- iärer basis. *Dtsch Z Nervenheilkd* 1893;**3**:427–70.
15. Hoffmann J. Weiterer Beiträge zur Lehre von der hereditären progressiven spinalen Muskelatrophie im Kindesalter. *Deut Zeitsch Nervenheilkd* 1897;**10**:292–320.
16. Hoffmann J. Dritter Eitrag zur Lehre von der hereditären progressiven spinalen Muskelatrophie im Kindesalter. *Deut Zeitsch Nervenheilkd* 1900;**18**:217–24.
17. Byers RK, Banker BQ. Infantile muscular atrophy. *Arch Neurol* 1961;**5**(2):140–64.
18. Sylvestre M. Paralysie flasque de quatre membres et des muscles du tronc (sauf le diaphragme) chez un nouveau-ne. *Bull Soc Pediatry Paris* 1899;**1**:3–10.
19. Beevor CE. A case of congenital spinal muscular atrophy (family type) and a case of hemorrhage into the spinal cord at birth, giving similar symptoms. *Brain* 1902;**25**:85–108.
20. Wohlfart G, Fex J, Eliasson S. Hereditary proximal spinal muscular atrophy – a clinical entity simulating progressive muscular dystrophy. *Acta Psychiatry Neurol (Kjobenhavn)* 1955;**30**:395–406.
21. Kugelberg E, Welander L. Heredofamilial juvenile muscular atrophy simulating muscular dystrophy. *Arch Neurol Psychiatry (Chic)* 1986;**75**:500–9.
22. Thomson J, Bruce A. Progressive muscular atrophy in a child with a spinal lesion. *Edinb Hosp Rep* 1893;**1**:372.
23. Borkowsa J, Rudnik-Schoneborn S, Hausmanowa-Petrusewicz I, Zerres K. Early infantile form of spinal muscular atrophy (Werdnig-Hoffmann disease) with prolonged survival. *Folia Neuropathol* 2002;**40**(1):19–26.

24. Dubowitz V. Ramblings in the history of spinal muscular atrophy. *Neuromuscul Disord* 2009;**19**(1):69–73.

25. Dubowitz V. Chaos in the classification of SMA: a possible resolution. *Neuromuscul Disord* 1995;**5**(1):3–5.

26. Iannaccone ST, Russman BS, Browne RH, et al. *Prospective analysis of strength in spinal muscular atrophy*. DCN/Spinal Muscular Atrophy Group. *J Child Neurol* 2000;**15**:97–101.

27. Pearn JH, Hudgson P, Walton JN. A clinical and genetic study of spinal muscular atrophy of adult onset. *Brain* 1978;**101**:591–606.

28. Brahe C, Servidei S, Zappata S, Ricci E, Tonali P, Neri G. Genetic homogeneity between childhood-onset and adult-onset autosomal recessive spinal muscular atrophy. *Lancet* 1995;**346**:741–2.

29. MacLeod MJ, Taylor JE, Lunt PW, et al. Prenatal onset spinal muscular atrophy. *Eur J Paediatry Neurol* 1999;**3**:65–72.

30. Kong L, Wang X, Choe DW, Polley M, Burnett BG, Bosch-Marcé M, et al. *J Neurosci* 2009;**29**(3):842–51.

31. d'Ydewalle C, Sumner CJ. Spinal muscular atrophy therapeutics: where do we stand? *Neurotherapeutics* 2015;**12**(2):303–16.

32. Brzustowicz LM, et al. Genetic mapping of chronic childhood-onset spinal muscular atrophy to chromosome 5q11.2-13.3. *Nature* 1990;**344**(6266):540–1.

33. Gilliam TC, et al. Genetic homogeneity between acute and chronic forms of spinal muscular atrophy. *Nature* 1990;**345**(6278):823–5.

34. Wang CH, et al. Refinement of the spinal muscular atrophy locus by genetic and physical mapping. *Am J Hum Genet* 1995;**56**(1):202–9.

35. Clermont O, et al. Use of genetic and physical mapping to locate the spinal muscular atrophy locus between two new highly polymorphic DNA markers. *Am J Hum Genet* 1994;**54**(4):687–94.

36. Melki J, et al. De novo and inherited deletions of the 5q13 region in spinal muscular atrophies. *Science* 1994;**264**(5164):1474–7.

37. McLean MD, et al. Two 5q13 simple tandem repeat loci are in linkage disequilibrium with type 1 spinal muscular atrophy. *Hum Mol Genet* 1994;**3**(11):1951–6.

38. Lefebvre S, et al. Identification and characterization of a spinal muscular atrophy-determining gene. *Cell* 1995;**80**(1):155–65.

39. Lorson CL, et al. A single nucleotide in the SMN gene regulates splicing and is responsible for spinal muscular atrophy. *Proc Natl Acad Sci USA* 1999;**96**(11):6307–11.

40. Lorson CL, Androphy EJ. An exonic enhancer is required for inclusion of an essential exon in the SMA-determining gene SMN. *Hum Mol Genet* 2000;**9**(2):259–65.

41. Chang HC, et al. Degradation of survival motor neuron (SMN) protein is mediated via the ubiquitin/proteasome pathway. *Neurochem Int* 2004;**45**(7):1107–12.

42. Campbell L, et al. Genomic variation and gene conversion in spinal muscular atrophy: implications for disease process and clinical phenotype. *Am J Hum Genet* 1997;**61**(1):40–50.

43. Velasco E, et al. Molecular analysis of the SMN and NAIP genes in Spanish spinal muscular atrophy (SMA) families and correlation between number of copies of cBCD541 and SMA phenotype. *Hum Mol Genet* 1996;**5**(2):257–63.

44. Hahnen E, et al. Molecular analysis of candidate genes on chromosome 5q13 in autosomal recessive spinal muscular atrophy: evidence of homozygous deletions of the SMN gene in unaffected individuals. *Hum Mol Genet* 1995;**4**(10):1927–33.

45. McAndrew PE, et al. Identification of proximal spinal muscular atrophy carriers and patients by analysis of SMNT and SMNC gene copy number. *Am J Hum Genet* 1997;**60**(6):1411–22.

46. Schwartz M, et al. Quantification, by solid-phase minisequencing, of the telomeric and centromeric copies of the survival motor neuron gene in families with spinal muscular atrophy. *Hum Mol Genet* 1997;**6**(1):99–104.

47. DiDonato CJ, et al. Association between Ag1-CA alleles and severity of autosomal recessive proximal spinal muscular atrophy. *Am J Hum Genet* 1994;**55**(6):1218–29.
48. Wirth B, et al. Allelic association and deletions in autosomal recessive proximal spinal muscular atrophy: association of marker genotype with disease severity and candidate cDNAs. *Hum Mol Genet* 1995;**4**(8):1273–84.
49. Qu Association of copy numbers of survival motor neuron gene 2 and neuronal apoptosis inhibitory protein gene with the natural history in a Chinese spinal muscular atrophy cohort. *J Child Neurol* 2015;**30**(4):429–36.
50. Monani UR, et al. A single nucleotide difference that alters splicing patterns distinguishes the SMA gene SMN1 from the copy gene SMN2. *Hum Mol Genet* 1999;**8**(7):1177–83.
51. Gavrilov DK, et al. Differential SMN2 expression associated with SMA severity. *Nat Genet* 1998;**20**(3):230–1.
52. Battaglia G, et al. Expression of the SMN gene, the spinal muscular atrophy determining gene, in the mammalian central nervous system. *Hum Mol Genet* 1997;**6**(11): 1961–71.
53. Coovert DD, et al. The survival motor neuron protein in spinal muscular atrophy. *Hum Mol Genet* 1997;**6**(8):1205–14.
54. Liu Q, Dreyfuss G. A novel nuclear structure containing the survival of motor neurons protein. *EMBO J* 1996;**15**(14):3555–65.
55. Kesari A, et al. Prenatal diagnosis of spinal muscular atrophy: Indian scenario. *Prenat Diagn* 2005;**25**(8):641–4.
56. Wu T, et al. Prenatal diagnosis of spinal muscular atrophy in Chinese by genetic analysis of fetal cells. *Chin Med J (Engl)* 2005;**118**(15):1274–7.
57. Beroud C, et al. Prenatal diagnosis of spinal muscular atrophy by genetic analysis of circulating fetal cells. *Lancet* 2003;**361**(9362):1013–4.
58. van der Steege G, et al. PCR-based DNA test to confirm clinical diagnosis of autosomal recessive spinal muscular atrophy. *Lancet* 1995;**345**(8955):985–6.
59. Simsek M, et al. Allele-specific amplification of exon 7 in the survival motor neuron (SMN) genes for molecular diagnosis of spinal muscular atrophy. *Genet Test* 2003;**7**(4):325–7.
60. Feldkotter M, et al. Quantitative analyses of SMN1 and SMN2 based on real-time light-Cycler PCR: fast and highly reliable carrier testing and prediction of severity of spinal muscular atrophy. *Am J Hum Genet* 2002;**70**(2):358–68.
61. Prior TW, Professional P, Guidelines C. Carrier screening for spinal muscular atrophy. *Genet Med* 2008;**10**(11):840–2.
62. Burlet P, et al. Improved single-cell protocol for preimplantation genetic diagnosis of spinal muscular atrophy. *Fertil Steril* 2005;**84**(3):734–9.
63. Girardet A, Fernandez C, Claustres M. Efficient strategies for preimplantation genetic diagnosis of spinal muscular atrophy. *Fertil Steril* 2008;**90**(2):443.e7–443.e12.
64. Bladen CL, et al. Mapping the differences in care for 5,000 spinal muscular atrophy patients, a survey of 24 national registries in North America, Australasia and Europe. *J Neurol* 2014;**261**(1):152–63.
65. Mercuri E, Bertini E, Iannaccone ST. Childhood spinal muscular atrophy: controversies and challenges. *Lancet Neurol* 2012;**11**(5):443–52.
66. Wang CH, et al. Consensus statement for standard of care in spinal muscular atrophy. *J Child Neurol* 2007;**22**(8):1027–49.
67. Oskoui M, et al. The changing natural history of spinal muscular atrophy type 1. *Neurology* 2007;**69**(20):1931–6.
68. Dubowitz V. Very severe spinal muscular atrophy (SMA type 0): an expanding clinical phenotype. *Eur J Paediatry Neurol* 1999;**3**:49–51.
69. Markowitz JA, Singh P, Darras BT. Spinal muscular atrophy: a clinical and research update. *Pediatric Neurol* 2012;**46**:1–12.

70. Carre A, Empey C. Review of spinal muscular atrophy (SMA) for prenatal and pediatric genetic counselors. *J Genet Counsel* 2014http://dx.doi.org/10.1007/s10897-015-9859-z

71. Simic G. Pathogenesis of proximal autosomal recessive spinal muscular atrophy. *Acta Neuropathol* 2008;**116**(3):223–34.

72. Finkel RS, et al. Observational study of spinal muscular atrophy type I and implications for clinical trials. *Neurology* 2014;**83**(9):810–7.

73. Khirani S, et al. Longitudinal course of lung function and respiratory muscle strength in spinal muscular atrophy type 2 and 3. *Eur J Paediatry Neurol* 2013;**17**(6):552–60.

74. Kaufmann P, et al. Prospective cohort study of spinal muscular atrophy types 2 and 3. *Neurology* 2012;**79**(18):1889–97.

75. Farrar MA, Vucic S, Johnston HM, du Sart D, Kiernan MC. Pathophysiological insights derived by natural history and motor function of spinal muscular atrophy. *J Pediatry* 2013;**162**(1):155–9.

76. Tiziano FD, Neri G, Brahe C. Biomarkers in rare disorders: the experience with spinal muscular atrophy. *Int J Mol Sci* 2010;**12**(1):24–38.

77. Lorson CL, Rindt H, Shababi M. *Spinal muscular atrophy: mechanisms and* therapeutic strategies. *Hum Mol Genet* 2010;**19**(R1):R111–8.

78. Barois A, et al. Spinal muscular atrophy. *A 4-year prospective, multicenter, longitudinal study (168 cases).* Bull Acad Natl Med 2005;**189**(6):1181–98.

79. Swoboda KJ, et al. Natural history of denervation in SMA: relation to age, SMN2 copy number, and function. *Ann Neurol* 2005;**57**(5):704–12.

80. Foust KD, Wang X, McGovern VL, et al. Rescue of the spinal muscular atrophy phenotype in a mouse model by early postnatal delivery of SMN. *Nat Biotechnol* 2010;**28**(3):271–4.

81. Burd L, Short SK, Martsolf JT, Nelson RA. Prevalence of type I spinal muscular atrophy in North Dakota. *Am J Med Genet* 1991;**41**(2):212–5.

82. Ge X, et al. The natural history of infant spinal muscular atrophy in China: a study of 237 patients. *J Child Neurol* 2012;**27**(4):471–7.

83. Zerres K, Wirth B, Rudnik-Schoneborn S. Spinal muscular atrophy--clinical and genetic correlations. *Neuromuscul Disord* 1997;**7**(3):202–7.

84. Zerres K, Rudnik-Schoneborn S. Natural history in proximal spinal muscular atrophy. Clinical analysis of 445 patients and suggestions for a modification of existing classifications. *Arch Neurol* 1995;**52**(5):518–23.

85. Lewelt A, Newcomb TM, Swoboda KJ. New therapeutic approaches to spinal muscular atrophy. *Curr Neurol Neurosci Rep* 2012;**12**(1):42–53.

86. Lefebvre S, et al. The role of the SMN gene in proximal spinal muscular atrophy. *Hum Mol Genet* 1998;**7**(10):1531–6.

87. Dubowitz V. Infantile muscular atrophy. A prospective study with particular reference to a slowly progressive variety. *Brain* 1964;**87**:707–18.

88. Zerres K, et al. Genetic basis of adult-onset spinal muscular atrophy. *Lancet* 1995;**346**(8983):1162.

89. Lorson MA, Lorson CL. SMN-inducing compounds for the treatment of spinal muscular atrophy. *Future Med Chem* 2012;**4**(16):2067–84.

90. Farooq F, et al. Prolactin increases SMN expression and survival in a mouse model of severe spinal muscular atrophy via the STAT5 pathway. *J Clin Invest* 2011;**121**(8):3042–50.

91. Dominguez E, et al. Intravenous scAAV9 delivery of a codon-optimized SMN1 sequence rescues SMA mice. *Hum Mol Genet* 2011;**20**(4):681–93.

92. Valori CF, et al. Systemic delivery of scAAV9 expressing SMN prolongs survival in a model of spinal muscular atrophy. *Sci Transl Med* 2010;**2**(35):35ra42.

93. Andreassi C, et al. Aclarubicin treatment restores SMN levels to cells derived from type I spinal muscular atrophy patients. *Hum Mol Genet* 2001;**10**(24):2841–9.

94. Angelozzi C, et al. Salbutamol increases SMN mRNA and protein levels in spinal muscular atrophy cells. *J Med Genet* 2008;**45**(1):29–31.

95. Tiziano FD, et al. Salbutamol increases survival motor neuron (SMN) transcript levels in leucocytes of spinal muscular atrophy (SMA) patients: relevance for clinical trial design. *J Med Genet* 2010;**47**(12):856–8.
96. Chen PC, et al. Identification of a maleimide-based Glycogen Synthase Kinase-3 (GSK-3) Inhibitor, BIP-135, that Prolongs the Median Survival Time of Delta7 SMA KO mouse model of spinal muscular atrophy. *ACS Chem Neurosci* 2012;**3**(1):5–11.

6

Genetics of Spinal Muscular Atrophy

A.H.M. Burghes and V.L. McGovern

The Ohio State University, Columbus, OH, United States

INTRODUCTION

Spinal muscular atrophy (SMA) refers to a group of disorders that affect the lower motor neuron, and a number of genes have been defined that cause SMA. The most common SMA is proximal SMA that maps to chromosome 5q. In contrast, the late-onset distal SMAs (DSMAs), which show considerable overlap with Charcot–Marie–Tooth disease phenotype, are mostly non-5q SMAs. For these disorders many genes have been implicated, including: *GARS, DCTN1, HSPB8, HSPB1, BSCL2, SETX, HSPB3, DYNC1H1, REEP1*, and *SLC5A7*.[1,2] Dominant SMAs with proximal weakness have mutations in *VAPB, TRPV4, LMNA*, and, most recently, *BICD2*.[1-5] The recessive non-5q SMAs are caused by mutations in *IGHMP2* for DSMA1 or SMA with respiratory distress, mutations in *Gle1* for severe lethal congenital contracture syndrome 1, which results in loss of motor neurons in the fetus, and lastly the X-linked SMA caused by mutations in *Ube1*.[1-7] Overall, there is no indication of an overlapping function that is disrupted by other causative SMA genes with the exception that, in dominant proximal SMA, the genes *VAPB, TRPV4*, and *BICD2* may play a role in vesicle release. However, the phenotype of these proximal dominant SMAs is quite different from proximal recessive 5q SMA. For instance, cases with the *BICD2* mutation show symptoms at birth with a slow progression, upper motor neuron involvement, and sparing of axial muscle.[3-5] Therefore it is possible that mutations in different SMA causative genes represent alternate paths to motor neuron destruction. Here we focus on the genetics of 5q SMA.

Proximal 5q SMA is an autosomal recessive disorder and is the most common genetic cause of infant death. Proximal 5q SMA is caused by loss or mutation of the *survival motor neuron 1* gene (*SMN1*) and retention of the copy gene *SMN2*.[8] 5q SMA has a frequency of 1/11,000 new births[9,10] and carrier frequencies range from 1/47 to 1/72, depending on racial group.[11,12] The *SMN1* and *SMN2* genes lie in a complex region of the genome that shows a very high divergence among different individuals. Thus, any one particular arrangement of the *SMN1* and *SMN2* genes does not reflect the arrangement of the genes across the population.[13,14] Of particular relevance to SMA disease is the inherent copy number variation of *SMN1* and *SMN2* in the population.[12,15] The *SMN2* copy number varies in SMA individuals and, provided that the *SMN2* gene is intact, these extra copies modify the phenotype. An inverse correlation of phenotypic severity to copy number of *SMN2* is found in SMA patients.[13,15] Notably, for many SMAs the mutated gene is expressed in a wide range of tissues and expression is not confined to the nervous system. This is also the case with SMN and as such the reason for selective motor neuron or motor circuit deficit is not known.[8,16]

THE SPINAL MUSCULAR ATROPHY GENE REGION

The SMA region is often diagramed as an inverted duplication of *SMN1* and *SMN2*. This arrangement of genes was identified in a particular yeast artificial chromosome (YAC).[8,17] However, this region is somewhat more complicated and we now know the arrangement of genes is not the same in all individuals[13,14] (Fig. 6.1). The gene for SMA was first mapped by linkage analysis to position 5q for all SMA types.[21,22] Subsequently, as in most positional cloning, closer markers and physical maps of the region between the two flanking markers were then determined.[13] However, the more closely located markers often gave rise to multiple copies in an individual (i.e., greater than two alleles) indicating duplication of the marker.[17,21–24] Furthermore, the variation in copy number of these markers between individuals indicated different associated genomic DNA. Some markers indicated reduced gene copy number in many type 1 patients thus suggesting a large deletion in this area.[13,24] The SMA area was cloned and physical maps were produced utilizing YACs.[17,21–27] One difficulty encountered while building these maps was that the YAC clones from this region frequently underwent deletion or some form of rearrangement. This could be easily detected by genotyping of the original cell line and the YAC clones, which most often contained one allele instead of multiple alleles.[25] Thus maps of the SMA region produced from YAC clones with deletions were interpreted as if only one arrangement of multiple chromosomes was possible.[13] Further studies, using pulsed field

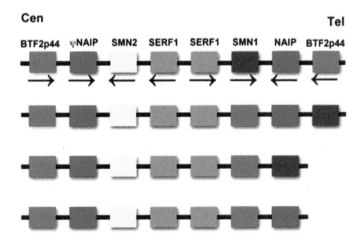

FIGURE 6.1   Diagram of the arrangement of genes in the spinal muscular atrophy (SMA) region. The survival motor neuron survival motor neuron (SMN)-gene-containing region is often depicted as a 500-kb inverted duplication based on the arrangement found in the original yeast artificial chromosome (YAC) clone. However, as depicted above, the gene can be arranged in a tandem duplication or with genes absent from one of the repeat units. Based on pulsed field gel analysis and mapping there are many possible organizations of these genes and not all organizations are depicted here. BTF2p44, *basic transcription factor 2 44-kDa subunit*[18]; ψNAIP, *Psi NLR family, apoptosis inhibitory protein* pseudo gene[19]; SMN2, *survival motor neuron 2*[8]; SERF1, *small EDRK-rich factor*[20]; SMN1, *survival motor neuron 1*[8]; NAIP, *NLR family, apoptosis inhibitory protein*[19]; Cen, centromeric; Tel, telomeric. An example of the direction of transcription in is indicated by arrows; however, other arrangements are possible.

gel electrophoresis to map the SMA region, revealed marked differences between individuals as different size restriction fragments were identified with probes from the SMA region.[14] Lastly, unequal crossing-over was found in some individuals indicating a noninverted duplication of the SMN genes.[13] Therefore, multiple alternative structures of the SMA region have been identified and as such this area is not always composed of an inverted duplication (Fig. 6.1). One consideration, which even today has not been addressed, is whether particular genomic structures are more prone to new mutations.[13] The rate of new mutation has been reported to be 2%.[28] However, considering the mutation rate by itself, then this should lead to an exceptionally high rate of disease alleles for an autosomal recessive disorder; yet, this is not the case. The most likely reason for this is the presence of an equal back-mutation rate of SMN2 exchange or conversion to SMN1. Due to the complex nature of this region, it is often incorrectly represented in sequence databases and caution should be applied when extracting data from these sources.

GENETICS OF 5q SPINAL MUSCULAR ATROPHY AND PHENOTYPE MODIFICATION IN MAN

Proximal 5q SMA is an autosomal recessive disorder and is caused by mutation or loss of the SMN1 gene and retention of the SMN2 gene.[8–17,21–29] The SMN1 and SMN2 genes essentially differ by a single C–T change in exon 7 of SMN2. This nucleotide change results in disruption of a splice modulator such that the majority of the transcript from SMN2 lacks exon 7.[30–34] The SMN lacking the amino acids encoded by exon 7 does not oligomerize efficiently and is rapidly degraded.[35,36] This results in SMN2 not producing as much full-length SMN protein as SMN1, and thus there is a deficiency of SMN in SMA.[37,38] The reduction of SMN occurs in all tissues; however, motor neurons are particularly affected.[16] It is clear that the amount of SMN produced is critical in determining the severity of the SMA phenotype.[16,29] Indeed, quantitative assays for the determination of SMN1 and SMN2 copy number permit both the detection of carriers and the correlation of phenotypic SMA severity with copy number of SMN2.[13,15]

SMA can be divided into five categories of severity from type 0 to type 4. Type 0 or 1a, which in general have one copy of SMN2, have the disease at birth. Type 1 SMA patients generally have two copies of SMN2 and develop SMA prior to 6 months of age and die before the age of 2 without ventilatory support. It should be noted that type 1 is often divided into subgroups 1b and 1c where 1c is slightly milder.[29] It has been reported that 85% of type 1 patients have one or two copies of SMN2 where the remaining 15% have three copies of SMN2.[39] In some but not all cases, type 1 patients with three copies of *SMN2* show a slightly milder progression.[39,40] In type 2 SMA, onset occurs between 6 and 18 months and patients can sit unaided but never stand or walk. In type 2 patients, approximately 82% have three copies of *SMN2*, 11% have two copies and 7% have four copies.[39] In type 3 and 4 SMA, onset is between 18 months and 30 years. In these cases, three or four copies of *SMN2* are most common; however, 4% of type 3 patients have two copies of *SMN2*.[39]

Not only does the copy number of *SMN2* influence phenotype, but the type of *SMN2* gene can also influence phenotype (Fig. 6.2). For example, the variant c.859G>C in exon 7 of *SMN2* results in an approximate 20% increase in full-length SMN RNA and a mild SMA phenotype.[41,42] Interestingly, the variant c.859G>C in exon 7 of *SMN2* does not occur in type 1 patients, is found in one copy in type 2 patients, and occurs in two copies in milder type 3b patients (the milder spectrum of type 3 patients).[43] Clearly the output of *SMN2* is critical to determination of phenotype. However one important consideration is that the ability to identify an intact *SMN2* gene is not possible with the current diagnostic

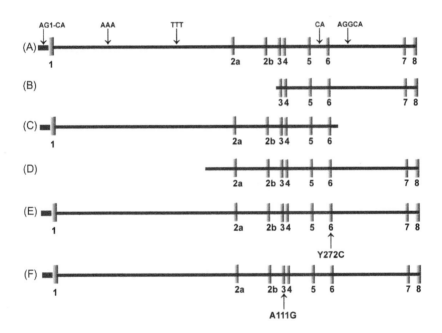

FIGURE 6.2 Diagram of *survival motor neuron 2 (SMN2)* alleles in SMA. (A) Intact *SMN2* with polymorphic regions and type indicated. (B) 5′-Truncated *SMN2* allele lacking exon1-2b. (C) 3′-Truncated allele reported as polymorphic variant. The exact breakpoint is not known. The allele shown here contains exon 1 but exons 2a–6 are questionable. (D) Alternative 5′-truncated *SMN2* allele with exon 1 missing. (E) Severe *SMN2* missense mutation. (F) Mild *SMN2* missense mutation. AG1-CA, AAA, TTT, CA, AGGCA: polymorphic markers identified in the intronic regions of SMN.

testing of SMN2. Some Multiplex Ligation-dependent Probe Amplification (MLPA) assays do look at the copy number of additional exons such as exon 1;[44] however, internal mutations of SMN2 will not be detected. Thus, with current screening procedures we cannot obtain precise information on whether these SMN2 genes are fully intact and functional (Fig. 6.2). Therefore, while type 1 patients with three copies of SMN2, and type 2 cases with four copies of SMN2 occur, it is not known if all the identified SMN2 genes are intact and producing full-length SMN.

Often there is discordance of the Ag1-CA/C272 marker with the SMN2 copy number[39] as well as results with MLPA assays. We identified Ag1-CA as a marker that showed allele loss in type 1 SMA and was in clear linkage disequilibrium with SMA.[24] It was subsequently found that Ag1-CA lies close to exon 1 of the SMN genes in the promoter (Fig. 6.2(A)).[8] The marker typing correlated to SMN2 copy number for The Ohio State University families[45] used in linkage analysis. Indeed, when SMN2 copy number and Ag1-CA genotype were examined in German families, there was clear evidence of SMN2 genes that lack the 5′-end marker and were therefore not intact functional SMN2 genes.[39]

Lastly, the SMN2 gene can have mutations in it just like SMN1. This has already been reported with the pSMNG278R missense mutation detected in exon 7 of SMN2.[39] It is predicted that SMN2 genes will have similar mutations to SMN1, including small deletions, duplications, and splice disrupting mutations. In this regard mild mutations could also occur in SMN2. We have shown that mild SMN1 mutations act by complementing the full-length SMN produced by SMN2[46] (Fig. 6.3). Therefore, a mild missense mutation occurring in an SMN2 gene is predicted to not complement as well as a wild-type SMN2 allele. Thus the presence of the mutated SMN2 gene may not give the same degree of alteration to the phenotype. We have diagramed in Fig. 6.2 potential SMN2 alleles that are likely to occur in SMA patients. Most are truncated alleles of SMN2 that have a 3′-end and thus test positive in the currently used diagnostic assays. In addition we indicated a reported polymorphic variant, which contains the 5′-end of either SMN1/2, but not the 3′-end that is used for copy number detection of SMN2.[47] The latter allele can confuse scoring of intact SMN2 alleles if you use a single site for indicating dosage. To fully analyze the SMN2 alleles in SMA requires a system that gives accurate copy number of all regions of the gene along with the complete sequence of the SMN2 genes present.

Although the form of SMN2 is critically important in determining the severity of the SMA phenotype there are certainly trans-acting modifiers. Evidence for trans-acting factors comes from the observation of haplo-identical siblings that have a marked difference in phenotype but identical SMN2 copy number. We and others have previously identified a number of these families where one sibling has type 1 and the other has type 2

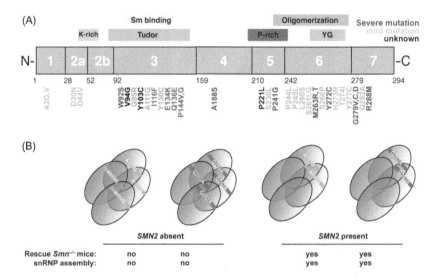

FIGURE 6.3   SMN missense mutations in complementation. (A) Diagram of the domains of SMN and the missense mutations that are found in SMA patient alleles. (B) Illustration of mild SMA causing missense mutations and their mode of action. Homomeric complexes of missense mutations do not rescue $Smn^{-/-}$ mice or perform snRNP assembly. In the presence of $SMN2$, however, small amounts of full-length SMN are produced and heteromeric complexes with SMN protein and the missense mutation are formed. These heteromeric complexes are functional, perform snRNP assembly, and rescue $Smn^{-/-}$ mice.

SMA, or more commonly type 2 and mild type 3 (usually 3b or milder) and finally, type 3 and unaffected.[45–50] Modifiers could act to ameliorate the phenotype by acting in critical pathways downstream of reduced SMN or they can act by altering the amount of SMN produced by $SMN2$. Currently we do not know the identity of the trans-acting modifier(s) or how they act. *Plastin 3* (*PLS3*), located on the X-chromosome, has been suggested to be a modifier of SMA.[51] This was based on the observation that lymphoblasts from the milder sibling had a higher level of *PLS3* expression. However, the reason for this increased expression is not clear as no DNA variant has been reported.[51] Furthermore, no variance in X inactivation or epigenetic regulation of *PLS3* has been demonstrated in these patients. Curiously, increased expression of *PLS3* also occurs in males but does not modify the SMA phenotype, and female patients with high *PLS3* expression with the more severe SMA phenotype in the sibling pair have been reported.[52] One possibility is the modifier *PLS3* is female dependent and not completely penetrant. An alternative explanation is that *PLS3* is in fact not the critical modifier of SMA phenotype. As no DNA change in *PLS3* or an activator of *PLS3* expression that segregates with the mild sibling has been reported, the role of *PLS3* overexpression in

SMA remains controversial. Finding a DNA change that accounts for the alteration in *PLS3* expression will allow a more straightforward approach to testing its importance in a larger patient population. In this regard studies on reoccurrence of colon cancer have clearly identified certain variants that alter *PLS3* expression. High *PLS3* levels have been found in Sezary Syndrome patients and this was associated with loss of CD26. In addition, *PLS3*-positive cells showed hypomethylation of the *PLS3* CpG island at sites 95–99.[53,54] Interestingly, the polymorphism SNP *PLS3* rs871773T allele is associated with a higher protein expression of the *PLS3* gene in colon cancer and an increased risk of recurrence of colon cancer.[55] If *PLS3* does alter severity of SMA, defining the role of both *PLS3* rs871773 and the hypomethylation of sites 95–99 will be important as it gives a mechanism of *PLS3* activation and may even result in a DNA marker that could be followed in patient material.

SMN1 loss accounts for 95% of SMA cases. In the remaining 5% of cases the *SMN1* gene contains a small mutation. Most of the mutations disrupt the SMN reading frame or disrupt a splice site in the gene and are thus similar to the loss of *SMN1*. However, many missense mutations that disrupt SMN function have been reported. These mutations are diagramed Fig. 6.3(A) along with the domains they disrupt and the severity of each mutation.[16,56] Insights can be gained by studying how these missense mutations function upon disruption of SMN domains. At least some of the severe missense mutations disrupt the ability of SMN to efficiently oligomerize and therefore act much like SMN lacking the sequence encoded by exon 7. This form of SMN is rapidly degraded and results in minimal amounts of SMN protein.[16,36] An example of a missense mutation like this is p.SMNY272C, which results in a severe phenotype in the presence of two copies of *SMN2*. Patient lymphoblasts containing the p.SMNY272C mutation have the same amount of SMN protein as type 1 SMA patients with two copies of *SMN2*.[38] This indicates that minimal amounts of SMN are produced by the p.SMNY272C allele. Other severe mutations result in unfolding of protein domains including the Tudor domain. Perhaps one of the most interesting severe missense mutations is p.SMNE134K, which appears to disrupt the binding of SmD1/HuD proteins.[57,58] This mutation could be used to dissect the contributions of different SMN assembly[59] reactions to the phenotype (see below for SMN function).

We then come to the mild SMN missense mutations. These mutations occur in similar domains (Fig. 6.3) yet produce a mild SMA phenotype. This raises the question: is a homomeric complex just composed of mutant SMNs functional, or is the small amount of full-length SMN protein produced by *SMN2* required to complement the missense mutations? We have found that two mutations p.SMNA2G and p.SMNA111G, both mild mutations, can rescue the survival of *Smn*$^{-/-}$ mice only in the presence of full-length SMN and therefore do not have function on their own.

Furthermore, these two amino-terminal mutations also do not complement each other in the absence of full-length SMN.[46] The full-length SMN protein from *SMN2* oligomerizes with the SMN missense mutation protein making a functional SMN complex. In summary, heteromeric SMN complexes are functional and homomeric complexes are not[46] (Fig. 6.3). The amount of full-length SMN and the amount of missense SMN expressed then becomes critical in determining phenotype as it appears likely that only one monomer of wild-type full-length SMN per complex is required for function. While this finding was first demonstrated with p.SMNA2G and p.SMNA111G, it also applies to the mutations p.SMND44V, p.SMNT274I, and p.SMNQ282A as well, and could be a general phenomenon of all mild missense alleles. Indeed, while missense alleles are rare, no consanguineous cases with the absence of *SMN2* have ever been reported.

One current conundrum in the field is the behavior of equivalent missense mutations in nonmammalian species. In *Caenorhabditis elegans* the equivalent of p.SMND44V (Smn.D27N) has been examined. Animals with this mutation can survive and offspring can be obtained in the absence of wild-type SMN.[60] Homozygous mutants show various phenotypes including a motility defect. However this invertebrate finding is different than the observation in mammals where p.SMND44V does not rescue $Smn^{-/-}$ mice in the absence of full-length SMN from *SMN2*. At least some of the mild SMN alleles have been analyzed in Drosophila and appear to complement the maternal SMN present in these mutants.[61] Although some severe alleles behave in the expected manner, others such as Smn-pSMN.G210V (equivalent to p.SMNG279V, a severe allele in humans) do not. The exact reasons for these differences among species remain unclear.

THE BIOCHEMICAL FUNCTION OF SURVIVAL MOTOR NEURON

SMN forms a complex with many other proteins called gemins.[59] The SMN complex, consisting of gemins 2–8 and unrip, is found in the cytoplasm.[59–66] The SMN complex functions in placing the Sm protein ring onto small nuclear RNAs (snRNA) to form a small nuclear ribonuclear protein (snRNP) that together with the SMN complex is transported into the nucleus. snRNPs are critical for the correct splicing of all genes.[63,65] Thus complete loss of SMN results in lethality as snRNPs are essential to the survival of any cell. Apart from SMN's function in assembly of snRNA,[59] SMN also functions in the assembly of the Lsm10, Lsm11, and Sm complex onto U7 snRNA.[67–69] The U7snRNP processes the 3'-end of the histone message. Both of these functions have been clearly demonstrated and have biochemical assays that can be used to determine activity in various samples. In addition to this assembly function, SMN has been proposed to potentially function in assembly of other RNP complexes.[59] SMN has been reported to bind many different proteins.[16,58–60,64–70] Yet the question remains, how many of these reported interactions give rise to a SMN complex with a specific function?[16] In a large number of cases the SMN colocalization studies only show a small proportion of the particular protein overlaps. In addition, coimmunoprecipitation (Co-IP) could be due to the fact that SMN is a sticky protein and is limited by the amount of the input in Co-IPs. It is also hard to eliminate the possibility that the interaction only occurs upon lysis of cells. It becomes key to know whether a function is associated with these putative other SMN complexes. The SMN protein in the axon has been associated with HuD, yet we do not know if SMN can assemble the HuD complex onto mRNA.[58,69,70] Furthermore, the function of an SMN axonal complex must be determined with a rigorous biochemical assay. The lack of any abnormality of axons in vivo in mammalian models of axon patterning, along with no clear biochemical assay, does give concern. For example, overexpression of HuD corrected axonal defects in vitro in motor-neuron-like cells with reduced SMN but it is not known whether HuD overexpression can correct SMA motor neuron physiology in the mouse in vivo.[58] However, the demonstration of SMN's importance in axon repair later in development indicates other functions of SMN could be critical.[71] Thus while the essential snRNP assembly function of SMN is well defined, the contribution of that function versus other proposed functions of SMN in SMA disease development is not. Therefore, there is an urgent need to define the parameters that cause SMA.

SPINAL MUSCULAR ATROPHY MODELS AND GENETIC SUPPRESSION

Many animal models of SMA have been created and are widely used. There are certain features that are important to realize when discussing these models for use in genetic screens to identify critical pathways in SMA. First, in species such as Drosophila there is a very large contribution of maternal SMN and a short life span. When a particular gene is zygotically deleted or mutated Drosophila can still survive to late larval stages based on the maternal RNA loaded into the embryo. This is also true in Zebrafish which have a large maternal contribution of RNA from the yolk. This situation of course is very different in mammals, where embryo development times are generally longer with a limited contribution of maternal RNA. Thus in the mouse, loss of the single functional *Smn* gene results in death of cells at an extremely early stage before any organs or tissues have developed.[72] Any animal model is further complicated by the presence of *SMN2* in humans. In humans, the loss of *SMN1* is compensated by the production of SMN from *SMN2* throughout the patient's life. In other words, SMA in humans is a condition of low levels of continuously produced SMN protein. In order to more closely model SMA in lower species or under specific circumstances, either siRNA or shRNA can be used. Invertebrates and nonmammalian vertebrates should be used to carry out experiments that cannot be done efficiently in mammals.[16] We would advocate exploiting the strengths of each species. Genetic screens in *C. elegans* and Drosophila can be used to quickly identify suppressors and genetic interactors that can then be confirmed in both Zebrafish and SMA mice.[16] In Drosophila a genetic screen was performed using RNAi lines to knock down SMN expression under a universal driver. This study resulted in the identification of 302 modifiers of SMN deficiency where survival of the organism was slightly altered.[73] None of the suppressors identified could be classified as a strong suppressor of the phenotype. Further homolog searches resulted in the identification of 322 human genes. The potential human genetic modifiers were then overlapped with the human interactome. Interestingly, one of the modifiers found in this screen was SmD1 as well as modifiers in other RNA pathways. Certainly this data can be used to generate candidate genes that can then be further investigated in mammalian systems. Indeed it is important to test both these candidates as well as the proposed biochemical pathways in vertebrates for confirmation. Even if survival of an organism is not corrected, the measurement of critical physiological parameters can indicate how a particular pathway contributes to SMA. The absence of an effect might well indicate that this is not critical in mammals.

In addition, it is important to consider that SMN functions in snRNP biogenesis. Altered snRNP production results in altered splicing of genes. However, intron structure is not well conserved across species; thus, sensitivity to SMN reduction can vary. As such, determining altered splicing patterns across species can be particularly informative in SMA. Currently there is limited information on these changes and their affects. Although some splicing targets of SMN deficiency have been described, the list is far from comprehensive, and in some cases the changes appear variable between SMA samples.[74,75] The question that remains is which changes contribute significantly to the SMA phenotype? Some propose that multiple small changes all contribute collectively to the SMA phenotype. However this appears unlikely as the SMA phenotype is relatively specific to the motor circuit, at least in humans. It should be possible to link the physiological changes in the motor circuit to specific SMN functions. Lastly, others have identified a role of SMN in mRNA translation and transport.[16] If these SMN functions are critical in SMA then specific rescue of these functions, apart from snRNP assembly, should have an impact on the SMA phenotype. It is also possible that more than one function contributes to different aspects of the SMA phenotype. Again, rescue using specific mutations that suppress a particular function can be informative about its role in SMA. As a large number of expression and splicing changes in SMA tissues are likely not primary but secondary to SMN deficiency in animals, it is very important to not just invoke all changes detected as having a role in SMA but to garner specific genetic information that shows that a particular biochemical pathway is involved.

DEVELOPMENT OF THERAPIES BASED ON THE GENETICS OF SPINAL MUSCULAR ATROPHY

The identification of *SMN1* and *SMN2* and their roles in SMA led directly to the development of therapeutics for SMA. *SMN2* is always present in SMA patients and produces at least some full-length SMN. Furthermore, increased copy number of *SMN2* results in a milder SMA phenotype.[29] Since SMN is always present in humans, immune reactions to SMN replacement in gene therapy strategies are not an issue. To date three major therapeutic approaches have been developed: antisense oligonucleotides (ASO), small molecules, and gene therapy. All three strategies have been shown to be highly effective in mouse models of SMA and are now in clinical trials.[29]

SMN2 possess both negative and positive elements that control the inclusion of SMN exon 7. Thus, enhancing the positive elements or blocking the negative elements can make *SMN2* behave like *SMN1* and produce a large amount of full-length SMN.[29] Indeed, this has been accomplished in mice with ASO that block the negative regulatory ISS-N1 sequence. In particular, two chemistries of ASO have been successfully used: the 2'-O-methoxyethyl (MOE) and the morpholino oligonucleotides (MO).[76,77] MOs can be administered at high concentrations without side effects and have shown marked impact when given in the central nervous system (CNS) in mice. More recently it has also become possible to obtain widespread distribution in the CNS of adult mice.[78] The MOE against ISS-N1, known as IONIS-SMNRx, is currently in clinical trials and has been safely administered with multiple injections (NCT02193074, Clinicaltrials.gov). The degree of efficacy at different stages of disease progression is determined by the response of biomarkers, electrophysiological measures, and muscle strength in the patient.

The second strategy has been to identify small molecules or compounds that can make *SMN2* produce more full-length SMN protein. A number of screens have resulted in small molecules that activated the SMN promoter or stabilized SMN mRNA and/or protein. Some compounds were identified in these screens that somewhat increased SMN, but in general not to the level required for therapy.[79] Of note is the fact that a number of these screens should have detected molecules that altered the incorporation of exon 7 by *SMN2* but did not. This is likely due to the fact that the libraries of small molecules screened did not cover the required chemical space for molecules that could or would alter the *SMN2* splicing or that weak hits that did alter splicing were not optimized further chemically. As such, the NIH-based libraries in particular appear to be missing critical molecules that can alter splicing. In the future the publically accessible libraries will need to be improved. PTC/Roche and Novartis have performed screens and identified compounds that alter the splicing of *SMN2*.[80,81] Indeed these compounds have a major impact on SMA mice and are in clinical trials (NCT02240355, NCT02268552). Interestingly, the compounds developed by Roche/PTC appear to affect the splicing of a limited set of genes, thus giving hope that there will not be significant off-target effects of these drug compounds.[80] As such, when searching for molecules that alter splicing in SMA or other disorders, the nature of the chemical library should be considered.

Lastly, gene therapy using scAAV9-SMN has shown remarkable improvement of SMA mice when delivered into the vascular system or delivered into the CNS via the cerebrospinal fluid.[82–85] Recent studies in nonhuman primates have shown no toxicity of the scAAV9 upon intrathecal injection.[86] Greater than 50% of motor neurons were targeted in all spinal cord sections when using the Trendelenburg position.[86] Upon intrathecal injection, similar transduction of motor neurons was demonstrated with a dose of virus that was 30-times lower than that used for intravascular delivery.[86] The scAAV9-SMN gene therapy is currently in Phase 1 clinical trials for type 1 patients (NCT02122952).

Until recently, a large animal model of SMA to assess therapy has not been available. The mouse can give information on the activation of *SMN2*; however, the drawbacks of the mice need to be considered. First, unlike humans, the blood–brain barrier is open in the early developmental stages of the mouse. Second, the SMA mouse model has heart abnormalities and necrosis, neither of which is typically present in human SMA. Third, the introns in mouse can have a different sensitivity to SMN depletion. Finally, there have been reports of a peripheral requirement of SMN in the mouse, whereas there is little indication even in older SMA patients, of abnormalities in peripheral organs. Thus investigation of SMN restoration to the motor neuron at different stages in SMA disease progression in a large animal is critical to effective therapeutic development. We have shown that reduction of SMN in motor neurons of the pig results in SMA with fibrillations on electromyography, reduction of compound muscle action potential (CMAP), reduction of motor unit number estimation (MUNE), and pathological loss of motor neuron roots and motor neuron cell bodies.[87] Administration of scAAV9-SMN early in disease progression resulted in correction of the motor neuron showing the SMN dependence of these changes. Furthermore, symptomatic correction resulted in stabilization of the phenotype, increased CMAP, partial correction of MUNE, and motor neuron loss.[87] While the phenotype did improve, the results indicate that early introduction of SMN will be necessary in order for the greatest rescue of the motor neuron, regardless of the particular therapeutic used.

SUMMARY AND FUTURE DIRECTIONS

The genetic cause of proximal SMA is well defined with the loss or mutation of the *SMN1* gene and retention of *SMN2*. The *SMN2* gene is clearly a modifier of phenotype; the more SMN protein produced, the milder the phenotype. However, apart from *SMN2*, modifiers in trans also exist. Once effective SMN-inducing therapies are in place then newborn screening for SMA will be implemented. Thus the identification of the DNA changes associated with these modifiers of phenotype, as well as rapid analysis of which *SMN2* genes are fully functional will become essential for accurate diagnostic testing. The early introduction of SMN restorative therapy is essential to protect against the loss of motor neurons. Early detection of SMA followed by analysis of *SMN2* and modifiers could give a prediction of when to introduce SMN induction therapy based on estimated severity of phenotype. Thus the precise genetic details of what fully determines SMA severity of phenotype must be determined. In addition, while we know that SMA is caused by reduced SMN we do not yet understand the downstream pathways affected that ultimately result in the development of SMA. While many biochemical changes might be detected, the important question is which changes can suppress or alter the SMA phenotype? Genetics can play a critical role in ensuring that the essential pathways are studied.

It should also be noted that with current newborn screening methodology approximately 7% of cases will not be detected. In approximately 5% of cases, one allele has a small SMN missense mutation instead of the loss of *SMN1* exon 7; 2% of SMA cases are caused by de novo mutations. With regard to either of these situations, or individuals that have had SMA for a number of years, it is likely that SMN inducers will still have an impact in stabilization of the motor neurons that are still present when the therapy is administered. A promising feature of SMN-inducing therapies is that they function through different modalities and can therefore be combined if necessary to give further increases in SMN. A question for the future is whether stabilizing SMN followed by additional therapies that either enhance the remaining motor neurons or enhance muscle function would be beneficial in these instances? It will be interesting over the next few years to see how these targeted therapies behave in clinical trials.

References

1. Darras BT. Non-5q spinal muscular atrophies the alphanumeric soup thickens. *Neurology* 2011;**77**:312–4.
2. Hoffman EP, Talbot K. A calm before the exome storm: coming together of dSMA and CMT2. *Neurology* 2012;**78**(22):1706–7.
3. Neveling K, Martinez-Carrera LA, Holker I, et al. Mutations in BICD2, which encodes a golgin and important motor adaptor, cause congenital autosomal-dominant spinal muscular atrophy. *Am J Hum Genet* 2013;**92**(6):946–54.
4. Oates EC, Rossor AM, Hafezparast M, et al. Mutations in BICD2 cause dominant congenital spinal muscular atrophy and hereditary spastic paraplegia. *Am J Hum Genet* 2013;**92**(6):965–73.
5. Peeters K, Litvinenko I, Asselbergh B, et al. Molecular defects in the motor adaptor BICD2 cause proximal spinal muscular atrophy with autosomal-dominant inheritance. *Am J Hum Genet* 2013;**92**(6):955–64.
6. Ramser J, Ahearn ME, Lenski C, et al. Rare missense and synonymous variants in UBE1 are associated with X-linked infantile spinal muscular atrophy. *Am J Hum Genet* 2008;**82**(1):188–93.
7. Nousiainen HO, Kestila M, Pakkasjarvi N, et al. Mutations in mRNA export mediator GLE1 result in a fetal motoneuron disease. *Nat Genet* 2008;**40**(2):155–7.
8. Lefebvre S, Burglen L, Reboullet S, et al. Identification and characterization of a spinal muscular atrophy-determining gene. *Cell* 1995;**80**(1):155–65.
9. Pearn J. Incidence, prevalence, and gene frequency studies of chronic childhood spinal muscular atrophy. *J Med Genet* 1978;**15**(6):409–13.
10. Prior TW, Snyder PJ, Rink BD, et al. Newborn and carrier screening for spinal muscular atrophy. *Am J Med Genet A* 2010;**152A**(7):1608–16.
11. Hendrickson BC, Donohoe C, Akmaev VR, et al. Differences in SMN1 allele frequencies among ethnic groups within North America. *J Med Genet* 2009;**46**(9):641–4.
12. Sugarman EA, Nagan N, Zhu H, et al. Pan-ethnic carrier screening and prenatal diagnosis for spinal muscular atrophy: clinical laboratory analysis of > 72,400 specimens. *Eur J Hum Genet* 2012;**20**(1):27–32.
13. Burghes AH. When is a deletion not a deletion? When it is converted. *Am J Hum Genet* 1997;**61**(1):9–15.
14. Campbell L, Potter A, Ignatius J, Dubowitz V, Davies K. Genomic variation and gene conversion in spinal muscular atrophy: implications for disease process and clinical phenotype. *Am J Hum Genet* 1997;**61**(1):40–50.
15. McAndrew PE, Parsons DW, Simard LR, et al. Identification of proximal spinal muscular atrophy carriers and patients by analysis of SMNT and SMNC gene copy number. *Am J Hum Genet* 1997;**60**(6):1411–22.
16. Burghes AH, Beattie CE. Spinal muscular atrophy: why do low levels of survival motor neuron protein make motor neurons sick? *Nat Rev Neurosci* 2009;**10**(8):597–609.
17. Melki J, Lefebvre S, Burglen L, et al. De novo and inherited deletions of the 5q13 region in spinal muscular atrophies. *Science* 1994;**264**(5164):1474–7.
18. Carter TA, Bonnemann CG, Wang CH, et al. A multicopy transcription-repair gene, BTF2p44, maps to the SMA region and demonstrates SMA associated deletions. *Hum Mol Genet* 1997;**6**(2):229–36.
19. Roy N, Mahadevan MS, McLean M, et al. The gene for neuronal apoptosis inhibitory protein is partially deleted in individuals with spinal muscular atrophy. *Cell* 1995;**80**(1):167–78.
20. Scharf JM, Endrizzi MG, Wetter A, et al. Identification of a candidate modifying gene for spinal muscular atrophy by comparative genomics. *Nat Genet* 1998;**20**(1):83–6.

21. Gilliam TC, Brzustowicz LM, Castilla LH, et al. Genetic homogeneity between acute and chronic forms of spinal muscular atrophy. *Nature* 1990;**345**(6278):823–5.
22. Melki J, Abdelhak S, Sheth P, et al. Gene for chronic proximal spinal muscular atrophies maps to chromosome 5q. *Nature* 1990;**344**(6268):767–8.
23. Burghes AH, Ingraham SE, McLean M, et al. A multicopy dinucleotide marker that maps close to the spinal muscular atrophy gene. *Genomics* 1994;**21**(2):394–402.
24. DiDonato CJ, Morgan K, Carpten JD, et al. Association between Ag1-CA alleles and severity of autosomal recessive proximal spinal muscular atrophy. *Am J Hum Genet* 1994;**55**(6):1218–29.
25. Carpten JD, DiDonato CJ, Ingraham SE, et al. A YAC contig of the region containing the spinal muscular atrophy gene (SMA): identification of an unstable region. *Genomics* 1994;**24**(2):351–6.
26. Kleyn PW, Wang CH, Lien LL, et al. Construction of a yeast artificial chromosome contig spanning the spinal muscular atrophy disease gene region. *Proc Natl Acad Sci U S A* 1993;**90**(14):6801–5.
27. Roy N, McLean MD, Besner-Johnston A, et al. Refined physical map of the spinal muscular atrophy gene (SMA) region at 5q13 based on YAC and cosmid contiguous arrays. *Genomics* 1995;**26**(3):451–60.
28. Wirth B, Schmidt T, Hahnen E, et al. De novo rearrangements found in 2% of index patients with spinal muscular atrophy: mutational mechanisms, parental origin, mutation rate, and implications for genetic counseling. *Am J Hum Genet* 1997;**61**(5):1102–11.
29. Arnold WD, Burghes AH. Spinal muscular atrophy: development and implementation of potential treatments. *Ann Neurol* 2013;**74**(3):348–62.
30. Lorson CL, Hahnen E, Androphy EJ, Wirth B. A single nucleotide in the SMN gene regulates splicing and is responsible for spinal muscular atrophy. *Proc Natl Acad Sci U S A* 1999;**96**(11):6307–11.
31. Monani UR, Lorson CL, Parsons DW, et al. A single nucleotide difference that alters splicing patterns distinguishes the SMA gene SMN1 from the copy gene SMN2. *Hum Mol Genet* 1999;**8**(7):1177–83.
32. Gennarelli M, Lucarelli M, Capon F, et al. Survival motor neuron gene transcript analysis in muscles from spinal muscular atrophy patients. *Biochem Biophys Res Commun* 1995;**213**(1):342–8.
33. Cartegni L, Krainer AR. Disruption of an SF2/ASF-dependent exonic splicing enhancer in SMN2 causes spinal muscular atrophy in the absence of SMN1. *Nat Genet* 2002;**30**(4):377–84.
34. Kashima T, Manley JL. A negative element in SMN2 exon 7 inhibits splicing in spinal muscular atrophy. *Nat Genet* 2003;**34**(4):460–3.
35. Lorson CL, Strasswimmer J, Yao JM, et al. SMN oligomerization defect correlates with spinal muscular atrophy severity. *Nat Genet* 1998;**19**(1):63–6.
36. Burnett BG, Munoz E, Tandon A, Kwon DY, Sumner CJ, Fischbeck KH. Regulation of SMN protein stability. *Mol Cell Biol* 2009;**29**(5):1107–15.
37. Coovert DD, Le TT, McAndrew PE, et al. The survival motor neuron protein in spinal muscular atrophy. *Hum Mol Genet* 1997;**6**(8):1205–14.
38. Lefebvre S, Burlet P, Liu Q, et al. Correlation between severity and SMN protein level in spinal muscular atrophy. *Nat Genet* 1997;**16**(3):265–9.
39. Feldkotter M, Schwarzer V, Wirth R, Wienker TF, Wirth B. Quantitative analyses of SMN1 and SMN2 based on real-time lightCycler PCR: fast and highly reliable carrier testing and prediction of severity of spinal muscular atrophy. *Am J Hum Genet* 2002;**70**(2):358–68.
40. Rudnik-Schoneborn S, Berg C, Zerres K, et al. Genotype-phenotype studies in infantile spinal muscular atrophy (SMA) type I in Germany: implications for clinical trials and genetic counselling. *Clin Genet* 2009;**76**(2):168–78.

41. Prior TW, Krainer AR, Hua Y, et al. A positive modifier of spinal muscular atrophy in the SMN2 gene. *Am J Hum Genet* 2009;**85**(3):408–13.
42. Vezain M, Saugier-Veber P, Goina E, et al. A rare SMN2 variant in a previously unrecognized composite splicing regulatory element induces exon 7 inclusion and reduces the clinical severity of spinal muscular atrophy. *Hum Mutat* 2010;**31**(1):E1110–25.
43. Bernal S, Alias L, Barcelo MJ, et al. The c.859G> C variant in the SMN2 gene is associated with types II and III SMA and originates from a common ancestor. *J Med Genet* 2010;**47**(9):640–2.
44. Arkblad EL, Darin N, Berg K, et al. Multiplex ligation-dependent probe amplification improves diagnostics in spinal muscular atrophy. *Neuromuscul Disord* 2006;**16**(12):830–8.
45. Burghes AH, Ingraham SE, Kote-Jarai Z, et al. Linkage mapping of the spinal muscular atrophy gene. *Hum Genet* 1994;**93**(3):305–12.
46. Workman E, Saieva L, Carrel TL, et al. A SMN missense mutation complements SMN2 restoring snRNPs and rescuing SMA mice. *Hum Mol Genet* 2009;**18**(12):2215–29.
47. Alias L, Bernal S, Barcelo MJ, et al. Accuracy of marker analysis, quantitative real-time polymerase chain reaction, and multiple ligation-dependent probe amplification to determine SMN2 copy number in patients with spinal muscular atrophy. *Genet Test Mol Biomarkers* 2011;**15**(9):587–94.
48. DiDonato CJ, Ingraham SE, Mendell JR, et al. Deletion and conversion in spinal muscular atrophy patients: is there a relationship to severity? *Ann Neurol* 1997;**41**(2):230–7.
49. Wirth B, Tessarolo D, Hahnen E, et al. Different entities of proximal spinal muscular atrophy within one family. *Hum Genet* 1997;**100**(5–6):676–80.
50. Cobben JM, van der Steege G, Grootscholten P, de Visser M, Scheffer H, Buys CH. Deletions of the survival motor neuron gene in unaffected siblings of patients with spinal muscular atrophy. *Am J Hum Genet* 1995;**57**(4):805–8.
51. Oprea GE, Krober S, McWhorter ML, et al. Plastin 3 is a protective modifier of autosomal recessive spinal muscular atrophy. *Science* 2008;**320**(5875):524–7.
52. Bernal S, Also-Rallo E, Martinez-Hernandez R, et al. Plastin 3 expression in discordant spinal muscular atrophy (SMA) siblings. *Neuromuscul Disord* 2011;**21**(6):413–9.
53. Jones CL, Ferreira S, McKenzie RC, et al. Regulation of T-plastin expression by promoter hypomethylation in primary cutaneous T-cell lymphoma. *J Invest Dermatol* 2012;**132**(8):2042–9.
54. Begue E, Michel L, Jean-Louis F, Bagot M, Bensussan A. Promoter hypomethylation and expression of PLS3 in human sezary lymphoma cells. *SOJ Immunol* 2013;**1**(1):4.
55. Szkandera J, Winder T, Stotz M, et al. A common gene variant in PLS3 predicts colon cancer recurrence in women. *Tumour Biol* 2013;**34**(4):2183–8.
56. Jedrzejowska M, Gos M, Zimowski JG, Kostera-Pruszczyk A, Ryniewicz B, Hausmanowa-Petrusewicz I. Novel point mutations in survival motor neuron 1 gene expand the spectrum of phenotypes observed in spinal muscular atrophy patients. *Neuromuscul Disord* 2014;**24**(7):617–23.
57. Tripsianes K, Madl T, Machyna M, et al. Structural basis for dimethylarginine recognition by the Tudor domains of human SMN and SPF30 proteins. *Nat Struct Mol Biol* 2011;**18**(12):1414–20.
58. Hubers L, Valderrama-Carvajal H, Laframboise J, Timbers J, Sanchez G, Cote J. HuD interacts with survival motor neuron protein and can rescue spinal muscular atrophy-like neuronal defects. *Hum Mol Genet* 2011;**20**(3):553–79.
59. Li DK, Tisdale S, Lotti F, Pellizzoni L. SMN control of RNP assembly: from post-transcriptional gene regulation to motor neuron disease. *Semin Cell Dev Biol* 2014;**32**:22–9.
60. Sleigh JN, Buckingham SD, Esmaeili B, et al. A novel *Caenorhabditis elegans* allele, smn-1(cb131), mimicking a mild form of spinal muscular atrophy, provides a convenient drug screening platform highlighting new and pre-approved compounds. *Hum Mol Genet* 2011;**20**(2):245–60.

61. Praveen K, Wen Y, Gray KM, et al. SMA-causing missense mutations in survival motor neuron (Smn) display a wide range of phenotypes when modeled in Drosophila. *PLoS Genet* 2014;**10**(8):e1004489.

62. Pellizzoni L, Baccon J, Rappsilber J, Mann M, Dreyfuss G. Purification of native survival of motor neurons complexes and identification of Gemin6 as a novel component. *J Biol Chem* 2002;**277**(9):7540–5.

63. Pellizzoni L, Yong J, Dreyfuss G. Essential role for the SMN complex in the specificity of snRNP assembly. *Science* 2002;**298**(5599):1775–9.

64. Meister G, Buhler D, Laggerbauer B, Zobawa M, Lottspeich F, Fischer U. Characterization of a nuclear 20S complex containing the survival of motor neurons (SMN) protein and a specific subset of spliceosomal Sm proteins. *Hum Mol Genet* 2000;**9**(13):1977–86.

65. Meister G, Buhler D, Pillai R, Lottspeich F, Fischer U. A multiprotein complex mediates the ATP-dependent assembly of spliceosomal U snRNPs. *Nat Cell Biol* 2001;**3**(11):945–9.

66. Otter S, Grimmler M, Neuenkirchen N, Chari A, Sickmann A, Fischer U. A comprehensive interaction map of the human survival of motor neuron (SMN) complex. *J Biol Chem* 2007;**282**(8):5825–33.

67. Tisdale S, Lotti F, Saieva L, et al. SMN is essential for the biogenesis of U7 small nuclear ribonucleoprotein and 3'-end formation of histone mRNAs. *Cell Rep* 2013;**5**(5):1187–95.

68. Pillai RS, Grimmler M, Meister G, et al. Unique Sm core structure of U7 snRNPs: assembly by a specialized SMN complex and the role of a new component, Lsm11, in histone RNA processing. *Genes Dev* 2003;**17**(18):2321–33.

69. Akten B, Kye MJ, Hao le T, et al. Interaction of survival of motor neuron (SMN) and HuD proteins with mRNA cpg15 rescues motor neuron axonal deficits. *Proc Natl Acad Sci U S A* 2011;**108**(25):10337–42.

70. Fallini C, Zhang H, Su Y, et al. The survival of motor neuron (SMN) protein interacts with the mRNA-binding protein HuD and regulates localization of poly(A) mRNA in primary motor neuron axons. *J Neurosci* 2011;**31**(10):3914–25.

71. Kariya S, Obis T, Garone C, et al. Requirement of enhanced Survival Motoneuron protein imposed during neuromuscular junction maturation. *J Clin Invest* 2014;**124**(2):785–800.

72. Schrank B, Gotz R, Gunnersen JM, et al. Inactivation of the survival motor neuron gene, a candidate gene for human spinal muscular atrophy, leads to massive cell death in early mouse embryos. *Proc Natl Acad Sci U S A* 1997;**94**(18):9920–5.

73. Sen A, Dimlich DN, Guruharsha KG, et al. Genetic circuitry of Survival motor neuron, the gene underlying spinal muscular atrophy. *Proc Natl Acad Sci U S A* 2013;**110**(26):E2371–80.

74. Lotti F, Imlach WL, Saieva L, et al. An SMN-dependent U12 splicing event essential for motor circuit function. *Cell* 2012;**151**(2):440–54.

75. Zhang Z, Pinto AM, Wan L, et al. Dysregulation of synaptogenesis genes antecedes motor neuron pathology in spinal muscular atrophy. *Proc Natl Acad Sci U S A* 2013;**110**(48):19348–53.

76. Hua Y, Sahashi K, Rigo F, et al. Peripheral SMN restoration is essential for long-term rescue of a severe spinal muscular atrophy mouse model. *Nature* 2011;**478**(7367):123–6.

77. Porensky PN, Mitrpant C, McGovern VL, et al. A single administration of morpholino antisense oligomer rescues spinal muscular atrophy in mouse. *Hum Mol Genet* 2012;**21**(7):1625–38.

78. Porensky PN, Burghes AH. Antisense oligonucleotides for the treatment of spinal muscular atrophy. *Hum Gene Ther* 2013;**24**(5):489–98.

79. Cherry JJ, Kobayashi DT, Lynes MM, et al. Assays for the identification and prioritization of drug candidates for spinal muscular atrophy. *Assay Drug Dev Technol* 2014;**12**(6):315–41.

80. Naryshkin NA, Weetall M, Dakka A, et al. Motor neuron disease. SMN2 splicing modifiers improve motor function and longevity in mice with spinal muscular atrophy. *Science* 2014;**345**(6197):688–93.

81. Palacino J, Swalley SE, Song C, et al. SMN2 splice modulators enhance U1-pre-mRNA association and rescue SMA mice. *Nat Chem Biol* 2015;**11**(7):511–7.

82. Foust KD, Wang X, McGovern VL, et al. Rescue of the spinal muscular atrophy phenotype in a mouse model by early postnatal delivery of SMN. *Nat Biotechnol* 2010;**28**(3):271–4.

83. Dominguez E, Marais T, Chatauret N, et al. Intravenous scAAV9 delivery of a codon-optimized SMN1 sequence rescues SMA mice. *Hum Mol Genet* 2011;**20**(4):681–93.

84. Valori CF, Ning K, Wyles M, et al. Systemic delivery of scAAV9 expressing SMN prolongs survival in a model of spinal muscular atrophy. *Sci Transl Med* 2010;**2**(35):35ra42.

85. Passini MA, Bu J, Richards AM, et al. Translational fidelity of intrathecal delivery of self-complementary AAV9-survival motor neuron 1 for spinal muscular atrophy. *Hum Gene Ther* 2014;**25**(7):619–30.

86. Meyer K, Ferraiuolo L, Schmelzer L, et al. Improving single injection CSF delivery of AAV9-mediated gene therapy for SMA - a dose response study in mice and nonhuman primates. *Mol Ther* 2014;**23**(3):477–87.

87. Duque SI, Arnold WD, Odermatt P, et al. A large animal model of spinal muscular atrophy and correction of phenotype. *Ann Neurol* 2015;**77**(3):399–414.

7

Introduction to Gene and Stem-Cell Therapy

D.M. O'Connor

Emory University, Atlanta, GA, United States

INTRODUCTION

Motor neuron diseases, such as amyotrophic lateral sclerosis (ALS) and spinal muscular atrophy (SMA), are devastating neurodegenerative diseases that are progressive and fatal. There is no cure and the current standard of care is merely palliative. For example, riluzole is the only

FDA-approved treatment for ALS and extends life for 3–6 months. Due to the inability of conventional therapies to have an impact on these diseases, research in gene and stem cell therapies have come into focus.

Gene and stem cell therapies have the potential to make a significant impact on the treatment of ALS and SMA. For this potential to be realized the optimal gene therapy vector or stem cell needs to be selected. The various options have their pros and cons, which will be discussed in this chapter. It is necessary to take the time to assess and match the disease with the appropriate therapy.

GENE THERAPY

Gene therapy proposes to replace a defective gene with the correct copy or, to introduce a therapeutic gene that will support and improve the disease environment. Vectors, both viral and nonviral, are used to deliver the therapeutic gene to the target tissue and cells.

Nonviral Vectors

Nonviral vectors consist of a nucleic acid, either naked or complexed with a carrier that aids its passage into the target tissue.[1] They have the advantage of a low risk factor due to the absence of a viral component. Also, there is unlikely to be an immune response caused by the vector, which increases their safety profile. Their capacity for delivering nucleic acid is large and they are easier to synthesize than viral vectors. However, their efficacy is low as they lack the efficient means to reach the nucleus that viral vectors possess. This limits their usefulness as a tool for treating diseases that are systemic or affect more than one organ type.[1] Advances in developing compounds that package the nucleic acid, to prevent degradation and enhance delivery, has resulted in improvements in the effect of nonviral gene therapy. These developments include cationic polymers, some of which have the ability to be retrogradely transported.[2] This would be useful for peripheral neuronal delivery. Some proteins have also been incorporated into nucleic acid carriers due to their ability to bind to cell-surface receptors. For example, tetanus toxin has the ability to target neuronal cells. Fragments of the tetanus toxin have the ability to bind to receptors, be taken up into cells and be transported in a manner similar to the full length toxin.[3–5] The fusion of tetanus-like peptides with nucleic acid may assist targeting specific cell populations of interest in ALS and SMA.[4] Another option for targeting of nonviral vectors are synthetic nanoparticles. Nanoparticles can be made from a variety of different materials such as lipids and polymers. For example, silica nanoparticles have been demonstrated to result in targeted expression in neuronal cells.

This expression was on a par with expression from a herpes simplex virus (HSV) vector and had the advantage of having none of the vector-associated toxicity.[6] It is interesting to note that the majority of the data for nonviral vectors in neuronal cells comes from in vitro studies. Further research is required in vivo to thoroughly evaluate the potential of these nonviral vector options.[7]

Viral Vectors

Viral vectors use the natural ability of viruses to infect host cells and then use the cells' machinery to express the transgene. They have the genes that allow them to replicate removed from their genome, which renders them safer to use in a gene therapy. Viral vectors can carry either DNA or RNA, can integrate their genetic material into the host genome or exist episomally, and have different tropisms for different tissue and cell types. They also vary in the size of the transgene they can encode. All of these factors have to be taken into account when matching a viral vector with the appropriate disease. The most common viral vectors used include Adenovirus (Ad), Adeno-associated virus (AAV), Lentivirus (LV) and HSV (Fig. 7.1).

Adenovirus

Ad is a nonenveloped, double-stranded DNA virus and has the ability to transduce both dividing and nondividing cells. It most commonly causes respiratory infections.[8,9] When Ad infects cells, it binds to cell-surface receptors via proteins on its capsid coat. Most Ad serotypes bind to the coxsackie and adenovirus receptor.[7] The virus is internalized by endocytosis prompted by interaction between the viral capsid and integrin receptors on the cell surface.[10,11] After cell entry, the virus is released from the endosome, interacts with cytoplasmic dynein and microtubules, and is moved toward the nucleus. The Ad capsid docks with the nuclear pore complex protein. The nuclear protein histone H1 protein attaches to the viral capsid and microtubule kinesin-1 disrupts nuclear-pore-complex-docked capsids and the nuclear pore complex. This allows for access of the viral genome to the nucleus.[12,13] Ad does not integrate into the host genome but exists as a linear episome in the cell nucleus.[14] The transcription and assembly of the progeny virus takes place in the cell nucleus. The progeny virus particles are released from the cell by virus-induced cell lysis.[15]

The development of the Ad vector has gone through several iterations as genes responsible for replication have been deleted from the original viral genome. The first-generation Ad vectors had deletions in the E1 region and/or in the E3 region of the viral genome. The E1 region is responsible for encoding proteins necessary for early gene expression.

Vector	Specifications	Application in the CNS
Adeno-associated virus (AAV) ~20 nm	Genome: ssDNA Capacity: ~4.7 kb (~ 2.2 kb with scAAV, ~8 kb with dual vectors) Forms circular and linear episomes; integrates with very low frequency Shown to infect neurons, astrocytes, glial and ependymal cells	Used extensively in clinical trials, including Parkinson's, Alzheimer's, batten, and canavan diseases. Preliminary studies suggest AAV vectors could also be used to treat mucopolysaccharidoses (MPS), Spinocerebellar ataxia, amyotrophic lateral sclerosis (ALS), epilepsy, and huntington's disease.
Retrovirus: human immunodeficiency virus (HIV) ~100 nm	Genome: ssRNA Capacity: ~ 8 kb NIL vectors form linear and circular episomes; integration is low. Other HIV vectors integrate with high efficiency Shown to infect neurons and astroglial cells	Used in clinical trials for treatment of Parkinson's and Alzheimer's diseases. Vectors are being developed for use with Huntington's and lysosomal storage diseases.
Adenovirus ~70–100 nm	Genome: dsDNA Capacity: ~36 kb Maintained as linear episomes; integration is minimal even with extensive homology to genome Shown to infect neural, astroglial and human glioma cells	Not in clinical use for gene therapy in the CNS due primarily to vector toxicity. Has been used for oncolytic potential as an anticancer agent.
Herpesvirus: Herpes simplex virus - 1 (HSV-1) ~186 nm	Genome: dsDNA Capacity: ~150 kb Genome circularizes upon entering nucleus and is maintained episomally; integration is minimal Shown to infect neurons	Not in clinical use for gene therapy in the CNS due to problems with vector toxicity and production. Vectors are being developed for use with Parkinson's disease. Has also been developed for anticancer therapy.

FIGURE 7.1   Viral vectors for gene delivery to the central nervous system. *Source: From Lentz T.B., Gray S. J., Samulski R.J. (2012) Viral vectors for gene delivery to the central nervous system. Neurobiol. Dis. 48(2):179–188.*

The E3 region is involved in replication and packaging of the virus.[16] Removal of these genes leaves a transgene capacity of approximately 8 kb. However, as this vector still expressed viral genes, it was found to elicit an immune response and was cytotoxic to cells.[17–19] This limited the expression of the transgene and, consequently, its usefulness as a vector. The second-generation Ad vectors had more of the viral genome deleted. The E2 and E4 regions were removed along with the original E1 and E3 regions.[20,21] The deletion of additional regions of the viral genome resulted in a reduced immune response but did not eliminate the problem completely. A side effect of the additional deletions left the vector unable to replicate by itself and it required a helper virus or a stably transfected cell line, expressing the deleted regions, to replicate. The third generation of Ad vectors had most of the viral genome deleted and have also been called gutted Ad vectors.[22,23] They contain the minimal genes for virus production and packaging and have a capacity of approximately 36 kb.[24,25]

Ad was first used in clinical trials in cystic fibrosis patients delivering the cystic fibrosis transmembrane receptor gene to the lung epithelium.[26] Subsequent studies demonstrated that the highly immunogenic response elicited by Ad vector administration restricted the duration of transgene expression to approximately 2 weeks. Repeated administration of the Ad vector resulted in reduced expression of the transgene with each additional administration.[27] Advances in Ad vector development toward third-generation Ad vectors have reduced the host immune response but not eliminated it entirely. This has had the effect of extending the expression of the transgene, with expression in nonhuman primate liver detected for up to 2 years.[28,29] However, to the best of our knowledge, third-generation Ad vectors have not been tested in clinical trials.

The immune response and the resulting limited gene expression have focused use of Ad vectors on cancer treatment. This includes treatment of cancers of the central nervous system (CNS).[30,31] For cancer treatment, the short-term expression of Ad is an advantage. A number of different strategies have been employed that utilize Ad vectors. They have been used to deliver cytotoxic genes directly to tumors. Another approach has been to deliver a gene that converts a prodrug to its active cytotoxic form.[32] Ad has also been used to express immune-related genes to activate or draw immune cells in the vicinity of the tumor.[33,34] Another option for Ad vector use is in the area of vaccine development. This has been achieved by raising antigens that recognize the transgene or by inserting the antigen into the capsid. This has been employed to develop vaccines against bioterrorism agents such as anthrax[35] and also against agents such as cocaine.[36]

Ad as viral vector has the capacity to be useful in specific areas such as cancer therapy. However, with its immunogenic profile, it is not a suitable choice for either systemic or organ-targeted gene therapy.

Adeno-Associated Virus

AAV is a small parvovirus virus, approximately 20 nm in size. It is classified as a helper-dependent virus, as it is unable to replicate without coinfection with either an Ad or a Herpes virus.[37] Its genomic material is packaged as a single-stranded DNA molecule. Upon infection, the viral genome is converted to a double-stranded episome with integration into the host genome a rare event.[38–40] The virion capsid consists of three proteins: VP1, 2, and 3. These proteins are responsible for the binding of the virus to the cell-surface receptors. Different serotypes, or subtypes, of AAV bind to different cell-surface receptors, determining the tropism of each serotype.[41] To package an AAV vector, the transgene cassette must simply be flanked by the inverted terminal repeats (ITRs), which function as an origin of replication for the vector. No viral genes are incorporated into these vectors.[42,43] AAV2 is the most extensively studied serotype and most AAV vectors use AAV2 ITRs. A limiting factor of AAVs is the cloning capacity of approximately 4.7 kb. Any greater than this and the packaging efficiency of the virus is adversely affected.[44] The first AAV vectors developed were single stranded.[44,45] The rate-limiting step in using these vectors was second-strand synthesis following cell transduction.[46] This resulted in delayed expression of the transgene. This issue was overcome by the development of double-stranded or self-complementary AAV (scAAV).[47] scAAV has a mutated ITR that is located between two copies of the transgene. The copies of the transgene are complementary to each other and yield a self-annealing double-stranded molecule. This eliminates the need for second-strand synthesis and the transgene is expressed earlier. The disadvantage of scAAV is that it reduces the cloning capacity of the vector to approximately 2.2 kb, limiting the number of genes that can be accommodated in this form of the vector.

As previously stated, AAV2 was one of the first serotypes identified and studied with a view to gene therapy. Since the initial identification of AAV2, a large number of serotypes of AAV have been identified, with each serotype having a different tropism. Along with naturally occurring serotypes, directed evolution has been utilized to develop serotypes with specific tropisms.[48] Directed evolution involves shuffling the capsid genes of naturally occurring AAV to create random chimeras of AAV capsid genes. These are then packaged and screened through either in vitro or in vivo methods. Vectors that transduce target tissues are purified. There may be several rounds of screening to optimize tissue targeting.[48] Other options include engineering the capsid proteins to have a mixture of proteins from two different serotypes to create a mosaic virus[49] or to incorporate targeting proteins into the capsid to direct the tropism of the virus.[50] However, it needs to be borne in mind that engineering capsids can affect the packing of the virus and, consequently, the titer.

Although AAV does not cause disease in humans and does not elicit the immunogenic response of Ad, it does result in the production of anti-AAV antibodies (Abs).[51] These Abs could eliminate the viral vector before the transgene had an effect. This is an important issue for any AAV-mediated therapy that is being translated to the clinic. Suppression of this response with steroids has been sufficient to suppress the loss of the transgene. In addition, Abs can also pose a hurdle at the initiation of treatment, too.

A large fraction of the population has already been exposed to naturally occurring AAVs, resulting in the production of neutralizing Abs that can block therapeutic vectors. Currently, the best option to address this problem is to screen prospective patients before they are enrolled in AAV gene therapy trials. Patients with an Ab titer that is too high would not be enrolled. This strategy has been employed in a number of clinical trials, for example the SMA trial that is delivering AAV-9 SMN intravenously (NCT02122952).

AAV serotypes have been demonstrated to be particularly effective in transducing neural tissue. Different serotypes have been tested in several neurodegenerative diseases and have progressed to clinical trials. These include Parkinson's disease,[52–54] Alzheimer's,[55] SMA (NCT02122952), and Canavan's disease.[56] These trials have repeatedly shown that AAV vectors are safe for clinical use for CNS gene therapies.

Lentiviral Vectors

Lentivirus is a member of the retrovirus family, which also includes HIV-1. They are 80–100 nm in diameter and contain two copies of single-stranded RNA. Following entry into the cell, the RNA is reverse transcribed into linear double-stranded DNA in the cytoplasm. The proviral DNA is then translocated to the nucleus where it stably integrates into the host's DNA. The integration sites within the host genome are believed to be random within transcriptionally active sites.[57] The viral DNA replicates with the host's own genome and is passed onto daughter cells. The proviral DNA is then transcribed to produce both the RNA and the viral proteins necessary to produce progeny virus. To acquire its lipid membrane, the progeny lentiviruses bud through the plasma membrane of the host cell.[57]

The first lentiviral vectors (LV) produced were used more for HIV research than for gene therapy. The main aim was to prevent the production of a replication-competent lentivirus. The safety profile of LV has been improved through several generations. The first generation of vectors used a three plasmid system, splitting the trangene and the viral genes, to limit the possibility of generating a replication-competent vector.[58] One of the plasmids contained the gene encoding the vesicular stomatitis virus glycoprotein (VSV-G) envelope protein, replacing the existing HIV envelope protein with that from another virus in a process termed "pseudotyping." This increases the safety profile of the vector and can also alter the tropism of the vector. Pseudotyping can target the vector toward a specific cell type.[58,59] The second generation of LV removed accessory genes from the viral genome which reduced the potential pathogenicity of the vector.[60,61] Further refinement in the third-generation of lentiviral vectors resulted in removal of regulatory genes *rev* and *tat*. Rev allows unspliced mRNA to exit the nucleus intact. Tat functions in producing large amounts of mRNA transcripts which increases the transcription of the lentiviral genome.[57] To remove *tat*, the viral long terminal repeat (LTR) was replaced with an LTR from an avian retrovirus. The *rev* gene was placed in a separate plasmid. This further reduced the possibility of a replication-competent lentivirus being produced.[62,63] The fourth generation of LV are *rev*-independent vectors.[64] However, replacement of the *rev* response element results in vectors titers lower than vectors produced with rev.

LV have been tested in preclinical studies in Parkinson's disease where it has been used as a tricistronic vector. A single LV was used to deliver three genes that encode the rate-limiting enzymes involved in dopamine synthesis; tyrosine hydroxylase, aromatic amino acid dopa decarboxylase, and GTP cyclohydrolase I. This demonstrated efficacy in preclinical testing[65] and has advanced to clinical trial. Phase 1 (NCT00627588) of this trial has been completed and it has met its primary safety endpoint. There was some indication of efficacy, but this will be more rigorously tested in a Phase 2 trial.[66]

Herpes Simplex Virus

HSV is a large double-stranded DNA virus that is naturally neurotrophic. It is an enveloped virus with a subenvelope (tegument) and a viral capsid.[67] HSV-1 can cause cold sores, keratitis and, in rare cases, encephalitis.[68] HSV-2 is more associated with genital lesions and is considered more pathogenic. For development of viral vectors, HSV-1 has been used, as a less pathogenic virus was deemed more favorable.[69]

HSV infects epithelial cells of the skin and mucosa via a lytic pathway. This allows de-enveloped progeny virus particles to infect the axons of sensory neurons in the affected region. The viral components are efficiently delivered to the nerve cell nucleus by retrograde axonal transport. They are transported to the neuron cell nucleus of either the dorsal root ganglion (DRG) or the trigeminal ganglion. The viral genome remains in the nerve cell nucleus as a latent circular episome for life. HSV natural viral latency has been found to be limited to neurons, which explains its neurotrophic nature. The latent virus can be reactivated in response to a recurring illness or immunosuppression. If this occurs, progeny viruses are produced and transported in an anteretrograde manner back to the nerve terminal to establish an infection at the original site of infection. However, this does not spread in individuals with a normally functioning immune system.[68]

The development of HSV vectors has followed a similar pattern to other viral vectors, where the virus has been made replication deficient to reduce its pathogenicity. The first genes to be deleted were two of the immediate early (IE) genes. This allowed for the production of replication defective HSV vectors in cell lines that expressed the deleted genes.[70,71] The vector retained the ability to transduce sensory nerves after intradermal delivery.[72] A further advancement in HSV vector development was the identification of the latency gene (LAP2) which was positioned downstream of the latency promoter. The latency gene had independent promoter activity. LAP2 was found to express transgenes for extended periods of time.[72] However, these vectors have limited lifetime in non-neuronal tissue and were cytotoxic to cells in culture.[73,74] Production of HSV vectors that has the rest of the IE genes removed was achieved with creation of cell lines that expressed the deleted genes. This resulted in the viral genome being transcriptionally inactive and, therefore, the virus was nontoxic and persists in nondividing cells.[75] Problems with transgene shutdown were overcome with use of an insulator sequence (CTRL) and an extended LAP2 (LATP2) to result in gene expression without any of the virus IE genes.[76–78] This gave a vector capable of long-term gene expression that was not cytotoxic.

An alternative strategy was the creation of amplicon vectors. These are HSV vectors that contain an origin of replication and a packaging signal but have no lytic function. They are propagated using a replication-defective helper virus. Amplicon vectors can accommodate large transgenes (up to 100 kb) but are difficult to produce in high titers.[79–81] Another option developed for HSV vectors was replication-competent vectors. These were developed with a view to treating cancer by delivering transgenes that activated anticancer drugs, induced innate killing mechanisms, and improved distribution of anticancer metabolites.[82,83] Conditionally replicative vectors were developed that had genes contributing to neuropathogenesis removed.[84] HSV replication-conditional vectors have been tested for their use in the treatment of cancer. This has been translated to the clinic and been applied in the treatment of glioblastoma and other forms of cancer.[81,85,86]

HSV vectors have also been used in the treatment of chronic pain. They have been used to deliver proenkephalin, an antinociceptive gene, to the DRG in a preclinical pain model in rats. This resulted is alleviation of pain in these animals in response to a formalin injection.[87] In a subsequent study, HSV-proenkephalin enhanced the effect of morphine in a spinal nerve ligation model and provided an analgesic effect in neuropathic pain.[88] Further testing in a mouse model of bone cancer pain determined that HSV-enkephalin decreased pain-related behavior in these animals.[89] This vector proceeded to phase 1 and 2 clinical trials (NCT00804076 and NCT01291901). The trials met its primary safety endpoint but did not proceed beyond Phase 2 testing.

STEM CELLS

Stem cells are defined as cells that have the ability, under the correct conditions, to differentiate into a number of different cell types.

Depending on the type of stem cell, they can differentiate into a wide variety of cell types or can be restricted to a few lineages. Stem cells have been extensively researched due to their therapeutic potential in regenerative medicine. The types of stem cells covered in this chapter include embryonic stem cells (ESCs), neural progenitor cells (NPCs), mesenchymal stem cells (MSCs), hematopoietic stem cells (HSCs), and induced pluripotent stem cells (iPS cells) (Fig. 7.2).

Embryonic Stem Cells

ESCs are cells that differentiate into all of the somatic cells in an embryo. ESCs can be used to understand the mechanisms of development of cells and organ systems in the body. They also have great potential to be used therapeutically in regenerative medicine to replace damaged tissue and cells in patients.

Human ESCs are isolated from the inner cell mass of the human blastocyst. During normal development, they are present at day 4–7 post-fertilization. After day 7, they begin to form the three embryonic tissue layers (ectoderm, mesoderm, and endoderm).[90] Human ESCs were first grown successfully in a laboratory by Thompson et al.[91,92] ESCs are termed "multipotent," as they have the ability to become many different cell types. They express genes relating to transcription, chromatin remodeling, and DNA repair factors at higher levels when compared to differentiated

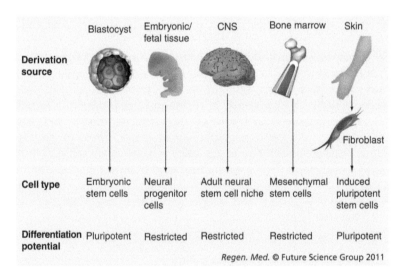

FIGURE 7.2 Sources of stem cells for therapeutic development. *Source: From Lunn J.S., Sakowski S.A., Federici T., Glass J.D., Boulis N.M., Feldman E.L. (2011) Stem cell technology for the study and treatment of motor neuron diseases. Regen. Med. 6(2): 201–213.*

cells.[93–97] Research has shown that differentiation of ESC in vitro results in the down-regulation of pluripotency-associated genes.[98]

There are a large number of studies that have differentiated ESCs into cells derived from the three germ layers. For example, endodermally derived cells include hepatocytes,[99,100] lung epithelium,[101,102] and insulin-producing β cells.[103,104] Cells derived from the mesoderm resulted in the differentiation of chondrocytes, osteocytes, and skeletal myoblasts.[105] Ectodermally derived cells include keratinocytes[106] and NPCs.[91] Functional motor neurons generated from ESCs in vitro were first reported by Li et al.[107]

The first cells derived from ESCs, which advanced to clinical trial, were retinal pigment epithelium (RPE) cells to treat Stargardt's macular dystrophy (NCT01345006) and dry age-related macular degeneration (NCT01344993). The RPE cells were administered via a submacular injection in a phase 1/2 open label trial. These trials were initiated in 2011 and first reported on in 2012. Initial follow-up on the trial participants reported no indication of rejection, tumorigenicity, hyper proliferation, or ectopic tissue formations at 4 months postsurgery. None of the patients' vision deteriorated in this time.[108] Longer-term follow-up reported safety in relation to RPE transplant and graft survival up to 22 months postsurgery. No adverse events were reported in relation to the transplanted cells. There were some improvements in vision but as this was an open label trial it is difficult to assess.[109] For other clinical trials using ESCs see www.clinicaltrials.gov.

Neural Progenitor Cells

NPCs are tissue-specific stem cells that have been obtained from a postimplantation embryo. These cells have the capacity to differentiate into the three major cell types of the CNS (neurons, astrocytes, and oligodendrocytes).[110] NPCs have some predetermined properties but their fate can be altered after exposure to certain external factors. Research has demonstrated that NPCs are spatially and temporally restricted in the profile of their gene expression. They are also limited in their ability to renew and differentiate.[111–113] However, other studies have determined that NSCs lose some of their region-specific characteristics in vitro.[114–116] External factors appear to play a significant role in determining what cell type the NSC will become. For example, basic fibroblast growth factor has been found to play a significant role in affecting the differentiating capacity of brain-derived NPCs and directing them toward a spinal motor neuron fate.[117]

The potential for NPCs to be differentiated into neuronal cells types, such as motor neurons, and to be used to treat neurodegenerative diseases such as ALS and SMA is significant. The other therapeutic possibility is to use undifferentiated NPCs therapeutically. In theory, these stem cells could differentiate into cells that have been lost to disease progression. Alternatively, they could differentiate into support cells, such as astrocytes. This has been tested in preclinical models of ALS where it showed therapeutic efficacy[118,119] and has been advanced to clinical trials.[120,121] These studies are discussed in more detail in subsequent chapters.

Mesenchymal Stem Cells

MSCs are adult-derived stem cells. They were first isolated from bone marrow[122,123] but have also been isolated from other adult tissues.[124,125] MSCs are defined as multipotent and have the ability to differentiate into a number of different lineages (i.e., adipogenic, osteogenic, and chondrogenic).[126,127]

Research has demonstrated that MSCs could be the foundation for an innate system of tissue repair.[128] MSCs have been tested for their ability to affect tissue repair in animal models of a variety of diseases such as lung injury,[129,130] kidney disease,[131] graft versus host disease (GVHD),[132] and myocardial infarction.[133] It was also demonstrated that MSCs had a therapeutic effect on tissue repair despite a low or transient level of engraftment. For example, human MSCs injected into immunodeficient mice, with acute myocardial infarction, were undetectable 3 weeks postinjection. However, these mice continued to show improvement in cardiac function beyond this timepoint.[134] This suggests that MSCs ability to secrete factors into the injured tissue may be a more important factor in tissue repair than their ability to differentiate into another cell type.[135] MSCs, also, have well-documented immunomodulatory properties, with human MSCs having been found to express intermediate levels of major histocompatibility complex class I proteins.[136] Other studies have found that MSCs have immunomodulatory effects on specific T-cell functions. This has been tested in experimental models of (GVHD).[137] MSCs have also been shown to inhibit B cell differentiation.[138]

One of the advantages of using MSCs as a cell therapy is the potential to isolate a patient's own MSCs, expand them in vitro and deliver them as an autologous therapy. This would avoid the whole issue of rejection and the necessity for immunosuppression. The problem associated with this is the possibility of the MSCs, especially in genetic conditions, carrying the targeted disease.

A large number of clinical trials have been initiated using MSCs. These have delivered both autologous and allogenic MSCs as a therapy to treat a wide variety of diseases including diabetes, cardiomyopathy, and spinal cord injury. These trials have primarily been phase I/II trials that have focused on safety as their primary outcome. Delivery of MSCs has been demonstrated to be safe with different delivery methods. Some studies have reported promising efficacy data but this needs to be tested in a placebo-controlled, blinded study to verify these findings. For a complete list see www.clinicaltrials.gov. A number of trials have been conducted using MSCs as a treatment for ALS. These are detailed in a subsequent chapter.

Hematopoietic Stem Cells

HSCs are stem cells isolated mostly from bone marrow. They proliferate and differentiate giving rise to leukocytes and erythrocytes.[139] Some HSCs circulate in the blood where they play a role in homeostasis and repopulate damaged bone marrow regions.[140] Alternative sources of HSCs include peripheral blood and umbilical cord blood. Isolation from peripheral blood requires that HSCs are mobilized from the bone marrow into circulation. The can be achieved by administering hematopoietic cytokines, such as granulocyte colony stimulating factor (G-CSF), to the donor.[141] Collecting HSCs from peripheral blood has the advantage of being less invasive that harvesting bone marrow, and this has become the preferred method of harvesting HSCs.[142,143] Umbilical cord blood was first reported as a source of HSCs in 1982.[144] The initial interest in cord blood was its potential to be used later in life as an autologous source of stem cells. However, with it being an easily accessed source of HSCs and the fact that it is tested and banked in advance, it is a valuable source for nonautologous transplants.

HSCs are primarily used in transplants to treat patients suffering from a range of hematologic diseases, particularly cancer. Patients receive a high dose of radiation or chemotherapy to ablate their endogenous HSCs. The HSCs are administered to repopulate the patient's bone marrow. The first transplants were between twins which avoided any immune incompatibility problems. When they were expanded to allogeneic transplants, the issue of graft rejection or GVHD became apparent. Graft rejection was due to the host immune system attacking the transplanted cells. GVHD resulted from the host-reactive donor T lymphocytes attacking tissue in the recipient. It can cause major damage and be fatal if not treated.[145] These problems led to a better understanding of the human leukocyte antigen complex[146] and the need for immunosuppression, even in matched donors and recipients.[147–149] T-lymphocyte depletion of the HSCs was found to reduce the risk of GVHD.[150,151] However, this increased the risk of rejection and prolonged the time it took to for the transplanted cells to reconstitute.[152,153] This resulted in these patients requiring additional support as they were vulnerable to infection for a prolonged period of time.

HSCs have also been used to treat diseases that do not have a hematopoietic origin. This includes diseases of the liver, heart, and brain.[154] For example, it was noted that marrow cells that were infused intravenously into adult mice migrated to the brain. These cells were found to differentiate and express neuronal markers.[155,156] These results were subsequently questioned due to conflicting data.[157] Despite the conflict in the data, HSCs have been translated to clinical trials for stroke/ischemic brain injury. These trials have shown that HSC administration is safe but there is not enough data to demonstrate efficacy (see www.clinicaltrials.gov).

Induced Pluripotent Stem Cells

iPS cells are adult-derived stem cells. They are developed by reprogramming somatic cells into a pluripotent state. This has been achieved in a number of different ways. One method reprogrammed somatic cells by transferring their nuclear content into oocytes.[158] Another method fused an ESC with a somatic cell.[159,160] These studies inferred that that certain factors in unfertilized eggs and ESCs were necessary for converting somatic cells into pluripotent stem cells. In 2006, Takahashi and Yamanaka conducted a study to identify these factors. They identified four factors that were necessary for induction of pluripotency. Two of the factors were found to be important in the maintenance of pluripotency in ESC and in early embryos, Oct3/4[161] and Sox2.[162] The other two factors functioned in maintaining the ESC phenotype long term and also played a role in the proliferation of ESCs in vitro, c-Myc[163] and Klf4.[164] Somatic cells, fibroblasts, were transduced with these factors and then tested for their pluripotency. The cells were found to exhibit the morphology and growth pattern of ESCs. When they were injected subcutaneously into nude mice, teratomas were formed that contained all three germ layers.[165] Researchers have continued to advance the protocol to develop iPS cells to use as alternatives to viral vectors. This includes episomal vectors, synthetic RNA replicons, human artificial chromosomes, transposon systems, and nanoparticle carriers. They have also looked for alternatives to the original four factors to induce pluripotency.[166]

iPS cells can be developed from a patient's fibroblast cells. One potential therapeutic use for iPS cells is to differentiate them into different cell types. These could then be transplanted back into the same patient, thereby circumventing the problem of rejection and the necessity for immunosuppression. iPS cells have been differentiated into a wide variety of cell types. This includes functional astrocytes,[167] dopaminergic neurons,[168] RPE cells,[169] hepatocytes,[170] cardiomyocytes,[171] and red blood cells.[172] A number of studies have been conducted to test cells in preclinical models of disease. One study transplanted NPCs, which had been derived from iPS cells, into the spinal cord of a rat model of ALS. These cells were found to survive for the duration of the study in both wild-type and in a rat model of ALS. The transplanted NPCs had differentiated toward a motor neuron fate and expressed motor neuron markers.[173] How these cells would affect disease progression and life span requires further study.

The main problem with the autologous derived iPS cell therapeutic approach is as the somatic cells are taken from a patient the cells derived from them may carry the disease. This could limit the transplanted cells' ability to be therapeutically effective. This is particularly true of genetic diseases. One way of exploiting iPS cells when dealing with a genetic disease is try and correct the defect in the stem cell. The corrected iPS cells can then either be administered or differentiated into the required cell type for treatment. This allows for use of autologous cells but eliminates the problem with genetic defects limiting therapeutic effectiveness. There are a number of genome-editing tools available that would allow for the correction of a mutation. These include zinc finger nuclease proteins, Clustered Regularly Interspaced Short Palindromic Repeats (CRISPR)—CRISPR associated (Cas9) (CRISPR/Cas9) system and Transcription Activator-Like Effector Nucleases.[174–176] These genome-editing proteins use endogenous mechanisms for DNA repair within the cell and can be engineered to target a specific mutation.

iPS cells can be used to model diseases in vitro. This has been applied across a number of different diseases. In diseases, such as ALS, where a number of different genes can be linked to the disease, iPS cells have been created from a broad spectrum of patients.[177] This allows for characterization of some of the underlying mechanisms of the different genetic causes of ALS and will assist is identifying where they overlap and diverge. This enhanced understanding will help in identifying new therapeutic targets. Another advantage of iPS cells is that they can be used to screen therapies in vitro before moving on to more expensive in vivo testing. This would be particularly useful for pharmacological agents.

ROUTES OF DELIVERY

Choosing a delivery route, targeting the CNS, for either a gene or stem cell therapy requires careful consideration. The delivery route has to be compatible with the therapy to ensure maximum therapeutic effect. One option is for a remote or systemic delivery. This can involve intravenous delivery or injection of the therapy into muscle to target the neuromuscular junctions for retrograde transport. Systemic delivery has the advantage of not requiring direct access to the CNS. This makes it a safer option. However it does require a much higher initial dose to ensure a therapeutic dose reaches the CNS. In the case of intravenous delivery, the therapy must also have the ability to cross the blood–brain barrier (BBB). One viral vector that has the ability to cross the BBB is AAV9.[178,179] This vector is being used in the clinical trial testing SMN gene therapy to treat SMA (NCT02122952), with the vector being delivered intravenously. Injection into the muscle targeting the neuromuscular junctions for retrograde delivery to the CNS is only feasible for certain vectors such as LV that has been pseudotyped with Rab G protein. However, to achieve a therapeutic dose in the CNS, a large number of injections would be required in muscle groups all over the body. This is not an easily translatable delivery option.

Direct delivery routes to the CNS have the advantage of needing a much smaller dose. Direct delivery also reduces the risk of off-target effects that can be seen with systemic delivery. However, direct delivery to the CNS carries a much higher risk. Delivery of a gene or stem cell therapy directly to the CNS can be achieved in a number of ways: direct intraparenchymal injection (brain or spinal cord), intracerebroventricular, or intrathecal. Direct intraparenchymal injection involves targeting specific areas of either the brain or spinal cord and injecting the therapy into that location. It has been tested and advanced to clinical trial for a number of neurodegenerative diseases such as Alzheimer's, Parkinson's, and ALS. This delivery route has been proven safe for administration to both the brain and spinal cord.[55,180–182] Intracerebroventricular delivery involves injecting the therapy into one of the cerebrospinal fluid filled ventricles of the brain. The gene or stem cell therapy will then travel within the cerebrospinal fluid to reach the entire CNS. This form of delivery has been tested in a phase 1 trial for ALS where VEGF protein is the therapy (NCT01384162). The phase 1 trial met its safety criteria and the delivery route was found to be safe. The third way of directly delivering a therapy to the CNS is intrathecal delivery. The mode of delivery accesses the intrathecal space which is between the dura and the spinal cord. This contains cerebrospinal fluid that bathes the spinal cord. Intrathecal delivery of gene and stem cells has been tested extensively in preclinical models of motor neuron disease. This method has been found to be effective in targeting motor neurons especially using AAV9.[183] Intrathecal delivery also has the advantage of targeting the entire CNS through the flow of cerebrospinal fluid.

CONCLUSION

ALS and SMA are devastating neurodegenerative diseases that currently have no cure. Gene and stem cell therapies have the potential to be effective in treating these diseases. Gene therapy vectors have been developed to improve targeting and to minimize an adverse immune response. Research into stem cells has improved our understanding of how they differentiate. This allows researchers to select the optimal cell type for specific diseases. Clinical trials have advanced to test potential gene and stem cell therapies for both of these diseases. Improvement in the development of vectors and stem cell therapies will assist in treatment, with the ultimate goal of finding a cure.

References

1. Yin H, Kanasty RL, et al. Non-viral vectors for gene-based therapy. *Nat Rev Genet* 2014;**15**(8):541–55.
2. Wang S, Ma N, et al. Transgene expression in the brain stem effected by intramuscular injection of polyethylenimine/DNA complexes. *Mol Ther* 2001;**3**(5 Pt 1):658–64.
3. Federici T, Liu JK, et al. Neuronal affinity of a C7C loop peptide identified through phage display. *J Drug Target* 2006;**14**(5):263–71.
4. Federici T, Liu JK, et al. A means for targeting therapeutics to peripheral nervous system neurons with axonal damage. *Neurosurgery* 2007;**60**(5):911–8. discussion 911-918.
5. Liu JK, Teng Q, et al. A novel peptide defined through phage display for therapeutic protein and vector neuronal targeting. *Neurobiol Dis* 2005;**19**(3):407–18.
6. Bharali DJ, Klejbor I, et al. Organically modified silica nanoparticles: a nonviral vector for in vivo gene delivery and expression in the brain. *Proc Natl Acad Sci USA* 2005;**102**(32):11539–44.
7. Bergelson JM, Cunningham JA, et al. Isolation of a common receptor for Coxsackie B viruses and adenoviruses 2 and 5. *Science* 1997;**275**(5304):1320–3.
8. Crystal RG. Adenovirus: the first effective in vivo gene delivery vector. *Hum Gene Ther* 2014;**25**(1):3–11.
9. Lentz TB, Gray SJ, et al. Viral vectors for gene delivery to the central nervous system. *Neurobiol Dis* 2012;**48**(2):179–88.
10. Davison E, Kirby I, et al. Adenovirus type 5 uptake by lung adenocarcinoma cells in culture correlates with Ad5 fibre binding is mediated by alpha(v)beta1 integrin and can be modulated by changes in beta1 integrin function. *J Gene Med* 2001;**3**(6):550–9.
11. Huang S, Kamata T, et al. Adenovirus interaction with distinct integrins mediates separate events in cell entry and gene delivery to hematopoietic cells. *J Virol* 1996;**70**(7):4502–8.
12. Strunze S, Engelke MF, et al. Kinesin-1-mediated capsid disassembly and disruption of the nuclear pore complex promote virus infection. *Cell Host Microbe* 2011;**10**(3):210–23.
13. Trotman LC, Mosberger N, et al. Import of adenovirus DNA involves the nuclear pore complex receptor CAN/Nup214 and histone H1. *Nat Cell Biol* 2001;**3**(12):1092–100.
14. Hillgenberg M, Tonnies H, et al. Chromosomal integration pattern of a helper-dependent minimal adenovirus vector with a selectable marker inserted into a 27.4-kilobase genomic stuffer. *J Virol* 2001;**75**(20):9896–908.
15. Chu RL, Post DE, et al. Use of replicating oncolytic adenoviruses in combination therapy for cancer. *Clin Cancer Res Off J Am Assoc Cancer Res* 2004;**10**(16):5299–312.
16. Bett AJ, Haddara W, et al. An efficient and flexible system for construction of adenovirus vectors with insertions or deletions in early regions 1 and 3. *Proc Natl Acad Sci USA* 1994;**91**(19):8802–6.
17. Tripathy S, Black H, et al. Immune responses to trangene-encoded proteins limit the stability of gene expression after injection of replication-defective adenovirus vectors. *Nat Med* 1996;**2**(5):545–50.
18. Wohlleber D, Kashkar H, et al. TNF-induced target cell killing by CTL activated through cross-presentation. *Cell Rep* 2012;**2**(3):478–87.
19. Yang Y, Jooss K, et al. Immune responses to viral antigens versus transgene product in the elimination of recombinant adenovirus vectors in mouse liver and lung tissues. *J Virol* 1996;**70**:6370–7.
20. Amalfitano A, Hauser MA, et al. Production and characterization of improved adenovirus vectors with the E1, E2b, and E3 genes deleted. *J Virol* 1998;**72**(2):926–33.
21. Armentano D, Sookdeo C, et al. Characterization of an adenovirus gene transfer vector containing an E4 deletion. *Hum Gene Ther* 1995;**6**(10):1343–53.
22. Parks R, Chen L, et al. A helper-dependent adenovirus vector system: removal of helper virus by Cre-mediated excision of the viral packaging signal. *Proc Natl Acad Sci USA* 1996;**93**:13565–70.

23. Volpers C, Kochanek S. Adenoviral vectors for gene transfer and therapy. *J Gene Med* 2004;**6**(Suppl. 1):S164–171.

24. O'Neal WK, Zhou H, et al. Toxicity associated with repeated administration of first-generation adenovirus vectors does not occur with a helper-dependent vector. *Mol Med* 2000;**6**(3):179–95.

25. Cregan SP, MacLaurin J, et al. Helper-dependent adenovirus vectors: their use as a gene delivery system to neurons. *Gene Ther* 2000;**7**(14):1200–9.

26. Rosenfeld MA, Yoshimura K, et al. In vivo transfer of the human cystic fibrosis trans-membrane conductance regulator gene to the airway epithelium. *Cell* 1992;**68**(1):143–55.

27. Harvey BG, Leopold PL, Hackett NR, Grasso TM, Williams PM, Tucker AL, et al. Airway epithelial CFTR mRNA expression in cystic fibrosis patients after repetitive administration of a recombinant adenovirus. *J Clin Invest* 1999;**104**(9):1245–55.

28. Morral N, O'Neal W, et al. Administration of helper-dependent adenoviral vectors and sequential delivery of different vector serotype for long-term liver-directed gene transfer in baboons. *Proc Natl Acad Sci U S A* 1999;**96**(22):12816–21.

29. Morsy MA, Gu M, et al. An adenoviral vector deleted for all viral coding sequences results in enhanced safety and extended expression of a leptin transgene. *Proc Natl Acad Sci U S A* 1998;**95**(14):7866–71.

30. Germano IM, Fable J, et al. Adenovirus/herpes simplex-thymidine kinase/ganciclovir complex: preliminary results of a phase I trial in patients with recurrent malignant gliomas. *J Neuro Oncol* 2003;**65**(3):279–89.

31. Immonen A, Vapalahti M, et al. AdvHSV-tk gene therapy with intravenous ganciclovir improves survival in human malignant glioma: a randomised, controlled study. *Mol Ther* 2004;**10**(5):967–72.

32. Wildner O, Blaese RM, et al. Therapy of colon cancer with oncolytic adenovirus is enhanced by the addition of herpes simplex virus-thymidine kinase. *Cancer Res* 1999;**59**(2):410–3.

33. Tagawa M, Kawamura K, et al. Cancer therapy with local oncolysis and topical cytokine secretion. *Front Biosci J Virtual Libr* 2008;**13**:2578–87.

34. Aguilar LK, Guzik BW, et al. Cytotoxic immunotherapy strategies for cancer: mechanisms and clinical development. *J Cell Biochem* 2011;**112**(8):1969–77.

35. Tan Y, Hackett NR, et al. Protective immunity evoked against anthrax lethal toxin after a single intramuscular administration of an adenovirus-based vaccine encoding humanized protective antigen. *Hum Gene Ther* 2003;**14**(17):1673–82.

36. Hicks MJ, De BP, et al. Cocaine analog coupled to disrupted adenovirus: a vaccine strategy to evoke high-titer immunity against addictive drugs. *Mol Ther J Am Soc Gene Ther* 2011;**19**(3):612–9.

37. Atchison RW, Casto BC, et al. Adenovirus-associated defective virus particles. *Science* 1965;**149**(3685):754–6.

38. McCarty DM, Young Jr. SM, et al. Integration of adeno-associated virus (AAV) and recombinant AAV vectors. *Ann Rev Genet* 2004;**38**:819–45.

39. Schnepp BC, Jensen RL, et al. Characterization of adeno-associated virus genomes isolated from human tissues. *J Virol* 2005;**79**(23):14793–803.

40. Duan D, Sharma P, et al. Circular intermediates of recombinant adeno-associated virus have defined structural characteristics responsible for long-term episomal persistence in muscle tissue. *J Virol* 1998;**72**(11):8568–77.

41. Rabinowitz JE, Rolling F, et al. Cross-packaging of a single adeno-associated virus (AAV) type 2 vector genome into multiple AAV serotypes enables transduction with broad specificity. *J Virol* 2002;**76**(2):791–801.

42. Samulski RJ, Srivastava A, et al. Rescue of adeno-associated virus from recombinant plasmids: gene correction within the terminal repeats of AAV. *Cell* 1983;**33**(1):135–43.

43. Hearing P, Samulski RJ, et al. Identification of a repeated sequence element required for efficient encapsidation of the adenovirus type 5 chromosome. *J Virol* 1987;**61**(8):2555–8.

44. Samulski RJ, Chang LS, et al. A recombinant plasmid from which an infectious adeno-associated virus genome can be excised in vitro and its use to study viral replication. *J Virol* 1987;**61**(10):3096–101.
45. Samulski RJ, Chang L, et al. Helper-free stocks of recombinant adeno-associated viruses: normal integration does not require viral gene expression. *J Virol* 1989;**63**:3822–8.
46. Ferrari FK, Samulski T, et al. Second-strand synthesis is a rate-limiting step for efficient transduction by recombinant adeno-associated virus vectors. *J Virol* 1996;**70**(5):3227–34.
47. McCarty DM, Monahan PE, et al. Self-complementary recombinant adeno-associated virus (scAAV) vectors promote efficient transduction independently of DNA synthesis. *Gene Ther* 2001;**8**(16):1248–54.
48. Kotterman MA, Schaffer DV. Engineering adeno-associated viruses for clinical gene therapy. *Nat Rev Genet* 2014;**15**(7):445–51.
49. Liu Y, Siriwon N, et al. Generation of targeted adeno-associated virus (AAV) vectors for human gene therapy. *Curr Pharm Design* 2015;**21**(22):3248–56.
50. Davis AS, Federici T, et al. Rational design and engineering of a modified adeno-associated virus (AAV1)-based vector system for enhanced retrograde gene delivery. *Neurosurgery* 2015;**76**(2):216–25. discussion 225.
51. Boutin S, Monteilhet V, et al. Prevalence of serum IgG and neutralizing factors against adeno-associated virus (AAV) types 1, 2, 5, 6, 8, and 9 in the healthy population: implications for gene therapy using AAV vectors. *Hum Gene Ther* 2010;**21**(6):704–12.
52. Marks Jr. WJ, Bartus RT, et al. Gene delivery of AAV2-neurturin for Parkinson's disease: a double-blind, randomised, controlled trial. *Lancet Neurol* 2010;**9**(12):1164–72.
53. Bartus RT, Weinberg MS, et al. Parkinson's disease gene therapy: success by design meets failure by efficacy. *Mol Ther J Am Soc Gene Ther* 2014;**22**(3):487–97.
54. LeWitt PA, Rezai AR, et al. AAV2-GAD gene therapy for advanced Parkinson's disease: a double-blind, sham-surgery controlled, randomised trial. *Lancet Neurol* 2011;**10**(4):309–19.
55. Rafii MS, Baumann TL, et al. A phase1 study of stereotactic gene delivery of AAV2-NGF for Alzheimer's disease. *Alzheimers Dement* 2014;**10**(5):571–81.
56. Leone P, Shera D, et al. Long-term follow-up after gene therapy for canavan disease. *Sci Trans Med* 2012;**4**(165) 165ra163.
57. Strauss JH, Strauss EG. *Viruses and human disease*. San Diego, Calif.; London: Academic Press; 2002.
58. Naldini L, Blomer U, et al. In vivo gene delivery and stable transduction of nondividing cells by a lentiviral vector. *Science* 1996;**272**(5259):263–7.
59. Bischof D, Cornetta K. Flexibility in cell targeting by pseudotyping lentiviral vectors. *Methods Mol Biol* 2010;**614**:53–68.
60. Zufferey R, Nagy D, et al. Multiply attenuated lentiviral vector achieves efficient gene delivery in vivo. *Nat Biotechnol* 1997;**15**(9):871–5.
61. Kim V, Mitrophanous K, et al. Minimal requirement for a lentivirus vector based on human immunodeficiency virus type 1. *J Virol* 1998;**72**(1):811–6.
62. Naldini L, Verma I. Lentiviral vectors. *Adv Virus Res* 2000;**55**:599–609.
63. Dull T, Zufferey R, et al. A third-generation lentivirus vector with a conditional packaging system. *J Virol* 1998;**72**(11):8463–71.
64. Pandya S, Boris-Lawrie K, et al. Development of an rev-independent, minimal simian immunodeficiency virus-derived vector system. *Hum Gene Ther* 2001;**12**(7):847–57.
65. Azzouz M, Martin-Rendon E, et al. Multicistronic lentiviral vector-mediated striatal gene transfer of aromatic L-amino acid decarboxylase, tyrosine hydroxylase, and GTP cyclohydrolase I induces sustained transgene expression, dopamine production, and functional improvement in a rat model of Parkinson's disease. *J Neurosci* 2002;**22**(23):10302–12.
66. Palfi S, Gurruchaga JM, et al. Long-term safety and tolerability of ProSavin, a lentiviral vector-based gene therapy for Parkinson's disease: a dose escalation, open-label, phase 1/2 trial. *Lancet* 2014;**383**(9923):1138–46.

67. Kantor B, Bailey RM, et al. Methods for gene transfer to the central nervous system. *Adv Genet* 2014;**87**:125–97.
68. Frampton Jr. AR, Goins WF, et al. HSV trafficking and development of gene therapy vectors with applications in the nervous system. *Gene Ther* 2005;**12**(11):891–901.
69. Smith JS, Robinson NJ. Age-specific prevalence of infection with herpes simplex virus types 2 and 1: a global review. *J Infect Dis* 2002;**186**(Suppl. 1):S3–28.
70. Shepard AA, DeLuca NA. Activities of heterodimers composed of DNA-binding- and transactivation-deficient subunits of the herpes simplex virus regulatory protein ICP4. *J Virol* 1991;**65**(1):299–307.
71. Marconi P, Krisky D, et al. Replication-defective herpes simplex virus vectors for gene transfer in vivo. *Proc Natl Acad Sci U S A* 1996;**93**(21):11319–20.
72. Goins WF, Sternberg LR, et al. A novel latency-active promoter is contained within the herpes simplex virus type 1 UL flanking repeats. *J Virol* 1994;**68**(4):2239–52.
73. Krisky DM, Wolfe D, Goins WF, Marconi PC, Ramakrishnan R, Mata M, et al. Deletion of multiple immediate-early genes from herpes simplex virus reduces cytotoxicity and permits long-term gene expression in neurons. *Gene Ther* 1998;**5**(12):1593–603.
74. Samaniego LA, Neiderhiser L, DeLuca NA. Persistence and expression of the herpes simplex virus genome in the absence of immediate-early proteins. *J Virol* 1998;**72**(4):3307–20.
75. Kaplitt MG, Makimura H. Defective viral vectors as agents for gene transfer in the nervous system. *J Neurosci Meth* 1997;**71**(1):125–32.
76. Palmer JA, Branston RH, et al. Development and optimization of herpes simplex virus vectors for multiple long-term gene delivery to the peripheral nervous system. *J Virol* 2000;**74**(12):5604–18.
77. Chattopadhyay M, Krisky D, et al. HSV-mediated gene transfer of vascular endothelial growth factor to dorsal root ganglia prevents diabetic neuropathy. *Gene Ther* 2005;**12**(18):1377–84.
78. Goins WF, Lee KA, et al. Herpes simplex virus type 1 vector-mediated expression of nerve growth factor protects dorsal root ganglion neurons from peroxide toxicity. *J Virol* 1999;**73**(1):519–32.
79. Epstein AL. Progress and prospects: biological properties and technological advances of herpes simplex virus type 1-based amplicon vectors. *Gene Ther* 2009;**16**(6):709–15.
80. Goins WF, Cohen JB, et al. Gene therapy for the treatment of chronic peripheral nervous system pain. *Neurobiol Dis* 2012;**48**(2):255–70.
81. Marconi P, Argnani R, et al. HSV as a vector in vaccine development and gene therapy. *Adv Exp Med Biol* 2009;**655**:118–44.
82. Moriuchi S, Oligino T, et al. Enhanced tumor cell killing in the presence of ganciclovir by herpes simplex virus type 1 vector-directed coexpression of human tumor necrosis factor-alpha and herpes simplex virus thymidine kinase. *Cancer Res* 1998;**58**(24):5731–7.
83. Marconi P, Tamura M, et al. Connexin 43-enhanced suicide gene therapy using herpesviral vectors. *Mol Ther J Am Soc Gene Ther* 2000;**1**(1):71–81.
84. Simonato M, Bennett J, et al. Progress in gene therapy for neurological disorders. *Nat Rev Neurol* 2013;**9**(5):277–91.
85. Markert JM, Gillespie GY, et al. Genetically engineered HSV in the treatment of glioma: a review. *Rev Med Virol* 2000;**10**(1):17–30.
86. Markert JM, Liechty PG, et al. Phase Ib trial of mutant herpes simplex virus G207 inoculated pre-and post-tumor resection for recurrent GBM. *Mol Ther J Am Soc Gene Ther* 2009;**17**(1):199–207.
87. Goss JR, Mata M, et al. Antinociceptive effect of a genomic herpes simplex virus-based vector expressing human proenkephalin in rat dorsal root ganglion. *Gene Ther* 2001;**8**(7):551–6.
88. Hao S, Mata M, et al. HSV-mediated gene transfer of the glial cell-derived neurotrophic factor provides an antiallodynic effect on neuropathic pain. *Mol Ther* 2003;**8**(3):367–75.

89. Goss JR, Harley CF, et al. Herpes vector-mediated expression of proenkephalin reduces bone cancer pain. *Ann Neurol* 2002;**52**(5):662–5.
90. *Stem cells and the future of regenerative medicine.* Washington, D.C.; [Great Britain]: National Academy Press; 2002.
91. Reubinoff BE, Pera MF, et al. Embryonic stem cell lines from human blastocysts: somatic differentiation in vitro. *Nat Biotechnol* 2000;**18**(4):399–404.
92. Thomson JA, Itskovitz-Eldor J, et al. Embryonic stem cell lines derived from human blastocysts. *Science* 1998;**282**(5391):1145–7.
93. Zhou Q, Chipperfield H, et al. A gene regulatory network in mouse embryonic stem cells. *Proc Natl Acad Sci U S A* 2007;**104**(42):16438–43.
94. Bhattacharya B, Miura T, et al. Gene expression in human embryonic stem cell lines: unique molecular signature. *Blood* 2004;**103**(8):2956–64.
95. Brandenberger R, Khrebtukova I, et al. MPSS profiling of human embryonic stem cells. *BMC Dev Biol* 2004;**4**:10.
96. Brandenberger R, Wei H, et al. Transcriptome characterization elucidates signaling networks that control human ES cell growth and differentiation. *Nat Biotechnol* 2004;**22**(6):707–16.
97. Liu Y, Shin S, et al. Genome wide profiling of human embryonic stem cells (hESCs), their derivatives and embryonal carcinoma cells to develop base profiles of U.S. Federal government approved hESC lines. *BMC Dev Biol* 2006;**6**:20.
98. Walker E, Ohishi M, et al. Prediction and testing of novel transcriptional networks regulating embryonic stem cell self-renewal and commitment. *Cell Stem Cell* 2007;**1**(1):71–86.
99. Cai J, Zhao Y, et al. Directed differentiation of human embryonic stem cells into functional hepatic cells. *Hepatology* 2007;**45**(5):1229–39.
100. Agarwal S, Holton KL, et al. Efficient differentiation of functional hepatocytes from human embryonic stem cells. *Stem Cells* 2008;**26**(5):1117–27.
101. Van Vranken BE, Rippon HJ, et al. The differentiation of distal lung epithelium from embryonic stem cells. *Curr Protocols Stem Cell Biol* 2007 **Chapter 1**: Unit 1G 1.
102. Wang D, Haviland DL, et al. A pure population of lung alveolar epithelial type II cells derived from human embryonic stem cells. *Proc Natl Acad Sci USA* 2007;**104**(11):4449–54.
103. Segev H, Fishman B, et al. Differentiation of human embryonic stem cells into insulin-producing clusters. *Stem Cells* 2004;**22**(3):265–74.
104. Brolen GK, Heins N, et al. Signals from the embryonic mouse pancreas induce differentiation of human embryonic stem cells into insulin-producing beta-cell-like cells. *Diabetes* 2005;**54**(10):2867–74.
105. Barberi T, Willis LM, et al. Derivation of multipotent mesenchymal precursors from human embryonic stem cells. *PLoS Med* 2005;**2**(6):e161.
106. Aberdam E, Barak E, et al. A pure population of ectodermal cells derived from human embryonic stem cells. *Stem Cells* 2008;**26**(2):440–4.
107. Li XJ, Du ZW, et al. Specification of motoneurons from human embryonic stem cells. *Nat Biotechnol* 2005;**23**(2):215–21.
108. Schwartz CM, Tavakoli T, et al. Stromal factors SDF1α, sFRP1, and VEGFD induce dopaminergic neuron differentiation of human pluripotent stem cells. *J Neurosci Res* 2012;**90**:1367–81.
109. Schwartz SD, Regillo CD, et al. Human embryonic stem cell-derived retinal pigment epithelium in patients with age-related macular degeneration and Stargardt's macular dystrophy: follow-up of two open-label phase 1/2 studies. *Lancet* 2015;**385**(9967):509–16.
110. Kim JK, Choi BH, et al. Effects of GM-CSF on the neural progenitor cells. *Neuroreport* 2004;**15**(14):2161–5.
111. Pearson BJ, Doe CQ. Specification of temporal identity in the developing nervous system. *Ann Rev Cell Dev Biol* 2004;**20**:619–47.

112. Seaberg RM, Smukler SR, et al. Intrinsic differences distinguish transiently neu-rogenic progenitors from neural stem cells in the early postnatal brain. *Dev Biol* 2005;**278**(1):71–85.
113. Abramova N, Charniga C, et al. Stage-specific changes in gene expression in acutely isolated mouse CNS progenitor cells. *Dev Biol* 2005;**283**(2):269–81.
114. Gabay L, Lowell S, et al. Deregulation of dorsoventral patterning by FGF con-fers trilineage differentiation capacity on CNS stem cells in vitro. *Neuron* 2003; **40**(3):485–99.
115. Hitoshi S, Alexson T, et al. Notch pathway molecules are essential for the maintenance, but not the generation, of mammalian neural stem cells. *Genes Dev* 2002;**16**(7):846–58.
116. Anderson D. Stem cells and pattern formation in the nervous system: the possible versus the actual. *Neuron* 2001;**30**:19–35.
117. Jordan PM, Ojeda LD, et al. Generation of spinal motor neurons from human fetal brain-derived neural stem cells: role of basic fibroblast growth factor. *J Neurosci Res* 2009;**87**(2):318–32.
118. Xu L, Yan J, et al. Human neural stem cell grafts ameliorate motor neuron disease in SOD-1 transgenic rats. *Transplantation* 2006;**82**(7):865–75.
119. Hefferan MP, Galik J, et al. Human neural stem cell replacement therapy for amyo-trophic lateral sclerosis by spinal transplantation. *PLoS One* 2012;**7**(8):e42614.
120. Feldman EL, Boulis NM, et al. Intraspinal neural stem cell transplantation in amyo-trophic lateral sclerosis: phase 1 trial outcomes. *Ann Neurol* 2014;**75**(3):363–73.
121. Glass JD, Boulis NM, et al. "Lumbar intraspinal injection of neural stem cells in patients with ALS: results of a phase I trial in 12 patients. *Stem Cells* 2012;**30**(6):1144–51.
122. Phinney DG. Building a consensus regarding the nature and origin of mesenchymal stem cells. *J Cell Biochem Suppl* 2002;**38**:7–12.
123. Bianco P, Riminucci M, et al. Bone marrow stromal stem cells: nature, biology, and potential applications. *Stem Cells* 2001;**19**(3):180–92.
124. Vayssade M, Nagel MD. Stromal cells. *Front Biosci* 2009;**14**:210–24.
125. Garcia-Castro J, Trigueros C, et al. Mesenchymal stem cells and their use as cell replace-ment therapy and disease modelling tool. *J Cell Mol Med* 2008;**12**(6B):2552–65.
126. Pittenger MF, Mackay AM, et al. Multilineage potential of adult human mesenchymal stem cells. *Science* 1999;**284**(5411):143–7.
127. Dennis JE, Merriam A, et al. A quadripotential mesenchymal progenitor cell isolated from the marrow of an adult mouse. *J Bone Miner Res Off J Am Soc Bone Miner Res* 1999;**14**(5):700–9.
128. Phinney DG, Prockop DJ. Concise review: mesenchymal stem/multipotent stromal cells: the state of transdifferentiation and modes of tissue repair--current views. *Stem Cells* 2007;**25**(11):2896–902.
129. Ortiz LA, Dutreil M, et al. Interleukin 1 receptor antagonist mediates the antiinflam-matory and antifibrotic effect of mesenchymal stem cells during lung injury. *Proc Natl Acad Sci U S A* 2007;**104**(26):11002–7.
130. Ortiz LA, Gambelli F, et al. Mesenchymal stem cell engraftment in lung is enhanced in response to bleomycin exposure and ameliorates its fibrotic effects. *Proc Natl Acad Sci U S A* 2003;**100**(14):8407–11.
131. Kunter U, Rong S, et al. Transplanted mesenchymal stem cells accelerate glomerular healing in experimental glomerulonephritis. *J Am Soc Nephrol JASN* 2006;**17**(8):2202–12.
132. Ringden O, Uzunel M, et al. Mesenchymal stem cells for treatment of therapy-resistant graft-versus-host disease. *Transplantation* 2006;**81**(10):1390–7.
133. Minguell JJ, Erices A. Mesenchymal stem cells and the treatment of cardiac disease. *Exp Biol Med* 2006;**231**(1):39–49.
134. Iso Y, Spees JL, et al. Multipotent human stromal cells improve cardiac function after myocardial infarction in mice without long-term engraftment. *Biochem Biophys Res Commun* 2007;**354**(3):700–6.

135. Prockop DJ. "Stemness" does not explain the repair of many tissues by mesenchymal stem/multipotent stromal cells (MSCs). *Clin Pharmacol Therap* 2007;**82**(3):241–3.
136. Le Blanc K, Ringden O. Mesenchymal stem cells: properties and role in clinical bone marrow transplantation. *Curr Opin Immunol* 2006;**18**(5):586–91.
137. Dazzi F, Marelli-Berg FM. Mesenchymal stem cells for graft-versus-host disease: close encounters with T cells. *Eur J Immunol* 2008;**38**(6):1479–82.
138. Corcione A, Benvenuto F, et al. Human mesenchymal stem cells modulate B-cell functions. *Blood* 2006;**107**(1):367–72.
139. Orkin SH, Zon LI. Hematopoiesis: an evolving paradigm for stem cell biology. *Cell* 2008;**132**(4):631–44.
140. Massberg S, Schaerli P, et al. Immunosurveillance by hematopoietic progenitor cells trafficking through blood, lymph, and peripheral tissues. *Cell* 2007;**131**(5):994–1008.
141. Drize N, Chertkov J, et al. Hematopoietic progenitor cell mobilization into the peripheral blood of mice using a combination of recombinant rat stem cell factor (rrSCF) and recombinant human granulocyte colony-stimulating factor (rhG-CSF). *Exp Hematol* 1995;**23**(11):1180–6.
142. Anasetti C, Logan BR, et al. Peripheral-blood stem cells versus bone marrow from unrelated donors. *New Engl J Med* 2012;**367**(16):1487–96.
143. Jantunen E, Varmavuo V. Plerixafor for mobilization of blood stem cells in autologous transplantation: an update. *Exp Opin Biol Ther* 2014;**14**(6):851–61.
144. Nakahata T, Ogawa M. Hemopoietic colony-forming cells in umbilical cord blood with extensive capability to generate mono- and multipotential hemopoietic progenitors. *J Clin Invest* 1982;**70**(6):1324–8.
145. Barriga F, Ramirez P, et al. Hematopoietic stem cell transplantation: clinical use and perspectives. *Biol Res* 2012;**45**(3):307–16.
146. Thomas JM, Owens L, et al. Correlation of in vitro immune reactivity and rejection in human transplant recipients. *Surg Forum* 1975;**26**:316–8.
147. Choi SW, Levine J. Indications for hematopoietic cell transplantation for children with severe congenital neutropenia. *Pediatr Transplant* 2010;**14**(8):937–9.
148. Deeg HJ. Allogeneic and autologous bone-marrow transplantation. *Can Fam Physician Medecin de famille canadien* 1988;**34**:2489–98.
149. Lochte Jr. HL, Levy AS, et al. Prevention of delayed foreign marrow reaction in lethally irradiated mice by early administration of methotrexate. *Nature* 1962;**196**:1110–1.
150. Prentice HG, Blacklock HA, et al. Depletion of T lymphocytes in donor marrow prevents significant graft-versus-host disease in matched allogeneic leukaemic marrow transplant recipients. *Lancet* 1984;**1**(8375):472–6.
151. Reisner Y, Kapoor N, et al. Transplantation for acute leukaemia with HLA-A and B nonidentical parental marrow cells fractionated with soybean agglutinin and sheep red blood cells. *Lancet* 1981;**2**(8242):327–31.
152. Marmont AM, Horowitz MM, et al. T-cell depletion of HLA-identical transplants in leukemia. *Blood* 1991;**78**(8):2120–30.
153. Martin PJ, Hansen JA, et al. Effects of in vitro depletion of T cells in HLA-identical allogeneic marrow grafts. *Blood* 1985;**66**(3):664–72.
154. Porada CD, Atala AJ, et al. The hematopoietic system in the context of regenerative medicine. *Methods* 2015;**99**:44–61.
155. Mezey E, Chandross KJ, et al. Turning blood into brain: cells bearing neuronal antigens generated in vivo from bone marrow. *Science* 2000;**290**(5497):1779–82.
156. Brazelton TR, Rossi FM, et al. From marrow to brain: expression of neuronal phenotypes in adult mice. *Science* 2000;**290**(5497):1775–9.
157. Roybon L, Ma Z, et al. Failure of transdifferentiation of adult hematopoietic stem cells into neurons. *Stem Cells* 2006;**24**(6):1594–604.
158. Wilmut I, Schnieke AE, et al. Viable offspring derived from fetal and adult mammalian cells. *Nature* 1997;**385**(6619):810–3.

159. Tada M, Takahama Y, et al. Nuclear reprogramming of somatic cells by in vitro hybridization with ES cells. *Curr Biol CB* 2001;**11**(19):1553–8.
160. Cowan CA, Atienza J, et al. Nuclear reprogramming of somatic cells after fusion with human embryonic stem cells. *Science* 2005;**309**(5739):1369–73.
161. Niwa H, Miyazaki J, et al. Quantitative expression of Oct-3/4 defines differentiation, dedifferentiation or self-renewal of ES cells. *Nat Genet* 2000;**24**(4):372–6.
162. Avilion AA, Nicolis SK, et al. Multipotent cell lineages in early mouse development depend on SOX2 function. *Genes Dev* 2003;**17**(1):126–40.
163. Cartwright P, McLean C, et al. LIF/STAT3 controls ES cell self-renewal and pluripotency by a Myc-dependent mechanism. *Development* 2005;**132**(5):885–96.
164. Li Y, McClintick J, et al. Murine embryonic stem cell differentiation is promoted by SOCS-3 and inhibited by the zinc finger transcription factor Klf4. *Blood* 2005;**105**(2):635–7.
165. Takahashi K, Yamanaka S. Induction of pluripotent stem cells from mouse embryonic and adult fibroblast cultures by defined factors. *Cell* 2006;**126**(4):663–76.
166. Rony IK, Baten A, et al. Inducing pluripotency in vitro: recent advances and highlights in induced pluripotent stem cells generation and pluripotency reprogramming. *Cell Proliferat* 2015;**48**(2):140–56.
167. Mormone E, D'Sousa S, et al. "Footprint-free" human induced pluripotent stem cell-derived astrocytes for in vivo cell-based therapy. *Stem Cells Dev* 2014;**23**(21):2626–36.
168. Theka I, Caiazzo M, et al. Rapid generation of functional dopaminergic neurons from human induced pluripotent stem cells through a single-step procedure using cell lineage transcription factors. *Stem Cells Transl Med* 2013;**2**(6):473–9.
169. Kamao H, Mandai M, et al. Characterization of human induced pluripotent stem cell-derived retinal pigment epithelium cell sheets aiming for clinical application. *Stem Cell Rep* 2014;**2**(2):205–18.
170. Ma X, Duan Y, et al. Highly efficient differentiation of functional hepatocytes from human induced pluripotent stem cells. *Stem Cells Transl Med* 2013;**2**(6):409–19.
171. Blazeski A, Zhu R, et al. Cardiomyocytes derived from human induced pluripotent stem cells as models for normal and diseased cardiac electrophysiology and contractility. *Prog Biophys Mol Biol* 2012;**110**(2–3):166–77.
172. Dias J, Gumenyuk M, et al. Generation of red blood cells from human induced pluripotent stem cells. *Stem Cells Dev* 2011;**20**(9):1639–47.
173. Popescu IR, Nicaise C, et al. Neural progenitors derived from human induced pluripotent stem cells survive and differentiate upon transplantation into a rat model of amyotrophic lateral sclerosis. *Stem Cells Transl Med* 2013;**2**(3):167–74.
174. Cai M, Yang Y. Targeted genome editing tools for disease modeling and gene therapy. *Curr Gene Ther* 2014;**14**(1):2–9.
175. Gaj T, Gersbach CA, et al. ZFN, TALEN, and CRISPR/Cas-based methods for genome engineering. *Trends Biotechnol* 2013;**31**(7):397–405.
176. Kim S, Kim D, et al. Highly efficient RNA-guided genome editing in human cells via delivery of purified Cas9 ribonucleoproteins. *Genome Res* 2014;**24**(6):1012–9.
177. Li Y, Balasubramanian U, et al. A comprehensive library of familial human amyotrophic lateral sclerosis induced pluripotent stem cells. *PLoS One* 2015;**10**(3):e0118266.
178. Bevan AK, Duque S, et al. Systemic gene delivery in large species for targeting spinal cord, brain, and peripheral tissues for pediatric disorders. *Mol Ther J Am Soc Gene Ther* 2011;**19**(11):1971–80.
179. Foust KD, Nurre E, et al. Intravascular AAV9 preferentially targets neonatal neurons and adult astrocytes. *Nat Biotechnol* 2009;**27**(1):59–65.
180. Riley J, Federici T, et al. Cervical spinal cord therapeutic delivery: preclinical safety validation of a stabilized microinjection platform. *Neurosurgery* 2009;**65**(4):754–62.

181. Riley J, Glass J, et al. Intraspinal stem cell transplantation in amyotrophic lateral sclerosis: a phase I trial, cervical microinjection, and final surgical safety outcomes. *Neurosurgery* 2014;**74**(1):77–87.

182. Bartus RT, Brown L, et al. Properly scaled and targeted AAV2-NRTN (neurturin) to the substantia nigra is safe, effective and causes no weight loss: support for nigral targeting in Parkinson's disease. *Neurobiol Dis* 2011;**44**(1):38–52.

183. Federici T, Taub JS, et al. Robust spinal motor neuron transduction following intrathecal delivery of AAV9 in pigs. *Gene Ther* 2012;**19**(8):852–9.

Gene Therapy for Amyotrophic Lateral Sclerosis: Therapeutic Transgenes

A. Donsante

Emory University, Atlanta, GA, United States

Molecular and Cellular Therapies for Motor Neuron Diseases.
DOI: http://dx.doi.org/10.1016/B978-0-12-802257-3.00008-0

167

INTRODUCTION

As discussed in Chapter 4, Molecular Mechanisms of Amyotrophic Lateral Sclerosis, amyotrophic lateral sclerosis (ALS) is a disease with a poorly understood etiology. There are a wide variety of abnormalities present, including mitochondrial dysfunction, glutamate excitotoxicity, oxidative stress, astrogliosis and microgliosis, loss of innervation of neuromuscular junctions (NMJs), and death of motor neurons (MNs). However, the relative contribution of these factors to disease initiation and progression is unclear at best. Therefore, a wide variety of approaches have been investigated to treat this disorder. This chapter will focus on the therapeutic transgenes that have been studied to date. These transgenes range from those that address a single abnormality, such as apoptosis, to those that are expected to alter several defects simultaneously. While this range of approaches has produced improvements in the ALS phenotype in animal models, none has become the silver bullet that will halt ALS in its tracks.

ANTIAPOPTOSIS

One of the defining characteristics of ALS is the loss of upper and lower MNs. Since the death of these cells could be a direct cause of NMJ denervation, one obvious therapeutic approach would be to prevent cell death from occurring. Cells can be killed by a number of mechanisms ranging from a tidy programmed cell death, where the cell essentially breaks itself down to be phagocytized by nearby cells, to messy necrosis, where the cell spills its contents into the extracellular space, inducing inflammation.[1] It is generally believed that MN death occurs via a process of programmed cell death known as apoptosis, or at the very least, something very similar to it.[2]

Apoptosis was first described by Kerr et al. in 1972, and it involves a highly regulated series of steps (Fig. 8.1).[3] There are three main pathways in apoptosis: death receptor-mediated, mitochondrial-mediated, and endoplasmic reticulum-mediated. Based on cell culture models, animal models, and postmortem tissue from ALS patients, it appears that the mitochondrial-mediated pathway is likely the most relevant to MN death in this disease.[2] In this pathway, the action of proteins like Bcl-2-associated X (Bax) and Bcl-2-homologous antagonist/killer (Bak) cause the mitochondria to release cytochrome *c* and other factors into the cytoplasm. This in turn activates a cascade of caspases, including caspases 3 and 9. This pathway is inhibited by a number of proteins, including X-linked inhibitors of apoptosis protein (XIAP), B-cell lymphoma 2 (Bcl-2), and B-cell lymphoma extra large (Bcl-X$_L$).

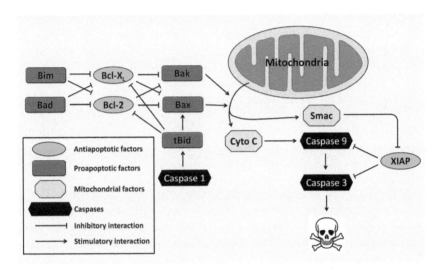

FIGURE 8.1   Apoptosis pathway. Mitochondrial-mediated apoptosis is believed to be the main route for cell death in amyotrophic lateral sclerosis (*ALS*). A simplified model of this pathway is illustrated. When they are not sequestered, proteins like Bcl-2-associated X (*Bax*) and Bcl-2-homologous antagonist/killer (*Bak*) are able to permeablize the mitochondrial outer membrane. These pores permit the release of several proteins, including cytochrome *c* (*Cyto c*) and second-mitochondrial-derived activator of caspases (*Smac*). Cytochrome *c* activates caspase 9 which then cleaves procaspase 3 into its active form, caspase 3, leading to cell death. Smac blocks the activity of caspase inhibitors, such as X-linked inhibitors of apoptosis protein (*XIAP*), which might stem the initiation of the cascade. Under normal conditions, Bax and Bak are sequestered by proteins that include B-cell lymphoma 2 (*Bcl-2*) and B-cell lymphoma extra large (*Bcl-X$_L$*), preventing the release of cytochrome *c* and Smac. However, upregulation of proteins like Bcl-2-associated death protein (*Bad*) and Bcl-2-interacting mediator of cell death (*Bim*) can allow Bax/Bak to be released and initiate apoptosis. In an additional route to initiation, caspase 1 can cleave BH3-interacting domain death agonist (Bid), producing truncated Bid (*tBid*). tBid activates Bax which induces permeabilization of the mitochondrial outer membrane.

During apoptosis, caspases break down the proteins of the cytoskeleton, causing cells to shrink and round. The nuclear chromatic condenses and endonucleases are activated, fragmenting the DNA. Finally, the plasma membrane blebs, breaking the cells into a host of small vesicles that are phagocytosed by macrophages.[4] Importantly, in contrast to necrotic cell death, the contents of the cytoplasm are not spilled into the extracellular space. Thus, apoptosis does not activate inflammatory pathways through intracellular milieu release.

There are a number of signs that apoptosis occurs in both animal models of ALS and ALS patients. In mice expressing mutant SOD1 (SOD1 mice), the antiapoptotic Bcl-2 and Bcl-X_L proteins exhibit reduced expression, while the proapoptotic proteins Bcl-2-associated death protein (Bad) and Bax are upregulated during the symptomatic phase of the disease.[5] Knocking out Bax in this model completely blocks MN death.[6] Cytoplasmic cytochrome c levels increase in the spinal cords of SOD1 mice as they age, peaking when the mice reach the early symptomatic stage of the disease.[7] Caspase 1 and caspase 3 mRNAs are also both upregulated, first in neurons and then in glial cells.[8] Sequential activation of several downstream caspases is also observed.[2]

In ALS patients, a similar pattern emerges. Bax and Bak are enriched in the mitochondrial membrane fraction, while Bcl-2 is decreased in this compartment. In addition, the binding of Bcl-2 to Bax, necessary for the suppression of Bax function, is also significantly reduced.[9] Similar to the animal model, ALS patients also exhibit elevated activity of caspase 1[10] and caspase 9.[11]

A number of studies have demonstrated that inhibiting apoptosis can slow or prevent some aspects of ALS. Inhibiting caspase activity is one potential approach. Overexpression of XIAP in neurons in a SOD1 mouse reduced caspase activity and extended life span.[12] Intracerebroventricular infusion of N-benzyloxycarbonyl-Val-Ala-Asp-fluoromethylketone (zVAD-fmk), a broad-spectrum caspase inhibitor, significantly extended the life span of the SOD1 mouse when delivered by an osmotic pump beginning at 60 days of age. Disease onset was delayed by 20 days and survival was increased by 27 days. MN death was reduced by the greatest amount in the cervical spinal cord, possibly suggesting poor delivery of the drug to the lumbar cord. It should be noted that this therapy significantly reduced the production of mature interleukin (IL)-1β, a proinflammatory factor. Thus, part of the effect of this treatment could be due to dampening of inflammation.[10]

Rather than targeting later steps in apoptosis, it may be more effective to target the initiating events. Overexpression of Bcl-2 in transgenic mice had previously been shown to block apoptosis due to a variety of factors, including the culling of excess neurons during normal development.[13]

This protein was also neuroprotective in other models of neuron death, including ischemia and axotomy.[13,14]

To determine whether overexpression of Bcl-2 could impact ALS, transgenic mice expressing Bcl-2 from the neuron-specific enolase (NSE) promoter were crossed with SOD1-G93A mice to create double transgenics (Bcl-2;SOD1). When compared to SOD1 mice, Bcl-2;SOD1 mice exhibited a 33-day delay in disease onset, and life span was increased by 35 days. MN loss was also delayed, but end-stage animals in both groups had similar levels of MN death. The loss of NMJ innervation and axons in the phrenic nerve exhibited a similar pattern. Thus, Bcl-2 delayed, but could not prevent, MN loss. While Bcl-2 has a critical role in apoptosis, the authors suggested that the effects that they observed may not simply be due to a direct effect on apoptosis. The Bcl-2;SOD1 mice also showed smaller numbers of ubiquitin-positive MNs relative to age-matched SOD1 mice, something unexpected if Bcl-2 was only affecting apoptosis. The authors speculate that antioxidant properties of Bcl-2 could also be contributing to the delayed disease onset.[15] It may be important to note that, while Bcl-2 can protect neurons from death, it may not prevent axon die back,[16] a process that is critical in ALS.

Another member of the Bcl-2 family of proteins with antiapoptotic activity is Bcl-X_L. When tested in a tissue culture model of glutamate-induced cytotoxicity, overexpression of Bcl-X_L substantially reduced TUNEL positivity and increased cell viability in primary MNs.[17] Ohta et al. studied a recombinant protein consisting of Bcl-X_L fused to the protein transduction domain (PTD) of the HIV TAT protein. The PTD is of interest because it can allow the coupled protein to cross cellular membranes, including those of the blood–brain barrier.[18] Thus, one can use intravenous, intraperitoneal, or intrathecal delivery to get a recombinant protein into the cytoplasm of cells. When injected intraperitoneally in an ischemic brain injury model, this fusion protein (TAT-Bcl-X_L) was able to reduce caspase-3 activation, substantially inhibit DNA fragmentation, and reduce neurological deficits.[19] In SOD1 mice, TAT-Bcl-X_L was delivered by intrathecal infusion over a period of 28 days, beginning at 91 days of age. This therapy delayed disease onset by 10 days and increased life span by 13 days. Motor function showed improvement, and MN death was reduced relative to controls. Markers of apoptosis, including caspase 3 and caspase 9, activation and TUNEL positivity were also significantly reduced.[20]

One other study should also be noted here. Kaspar et al. examined the effects of exercise in SOD1 mice. *Ad libitum* exercise increased life span by 24 days. Interestingly, exercise also elevated the expression of the anti-apoptotic genes Bcl-2 and Bcl-X_L, suggesting that exercise could function, in part, by increasing MN resistance to apoptosis.[21]

In the early stage of disease progression in SOD1 mice, activation of caspase 1, rather than caspase 8, leads to the cleavage of BH3-interacting domain death agonist (Bid), yielding truncated Bid (tBid), the active form of Bid that can initiate the mitochondrial apoptotic pathway.[22] Thus, targeting caspase 1 is a promising target for an antiapoptotic therapy. Friedlander et al. created a transgenic mouse expressing a dominant negative form of caspase 1 by mutating an active site cysteine residue to glycine.[23] Crossing this mouse with a SOD1 mouse substantially reduced the amount of cytochrome c release from mitochondria and delayed the activation of caspases 3 and 9.[22] The dominant negative caspase 1 did not delay disease onset, but it did significantly increase survival by 21 days.[23]

While there have been many successful studies using transgenes to inhibit apoptosis in animal models of ALS, only one study has evaluated a gene therapy approach. Azzouz et al. injected an adeno-associated virus (AAV) vector expressing Bcl-2 into the lumbar spinal cord of SOD1 mice. The therapy delayed spontaneous fibrillation potentials in the gastrocnemius muscle by approximately 10 days and increased MN survival in the injected region. However, no improvement in life span was detected.[24]

The use of parenchymal injections might explain why the effect of this therapy is so muted. In contrast to the previous studies, Bcl-2 was expressed only in the region injected. Thus, disease progression in the rest of the spinal cord was free to proceed unchecked. In addition, even near the injection sites, it is unlikely that all of the target cells were transduced. A broader delivery of the transgene would likely increase its effectiveness. It would be interesting to see how Bcl-2 would perform using an AAV9 vector, which can provide robust expression throughout the central nervous system (CNS) when delivered intravenously.[25]

GLUTAMATE SIGNALING (GLUTAMATE TRANSPORTER 1, ADAR2/AMPA)

Upper MNs, residing in the motor cortex, communicate with lower MNs in the spinal cord using the excitatory neurotransmitter glutamate. While it is the most prevalent neurotransmitter in the CNS, excess levels of glutamate can be toxic to neurons due to overstimulation of glutamate receptors.[26–28] This aberrant signaling leads to excessive calcium influx into the cell and, possibly, to activation of other signaling pathways.[29]

Glutamate-induced cytotoxicity is believed to contribute to the demise of MNs in ALS. In the spinal cord, excitatory amino acid transporter 2 (EAAT2, Glt1) is primarily responsible for glutamate clearance.[30] Studies of ALS patients have demonstrated abnormally spliced EAAT2 transcripts in the spinal cord,[31] loss of EAAT2 expression,[32] reduced glutamate transport,[33] and increased levels of glutamate in the cerebrospinal fluid,[34] all pointing to a possible role for glutamate-induced excitotoxicity in the development and/or progression of ALS. Thus, increasing EAAT2 expression was hoped to help mitigate the disease.

To test this hypothesis, Guo et al. generated transgenic mice that over-express EAAT2 in astrocytes. When this transgene was crossed with the SOD1 mouse to create EAAT2/SOD1 double transgenics, there was a small delay in the onset of motor decline, but the onset of paralysis was not delayed, and there was no improvement in life span. Interestingly, EAAT2/SOD1 mice did show increased MN survival relative to age-matched SOD1 mice. Astrogliosis appeared to be unaffected, but the double transgenic mice did exhibit lower levels of high-molecular-weight SOD1 aggregates.[35] Similar results were found in a study evaluating an AAV8 vector expressing EAAT2 from the astrocyte-specific promoter Gfa2. Despite transducing 83% of GFAP[+] cells, no improvement was seen in motor function and MN loss was not reduced.[36] The authors of both studies concluded that the loss of EAAT2 might contribute to, but probably does not cause, MN degeneration.

How glutamate contributes to the progression of ALS may be a more complicated story than a simple model of poor glutamate clearance. Glutamate signaling in MNs primarily occurs through α-amino-3-hydroxy-5-methyl-4-isoxazolepropionic acid (AMPA) receptors. In general, these receptors are heterotetramers, composed of four subunits. Four different AMPA receptor subunits have been described (GluA1–GluA4, also known as GluR1–GluR4). Depending on the cell type and other factors, different subunits are expressed and combine to form AMPA receptors. The GluA2 subunit is particularly interesting. The mRNA encoding this protein is the target of RNA editing by adenosine deaminase acting on RNA 2 (ADAR2). This enzyme induces a posttranscriptional modification, converting adenosine (A) to inosine (I) by deamination. During translation, inosine pairs with tRNAs as if it were guanosine. In the case of GluA2, this results in glutamine-to-arginine substitution in the protein. This change substantially alters the channel of AMPA receptors containing GluA2. When arginine is present (the edited form, GluA2(R)), the channel is permeable to sodium and potassium but not to calcium. However, in the absence of editing (GluA2(Q)), the presence of the glutamine residue also allows calcium to pass through the channel. In addition, the edited form of GluA2 affects assembly in the ER and slows trafficking to the plasma membrane. Thus, the presence of unedited GluA2 can increase conductance both through permitting calcium influx and increasing the number of channels on the cell surface (reviewed in refs 37 and 38).

Under normal conditions for most cells, including MNs, the vast majority of GluA2 subunits are of the edited variety. Interestingly, incomplete editing of GluA2 is present in some neurodegenerative disorders, including Huntington's disease,[39,40] Alzheimer's disease,[40] and schizophrenia.[40] In patients with sporadic ALS, a fraction of MNs exhibit loss of expression of edited GluA2.[41] In addition, a subset of spinal MNs from ALS patients lack expression of ADAR2, an abnormality not found in control cases. These ADAR2⁻ cells exhibit TDP-43-positive inclusions, a common finding in ALS.[42] These observations have led some to suggest that ADAR2 deficiency could lead to the development of TDP-43 inclusions and thus represents a possible therapeutic target. Alternatively, it is equally possible that TDP-43 inclusions lead to loss of ADAR2 expression and GluA2 editing. A number of mouse models have been created to begin testing the former hypothesis.

Kuner et al. investigated mice expressing GluA2 with an asparagine inserted in place of the edited codon (GluA2(N)) in addition to the endogenous GluA2 alleles. Like GluA2(Q), this subunit creates an AMPA receptor that is permeable to calcium. These mice began to show deficits in motor function around 35 weeks of age, although they did not progress to paralysis by 2 years of age. By 1 year of age, there was a statistically significant decrease in the number of neurons in the ventral horns. At 2 years of age, about 30% of these cells had been lost. Crossing the GluA2(Q) mouse with the SOD1 mouse led to a significant increase in disease severity.[43] In contrast, mice overexpressing GluA2(R) in MNs have decreased calcium influx into cells. When crossed with SOD1 mice, disease severity was reduced. These studies point to an important role for calcium influx through AMPA receptors in ALS disease progression.[44]

To directly determine whether the failure to edit GluA2 might contribute to ALS disease progression, Hideyama et al. created a conditional ADAR2 knock-out by crossing a floxed ADAR2 mouse with a mouse expressing Cre recombinase from the vesicular acetylcholine transporter.[45] This scheme significantly reduced ADAR2 expression in MNs and led to a significant fraction of MNs expressing unedited GluA2. This mouse exhibited a progressive loss of strength and motor function, although at a fairly protracted rate. Median life span was 82 weeks. The mice exhibited many symptoms of ALS, including fibrillation potentials and fasciculations, loss of NMJ enervation, astrogliosis and microgliosis, and MN loss. Importantly, crossing this mouse to one expressing a GluA2 mutated to only express the edited allele rescued the phenotype, demonstrating that the abnormalities observed in ADAR2-deficient mice were due to the loss of GluA2 editing.

Given the potential role that ADAR2 loss might play in the development or progression of ALS, Yamashita et al. examined gene therapy in the conditional ADAR2 knock-out mouse. They performed an intravenous injection of an AAV9 vector expressing ADAR2 into these mice before or at disease onset. This therapy significantly increased the amount of GluA2 that was correctly edited. Performance on the rotarod task, which deteriorated over time in untreated mice, stabilized with therapy for at least 20 weeks. At 39 weeks of age, MN survival and axon counts were higher in the treated group relative to controls.[46]

While it is not yet clear whether defects in GluA2 editing represent an early or late event in the progression of ALS, the data from the mouse models strongly suggest that altering calcium influx via modulation of the AMPA receptor can at least slow the progression of ALS. Since reduced ADAR2 expression in MNs seems to be common in sporadic ALS patients, gene therapy to restore its lost function represents a plausible approach to treat this aspect of the disorder. However, given the mild presentation of disease in the conditional knock-out of ADAR2, it is likely that ADAR2 gene therapy would need to be given in the context of additional therapies that target other aspects of ALS.

ANTIOXIDANT GENES

Oxidative stress may also contribute to ALS disease progression. Reactive oxygen species (ROS), including hydrogen peroxide and the superoxide radical anion, are generated as a byproduct of aerobic metabolism. Unneutralized ROS can react with a number of molecules. For instance, the superoxide radical anion can combine with a nitric oxide radical, producing peroxynitrite which can go on to react with tyrosine residues producing nitrotyrosine.[47] Other molecular changes include carbonylation[48] and oxidative damage to DNA.[47] A number of signs of oxidative damage have been found in postmortem tissue samples taken from ALS patients. These include higher 3-nitrotyrosine levels,[49–51] evidence for an increased carbonylation,[52,53] and elevated oxidative damage to DNA.[53,54] In an in vitro model of ALS, application of catalase can increase cell viability, suggesting that targeting oxidative stress could be beneficial.[55]

Metallothioneins (MTs) are a set of zinc binding proteins that likely have many functions, including protection from oxidative stress.[56] They can also serve as zinc chaperones for SOD1.[57] When zinc is depleted from SOD1, the enzyme more rapidly catalyzes the nitration reaction.[58–60] Interestingly, mutant SOD1 also has reduced zinc binding and increased nitration activity,[59,61] possibly pointing to one mechanism of its toxicity. Taken together, these observations suggest that MTs can play a role in slowing ALS disease progression.

Puttaparthi et al. hypothesized that inhibiting one or more MTs would exacerbate disease progression. Three MTs are expressed in the nervous system of mice: MT-I and -II are expressed in glial cells, while MT-III is expressed in neurons. When MT-I and MT-II were both knocked out in a SOD1 animals (MT-I$^-$;MT-2$^-$;SOD1 mice), mean survival time was reduced by 32 days (229 vs 261 days). Disease onset was approximately 2 months earlier in the MT-I⁻;MT-2⁻;SOD1 mice. Astrogliosis was more pronounced at 5.25 months in the MT-I⁻;MT-2⁻;SOD1 mice than in SOD1 mice, but MN counts were similar between both groups at all time points evaluated. This suggested the glial component of the disease could be altering the function of the remaining MNs, resulting in a faster decline in motor function than would be predicted from the MN counts. When MT-III was knocked out in SOD1 mice, survival was reduced by 51 days (202 vs 253 days). In contrast to the MT-I/II knockout, disease onset was not significantly affected by the loss of MT-III. Instead, the decline in motor function was substantially steeper. In addition, MN loss was significantly greater at 5.25 months in MT-III knockouts, suggesting that MT-III protects MNs from death.[62] These results suggest that altering the expression of MTs using gene therapy could also slow disease progression.

Hashimoto et al. investigated weekly injection of Ad5-MT-III into female SOD1 mice beginning at 20 weeks of age, just before the animals would become overtly symptomatic. The vector was injected into the lower limbs with the hope of getting retrograde transport of the vector to the MNs. Disease onset was not affected, which is not remarkable since treatment was so close to onset. MN counts were significantly higher at 160 days of age in Ad5-MT-III-treated animals. Life span was longer in treated mice as well. The authors found that the Ad vectors themselves were toxic and could accelerate aspects of the disease, and thus a different delivery system should be investigated.[63]

Two other antioxidant genes have been evaluated in the context of gene therapy. Peroxiredoxin 3 (PRDX3) is a mitochondrial thioredoxin-dependent hydroperoxidase and is upregulated in response to oxidative stress. Nuclear factor erythroid 2-related factor 2 (NRF2) is involved in antioxidant response element-mediated gene expression and regulates many phase 2 detoxifying enzymes. Many of the genes controlled by NRF2 are downregulated in ALS.[64] Overexpression of PRDX3 in NSC34 cells significantly increased cell survival under basal conditions and serum starvation, but did not have a significant effect in the face of oxidative stress induced by menadione. This was attributed to the severity of the menadione treatment and the susceptibility of the cells. Murine NRF2 showed protection against menadione in astrocytes and increased survival of NSC34 cells expressing mutant SOD1. These genes were evaluated in vivo using intramuscular injection of AAV6 vectors into 30-day-old SOD1 mice (facial muscles, tongue, intercostal muscles, diaphragm, and hindlimb). No improvement was seen with either vector. However, at endpoint, only 5% of the surviving neurons in green fluorescent protein controls were positive for the transgene. This suggested that poor transduction was to blame for the lack of efficacy.[65]

NEUROTROPHIC FACTORS

Neurotrophic factors are a set of signaling molecules that are involved in the development and maintenance of the nervous system. They bind to receptors on the cell surface, activating two major signaling cascades: the phosphoinositol-3 kinase (PI3K)/Akt pathway and the mitogen-activated protein kinase/extracellular signal-regulated kinase (MapK/Erk) pathway. In addition, IL-6-related cytokines and granulocyte-colony stimulating factor (G-CSF) can also activate the Janus kinase/signal transducer and activator of transcription (Jak/Stat) pathway (Fig. 8.2). Since space

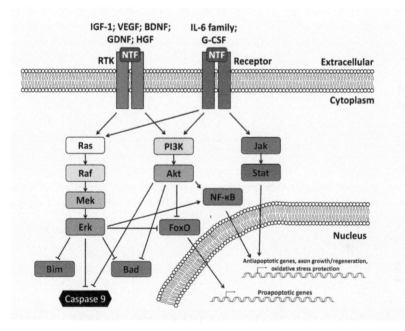

FIGURE 8.2   Neurotrophic factor signaling. Neurotrophic factors activate two or three signaling cascades. A simplified model of these pathways is illustrated. Signaling molecules, such as insulin-like growth factor 1 (*IGF-1*), vascular endothelial growth factor (*VEGF*), brain-derived neurotrophic factor (*BDNF*), glial-derived neurotrophic factor (*GDNF*), and hepatocyte growth factor (*HGF*), initiate signaling by binding to a receptor tyrosine kinase (*RTK*). This event causes the receptor to dimerize and trans-phosphorylate, activating a second kinase domain that initiates the phosphoinositol-3 kinase (*PI3K*)/Akt and mitogen-activated protein kinase/extracellular signal-regulated kinase (*MapK/Erk*) pathways. The interleukin-6 (*IL-6*) family of cytokines and granulocyte-colony stimulating factor (*G-CSF*) activates the Janus kinase/signal transducer and activator of transcription (*Jak/Stat*) pathway in addition to these cascades. Erk and Akt phosphorylate proapoptotic factors such as Bcl-2-interacting mediator of cell death (*Bim*), Bcl-2-associated death protein (*Bad*), and caspase 9, inhibiting their function (see Fig. 8.1). They also inactivate members of the forkhead transcription factor subtype O (*FoxO*) that can activate the transcription of proapoptotic genes. In addition, these two kinases activate NF-κB. Both NF-κB and Stat drive the expression of prosurvival genes.

does not permit a detailed description of the biology of each factor, the reader is directed to several excellent reviews: insulin-like growth factor 1 (IGF-1),[66] vascular endothelial growth factor (VEGF),[67] brain-derived neurotrophic factor,[68] glial-derived neurotrophic factor (GDNF),[69] G-CSF,[70] and the IL-6 family of cytokines.[71]

These signaling cascades allow neurotrophic factors to act on the ALS disease state in a number of ways. All three pathways play important roles in regulating axon growth and regeneration, maintaining and, possibly, restoring NMJ innervation.[72,73] They can also block apoptosis by inhibiting proapoptotic factors and upregulating antiapoptotic factors. Activated Akt phosphorylates serine residues on Bad, caspase 9, and some forkhead family transcription factors, inactivating these proteins and inhibiting apoptosis.[74,75] The Erk pathway similarly phosphorylates Bad, Bcl-2-interacting mediator of cell death (Bim), and caspase 9.[76] All three pathways upregulate expression of the antiapoptotic proteins Bcl-2 and Bcl-X_L.[77–79] Finally, both the Jak/Stat and MapK/Erk pathways have been implicated in resistance to ROS.[76,80] Thus, neurotrophic factors have the potential to attack ALS on a number of fronts.

Recombinant neurotrophic factors have been evaluated in ALS patients. However, the results of these trials have been generally disappointing, with little efficacy shown.[81–85] There are many potential causes for these failures, including subtherapeutic dosing, poor penetration into the CNS, and off-target effects. It is hoped that gene therapy approaches can overcome some or all of these hurdles.

Cardiotrophin-1

Cardiotrophin-1 (CT-1) is a member of the IL-6 family of cytokines which also includes leukemia inhibitory factor and ciliary neurotrophic factor (CNTF). These cytokines have been shown to have a potent neuroprotective effect in a variety of models, including cultured primary MNs,[86–88] peripheral nerve axotomy,[89] and mouse models of neurodegeneration.[90–95]

One gene therapy study has investigated the use of intramuscular injection of an Ad virus expressing CT-1. Neonatal SOD1 mice received bilateral injections of vector into the gastrocnemii, triceps brachii, and the long dorsal trunk muscles. Treatment delayed the onset of symptoms by 27 days and increased survival by 13 days. Neuromuscular function was improved, and there was both a reduction in the loss of muscle mass and improved muscle morphology. The authors hypothesized that the observed efficacy might have been achieved by direct protection of the NMJ and/or by myotrophic affects on the skeletal muscle.[96]

Although these molecules have been shown to be effective in animal models, their translation into the clinic has not been smooth. Chronic systemic infusion of CNTF had significant side effects in clinical trials, including activation of acute-phase proteins and elevated body temperature, anorexia, weight loss, and cough. These side effects were severe enough to be dose-limiting in about half of the patients.[81,97,98] Critically, the side effects limited the doses tested in these clinical trials to levels well below those evaluated in an animal model.[99] Thus, it might not be surprising that a double-blind trial saw no efficacy.[81] The possible side effects of these cytokines highlight the need to efficiently deliver the therapeutic to the spinal cord while minimizing exposure to the rest of the body. This is especially true for gene therapy approaches where discontinuing therapy may not be possible in the face of severe side effects. Employing fusion proteins to help target these cytokines to particular cell types may be one approach to consider.[100]

Granulocyte-Colony Stimulating Factor

G-CSF was originally described as a factor that could induce the differentiation of a myelomonocytic leukemic cell line into granulocytes.[101,102] While it has found many uses in the clinic targeting the hematopoietic compartment, this molecule also has roles in other tissues as well.[70] G-CSF has been shown to be neuroprotective in multiple models. For instance, it can block glutamate-induced death in cultured neurons,[103] and it can reduce the damage caused by stroke.[103,104]

The action of G-CSF in the CNS likely goes beyond direct effects on neurons. Yamasaki et al. investigated the effect that hypoglossal axotomy had on neuron death in wild-type and SOD1 mice. The procedure was performed at 12 weeks of age, well before the animals exhibited symptoms of ALS. Unsurprisingly, 40 days after hypoglossal axotomy, neuron death was higher in SOD1 mice than in wild-type controls. However, in the time frame of 3–20 days postaxotomy, staining for Iba1 showed more microglia in wild-type mice than in SOD1 mice, suggesting a protective role for microglia. These glia expressed the G-CSF receptor and were GDNF-positive. Treatment of these mice with G-CSF further increased the number and size of the microglia. Since the effect of G-CSF on microglia appeared to be neuroprotective, the authors evaluated recombinant G-CSF as a therapy for ALS.[105]

In SOD1 mice subcutaneous injection of G-CSF ($100\,\mu g/kg$ per day, 5 days a week) was initiated at 10 weeks of age. This therapy increased life span by 8 days. MN survival at the early symptomatic phase of the disease was not improved, but there was an increase in the number of large myelinated axons in the C5 ventral root. In vitro, microglia and peritoneal macrophages from SOD1 animals exhibited less migration when stimulated with monocyte chemotactic protein-1 compared to cells from wild-type animals. Looking for the basis for this finding, the authors found reduced levels of matrix metalloproteinase-9 in SOD1 microglia, an enzyme needed for transmigration. In addition, there were reduced levels of both phosphorylated p38 MAPK and NF-κB p65, suggesting that impaired signal transduction underlies the migration defect. Treatment of SOD1 microglia with G-CSF resulted in the nuclear translocation of phosphorylated Stat3 and significantly improved the ability of the cells to migrate.[105]

Given that chronic dosing will be required for ALS therapies and that G-CSF has functions outside the CNS, Henriques et al. investigated a CNS-directed gene therapy approach. AAV1-G-CSF was delivered by intraspinal injection into 10-week-old SOD1 mice. This vector produced high levels of G-CSF in the cord, but only mildly elevated levels in the serum. Neutrophil counts were elevated by threefold, but this was still in the normal range. Disease onset, as measured by a loss in body weight, was delayed by 2 weeks. MN counts and motor function showed some improvement, and NMJ innervation was improved by about 50%. Life span was increased by 15 days. Interestingly, microglial activation seemed unchanged,[106] a deviation from that which was observed in the axotomy study.

Vascular Endothelial Growth Factor

As the name implies, VEGF was originally discovered due to its ability to serve as a mitogen for vascular endothelial cells.[107,108] In the intervening years, it has been discovered that the first protein discovered, now termed VEGF-A, is only one member in a family of similar factors. The other members are VEGF-B, -C, -D, -E, -F, and placental growth factor.[67] Since VEGF-A is the only member that has been investigated as a therapeutic for ALS, this chapter will focus specifically on this gene and simply refer to it as VEGF. VEGF transcripts can undergo alternative splicing, yielding six proteins designated by their amino acid length: VEGF-121, VEGF-145, VEGF-165, VEGF-183, VEGF-189, and VEGF-206.[109,110] The majority of VEGF produced is generally the VEGF-165 form,[111] and the majority of ALS studies employing VEGF use this isoform.

In addition to its role in angiogenesis, VEGF plays important roles in the CNS as well. It can stimulate neurogenesis[112–114] and is involved in axon guidance.[115,116] VEGF is also neuroprotective in the face of a number of insults that may play a role in ALS. By activating both the ERK and PI3K/Akt pathways, VEGF significantly inhibits glutamate-induced cell death.[117] This effect may be due to the inhibition of excitotoxic processes rather than to caspase-dependent apoptosis.[118] VEGF also protects neurons from hypoxia via signaling through the VEGFR-2 receptor and the PI3K/Akt pathway.[119] It may also play a role in inhibiting apoptosis.[120]

A lot of the enthusiasm for VEGF as a therapeutic agent for ALS began with the study of a mouse knock-out of the hypoxia-response element in the VEGF promoter. This knockout has reduced expression of VEGF, primarily in neural tissue. Although 60% of knockouts die before or around birth, the remainder survive more than 23 months. The survivors develop significant muscle weakness by 5 months of age. Sensory systems, though, are not affected. Muscles atrophy, motor unit action potentials are significantly reduced, and the latency of compound muscle action potentials (CMAPs) is reduced in these mice. MN loss is detectable by 7 months of age, and 30% of MNs are lost by 17 months of age. Reactive astrocytosis is evident in the ventral horns, and phosphorylated neurofilaments are observed in older knock-out mice. Thus, these mice recapitulate many of the pathologic and behavioral aspects of ALS. Interestingly, resting neural blood flow rates are reduced by about 40% in the knock-out mice, while blood flow in other tissues is unaffected, supporting a role for hypoxia in the development of ALS.[120]

There have been a number of proof-of-principle studies examining the effects of VEGF on mutant SOD1. Using NSC34 cells transduced with an Ad vector expressing wild-type or mutant (G93A) SOD1, Li et al. showed that overexpression of mutant SOD1 increased oxidative stress and decreased cell viability. Pretreatment of cells with VEGF was effective at reducing the cytotoxic effects of mutant SOD1 expression. Since VEGF can signal through PI3K/Akt and MAPK, the contribution of these two pathways on the phenotype was examined. In an elegant series of experiments, they showed that PI3K was necessary to transmit the effect of VEGF. The authors hypothesized that Akt might inhibit apoptosis by inactivating glycogen synthase kinase 3, forkhead transcription factors, and/or caspase 9.[55]

Moving to an in vivo model, Wang et al. crossed a transgenic mouse that expressed VEGF-165 from the NSE promoter with the SOD1 mouse to see whether neuronal expression of VEGF-165 could alter the disease state. VEGF;SOD1 double transgenics had significantly reduced MN loss. Disease onset was delayed by 20 days. However, disease progression after the onset of motor deficits was similar to that of SOD1 mice. Thus, life span was similarly increased by 23 days. This demonstrated that neuronal expression of VEGF-165 can be protective in ALS.[121]

Two studies have examined the efficacy of recombinant VEGF in SOD1 rodent models. In the first study, female SOD1 mice received either $1\,\mu g/kg$ or $0.1\,\mu g/kg$ of murine VEGF once a week via an intraperitoneal injection. In the high-dose group, disease onset was delayed by ~12 days, while the low dose had no discernible effect.[122] Recognizing that continuous delivery of VEGF to the CNS might be more effective than daily or weekly systemic delivery, Storkebaum et al. evaluated recombinant VEGF delivered intracerebroventricularly by a micropump. Starting at 60 days of age, they infused $0.2\,\mu g/kg$ per day of VEGF-165. Life span was increased by 22 days, and disease onset was delayed by 17 days. At 110 days of age, MN survival was substantially greater in the brainstem and cervical spinal cord in the treated mice. Rotarod performance was also significantly improved, and NMJ enervation was doubled relative to controls. Delaying treatment until symptom onset reduced the efficacy of the therapy, increasing survival by only 10 days.[123]

Two studies have evaluated gene therapies for ALS by delivering VEGF-165. Azzouz et al. employed a rabies-G pseudotyped equine infectious anemia virus (EIAV) injected at either 21 days or 90 days, the latter corresponding to disease onset. Injection of the vector into the gastrocnemius, diaphragm, intercostal, facial, and tongue muscles at 21 days of age resulted in extensive transgene expression in the brainstem and spinal cord. Disease onset was delayed by 28 days and survival was increased by 38 days. MN survival was higher in EIAV-VEGF-treated mice at both 115 days and endpoint, although the increase at endpoint was smaller. Motor function was also improved. Similar to the use of recombinant VEGF-165, delivering the therapy at disease onset halved the improvement in life span with treated mice living an extra 19 days relative to controls. The authors noted that transduction was lower at 90 days which might be attributed to the death of MNs by that time. Since therapies for ALS will be initiated after disease onset for the vast majority of patients for the foreseeable future, this result suggests that delivery methods requiring retrograde transport of the vector will be less than ideal.[124]

A second approach examined was to use AAV4 to express VEGF-165 in choroid plexus epithelial cells and ependymal cells of the ventricles. The hope was that these cells would secrete the protein into the cerebrospinal fluid where it could be delivered throughout the CNS. SOD1 mice received bilateral intracerebroventricular injections at 85 days of age. Due to sex-related differences in disease presentation, males and females were analyzed separately. AAV4-treated female SOD1 mice had a 20-day increase in life span relative to controls, while the life span of males increased by 9 days. Measurements of motor function showed improvement for both sexes.[125]

As previously described, the VEGF gene produces at least six different isoforms, and studies have suggested that expressing combinations of these isoforms is more efficacious than expressing single isoforms.[126–128] This provides a challenge for current gene-therapy vectors, since their DNA and RNA packaging limits are generally large enough for cDNAs but insufficient to package entire genes. How might one then tackle this challenge for VEGF expression? One approach, taken first by Rebar et al., is to engineer a zinc finger protein (ZFP) transcription factor to increase VEGF expression from the endogenous gene. [129] This approach has the benefit of increasing the expression levels of the individual isoforms while preserving their expression levels relative to one another.

Sakowski et al. studied the use of this transcription factor in rats that underwent a recurrent laryngeal nerve crush injury that causes vocal cord paralysis. Delivery of an adenovirus expressing the transcription factor immediately after the nerve crush injury significantly improved recovery of vocal cord function, suggesting that this approach might be useful for treating diseases that exhibit loss of innervation by MNs.[130]

Kleim et al. attempted to extend this work to ALS. 80-day-old rats received weekly injections of a plasmid expressing the ZFP transcription factor unilaterally into the medial gastrocnemius muscle for six weeks. No significant improvements were observed for MN survival, NMJ innervation, and muscle fiber composition, but there was a hint of some improvement observed in hindlimb grip strength and rotarod performance.[131] The authors did not report any data on transgene expression. However, it is unlikely that they succeeded in transfecting MNs, and expression in the muscle was probably suboptimal relative to what could be achieved using viral vectors. It would be interesting to see how effective this approach would be using more efficient delivery methods.

Brain-Derived Neurotrophic Factor

BDNF is a member of the neurotrophin family, and it is expressed throughout the CNS. Like other neurotrophic factors, BDNF can reduce or prevent neuronal death caused by a wide variety of insults.[68] These include many possible contributors to ALS, such as glutamate toxicity,[132] toxic proteins,[133] and hypoxia.[134] In addition, BDNF can also affect synapses, increasing axonal branching,[135] which could be beneficial at the NMJ. One gene therapy study examined intramuscular injection of naked DNA expressing BDNF. No significant improvements in life span or motor function were found, but caspase activation was reduced.[136]

Some have argued that BDNF is a poor choice for the protection of MNs.[137] In vitro BDNF not only fails to protect MNs from glutamate-induced cytotoxicity, but it increases sensitivity through activation of the PI3K signaling pathway.[138] A lack of protection against cytotoxicity was also found in vivo.[139] Another potential hurdle is the MN itself. This cell type expresses a truncated form of the BDNF receptor, TrkB, that appears to reduce the efficacy of BDNF.[140] These findings suggest that BDNF alone will not be therapeutic.

Glial-Derived Neurotrophic Factor

GDNF is a neurotrophic factor that expressed in both CNS and non-CNS tissues.[141,142] Those locations that are likely relevant for ALS include the spinal cord,[142] Schwann cells, and skeletal muscle.[143] In the hope of identifying a therapeutic molecule for Parkinson's disease, GDNF was originally described as a neurotrophic factor that could protect midbrain dopaminergic neurons in cell culture. In vitro it increased the size of dopaminergic neurons and enhanced neurite outgrowth.[144] In vivo, GDNF could protect these neurons in a chemically induced model of Parkinson's disease.[69]

Could GDNF be neuroprotective for MNs as well? Interestingly, activation of GDNF's receptor, RET, is reduced in the MNs of ALS patients, suggesting that reduced GDNF signaling may contribute to disease progression.[145] GDNF is a potent protector of cultured embryonic MNs[146] and can protect MNs from serum deprivation.[147] In vivo it can protect fetal MNs from the usual apoptosis that occurs during development[148] and from a number of insults,[69] including excitotoxicity.[149]

A number of studies have evaluated delivery of GDNF to the muscle. Manabe et al. injected recombinant GDNF unilaterally into the gastrocnemius muscle of SOD1 mice three times a week beginning at 35 weeks of age. At 46 weeks of age, MN counts were higher in GDNF-treated SOD1 mice on the injected side, and there was a similar increase in the number of phosphorylated Akt$^+$ MNs. However, the activation of caspase 3 and caspase 9 was not suppressed, and phosphorylated ERK staining was unchanged.[150] Ciriza et al. injected GDNF fused to the tetanus toxin heavy chain C fragment (TTC) into each limb beginning at disease onset. This extended life span by 8 days and improved performance in an open-field test.[151]

To avoid the need for repeated injection of GDNF, Mohajeri et al. created myoblasts overexpressing GDNF using retroviral vectors. These cells were transplanted bilaterally into the medial and lateral gastrocnemius muscles of 6-week old SOD1 mice. This led to increased survival of large-diameter MNs, delayed disease onset and loss of motor function, and slowed muscle atrophy.[152]

Transducing the muscle directly may be preferable to ex vivo transduction of myocytes. Injection of an AAV2 vector expressing GDNF into the gastrocnemius muscles led to transduction of 30% of the muscle fibers. GDNF levels increased over the first 2 months and then remained fairly stable for at least 10 months. GDNF was concentrated in the sarcolemma, suggesting that the protein was secreted and concentrated at NMJs. GDNF was also retrogradely transported from the muscle to the cell bodies of MNs.[153]

To determine how this therapy impacts disease onset and progression, SOD1 animals were injected at 9 weeks of age into the gastrocnemius and triceps brachii muscles. Expression of GDNF was about 120-times normal. At 110 days of age, muscle atrophy was reduced by about 75%. Treatment preserved ~40% of the MNs that would have died by this time point. Rotorod testing showed improved motor function. Disease onset was delayed by 13 days, and survival was increased by 17 days.[154]

In a test of nonviral gene therapy, Yamamoto et al. electroporated a GDNF plasmid into tibialis anterior and gastrocnemius every 2 weeks beginning at 9 weeks of age. GDNF was present in both the muscle and

MNs. Survival was not improved, but motor function did respond to therapy.[155] Moreno-Igoa extended this line of study by delivering the therapeutic plasmid to both the forelimbs and hindlimbs. The plasmid was injected into the quadriceps and triceps of SOD1 mice at 8 weeks of age. Disease onset was delayed by 9 days and life span was increased by 15 days. Motor function was also improved relative to control mice. To begin to understand what aspects of the disease GDNF was targeting, the authors evaluated a number of molecular markers. Caspase-3 activation was reduced to levels similar to that of wild-type mice. Bax and Bcl2, both upregulated in SOD1 mice, were reduced. Phosphorylated Akt was elevated in treated mice, suggesting activation of survival signals. In SOD1 mice there was increased phosphorylation of ERK1/2; GDNF reduced this to almost normal levels.[156]

A few studies have targeted GDNF expression to the CNS. Foust et al. injected AAV5-GDNF into the red nucleus and motor cortex to get anterograde transport of GDNF to the spinal cord via the rubrospinal tract (RST) and the corticospinal tract, respectively. The RST path delivered more GDNF to the spinal cord. GDNF concentration dropped with increasing distance from the brain, which could have implications for clinical translation. However, no therapeutic application was measured in this study.[157]

Guillot et al. looked at the efficacy of lentiviral delivery of GDNF in SOD1 mice. Intraspinal injections were performed in 40-day-old SOD1 mice. Onset of disease and MN death was not affected. In contrast, when the vector was delivered to the facial nucleus, the therapy protected about 1/3 of the MNs that would have died 3 months after injection. The difference in efficacy for these two cell populations might suggest that there are differences in how these different populations respond to GDNF. Alternatively, it may simply reflect differences in disease severity, as MN loss is slower in the facial nucleus.[158]

Similar results were found when human cortical neural progenitor cells were transduced with a lentivirus expressing GDNF and transplanted into the spinal cord. There was wide distribution of transplanted cells throughout the lumbar cord on the injected side and high levels of GDNF expression.[159] MN loss was essentially prevented on the injected side, but the therapy failed to preserve innervation of NMJs. Consequently, no improvement in limb function was observed.[160]

The CNS-directed studies suggest that delivery of GDNF solely to the CNS will not be effective for the treatment of ALS. Instead, a GDNF approach may need to target both the MN soma and the NMJ. Ad vectors have the potential to transduce both MNs and muscle when injected intramuscularly. Acsadi et al. performed bilateral injections of Ad-GDNF into the anterior tibialis, gastrocnemius, quadriceps, and paraspinal muscles of neonatal SOD1 mice. Injecting neonates avoided the immune response that normally plagues Ad-mediated gene therapies. Expression was stable from 1 to 4 months in the muscle and present in the spinal cord for at least 4 months, although 72% of the expression there was lost over time. Disease onset was delayed by 9 days, and life span was increased by 13 days. Treatment improved rotarod performance and slowed MN loss.[161]

An attempt to initiate therapy closer to the time of disease onset failed to significantly alter the disease trajectory.[162] Due to the immunogenic nature of Ad, this therapy had to be given weekly when initiated in adult animals. Because of the immune complications, the results of this study may not be representative of the potential for combined CNS/muscle delivery of GDNF. However, as studies of VEGF have suggested, loss of innervation at disease onset may also make retrograde transport-dependent therapies difficult to translate.

Insulin-Like Growth Factor 1

IGF-1 was discovered far from the brain. This neurotrophic factor was originally identified as a mediator for the effects of the somatotrophic hormone.[163–165] Since that initial discovery, researchers have learned that the roles that this molecule plays go far beyond skeletal growth. IGF-1 and its receptor are expressed in several regions of the CNS, including on MNs.[66] When overexpressed in muscle, IGF-1 can accelerate the recovery of function following a nerve crush injury.[166] It can also block the death of MNs in vitro, inhibit MN death in embryos following axotomy,[167] and protect against excitotoxicity.[149] Thus, IGF-1 has been a strong candidate for ALS therapies.

Kaspar et al. investigated retrograde delivery of an AAV vector expressing IGF-1, comparing its efficacy with that of GDNF. When given at 60 days of age, IGF-1 increased median life span by 37 days, while GDNF extended life span by only 11 days. When given at disease onset, IGF-1 therapy increased median life span by 22 days, while GDNF extended it by 7 days. IGF-1 treatment delayed weight loss, improved motor function, reduced cellular vacuolization, and reduced MN loss. Astrogliosis was also reduced. Ubiquitin-positive aggregates in the spinal cord were ameliorated. TUNEL staining suggested that apoptosis was reduced as well. Using a lentivirus to deliver IGF-1 expression to only the muscle yielded a smaller improvement in life span (9 days), suggesting that expression of IGF-1 in MNs was critical for the therapy. Akt, a kinase activated by IGF-1, exhibited elevated phosphorylation with IGF-1 treatment. This likely led to the reduction in activation of caspases 3 and 9. Tumor necrosis factor alpha (TNF-α) was also reduced, suggesting a mechanism by which IGF-1 dampened microgliosis and astrogliosis. In unpublished data, the authors noted that ubiquitin-positive aggregates were smaller and more focalized in IGF-1-treated mice, suggesting that the therapy also enhanced degradation of the aggregates or slowed their production.[168]

Two studies evaluated parenchymal injection of AAV2-IGF-1 into the spinal cord. The first evaluated delivery of AAV2-IGF-1 into the lumbar cord of 60-day-old, presymptomatic SOD1 mice. This route resulted in neuronal transduction. MN survival was higher in IGF-1-treated mice at 110 days of age, but substantially lower than wild type. Microgliosis was unaffected. In males, disease onset was delayed, but once it started, the rate of motor function loss was similar to that of control animals. Survival in males was increased by 12 days. No improvement was observed in SOD1 females.[169] Franz et al. found similar results following unilateral parenchymal injection into the cervical spine.[170]

Dodge et al. have undertaken two approaches to try to deliver IGF-1 to a broader section of the CNS. In their first study, AAV1-IGF-1 was injected into the deep cerebellar nuclei in SOD1 mice at ~89 days of age. This site was chosen because it has multiple afferent and efferent connections to all levels of the spinal cord. The authors hoped that this would translate to vector delivery along the length of the cord via retrograde transport. Improvement was seen in MN survival at 110 days of age. Grip strength and rotarod performance was improved relative to mock-treated controls, and life span was increased by 14 days. Astrogliosis and microgliosis were dramatically reduced as were nitric oxide synthase and 3-nitrotyrosine. In vitro data suggested that IGF-1 acted both through MNs and through modulation of glial activation.[171] In their second study, Dodge et al. used an AAV4 to drive secretion of IGF-1 from the choroid plexus and ependymal cells as they did with VEGF. SOD1 mice received bilateral ICV injections at 85 days of age. Improvements were seen in survival and motor function.[125]

Like VEGF, IGF-1 can also generate an array of isoforms with distinct functions. IGF-1 has two major isoforms that arise from alternate splicing: circulating (class 2) and local (class 1). The former is expressed in the liver and is secreted. The latter is expressed in other tissues and generally stays within the tissue. One study crossed a transgenic mouse with muscle expression of the local isoform of IGF-1 (mIGF-1) with a SOD1 mouse. mIGF-1;SOD1 mice exhibited a delay in disease onset of 10 days and an increase in survival of 30 days. Muscle atrophy was attenuated, and markers of satellite cell activity suggested that these cells contributed to muscle maintenance. The mIGF-1;SOD1 mice also maintained fiber composition longer. NMJs showed more innervation in mIGF-1;SOD1 mice than in SOD1 mice, even at end-stage disease. MN survival was increased relative to SOD1 mice, and astrogliosis was delayed. The authors also looked for evidence of muscle regeneration. They had previously shown that the calcineurin-A mRNA was alternatively spliced during muscle regeneration. Although SOD1 mice and mIGF-1 transgenics did not express this isoform, mIGF-1;SOD1 mice expressed this protein after the onset of clinical symptoms. This could point to a mechanism by which IGF-1 could prolong muscle integrity after denervation began.[172]

Hepatocyte Growth Factor

Hepatocyte growth factor (HGF) was originally identified by its effects outside of the CNS, functioning as a mitogen for mature hepatocytes.[173,174] Like many signaling molecules, though, this factor plays roles in a variety of tissues, including the nervous system.[175] HGF and its receptor, Met, are expressed in the brain in both neuronal[176–179] and nonneuronal cells.[180,181] During development, HGF is necessary for MN neurite outgrowth to the limb, serving as a chemoattractant for MNs.[182] In vitro HGF promotes the survival of MNs,[182–184] while in vivo it protects them from axotomy.[185,186] In SOD1 rats, HGF and Met levels increase with disease progression, possibly pointing to a protective role for HGF signaling in ALS.[187]

To determine whether this growth factor might be therapeutic for ALS, Sun et al. created a transgenic mouse overexpressing rat HGF in the CNS using the NSE promoter. When crossed to the SOD1 mouse, this transgene delayed MN loss by more than a month and protected axons in the ventral roots. HGF;SOD1 animals also had improved motor function, and median survival was increased from 260 days to 287 days. Interestingly, the effect of HGF did not appear to be through the induction of the antiapoptotic gene Bcl-X_L. However, HGF did reduce induction of caspase-1 and iNOS. When studying astrocytes in HGF;SOD1 mice, they found that astrogliosis was attenuated and EAAT2 levels were substantially elevated.[188]

Ishigaki et al. studied the effects of continuous intrathecal administration of recombinant human HGF to SOD1 rats. MN death and disease onset were delayed in a dose-dependent manner. When given just before disease onset, the highest dose, 100 µg/week, increased survival by 11 days, and the decline in motor function was slowed relative to controls. There was a concomitant delay in MN loss 2 weeks after therapy was begun. Phosphorylated c-Met was increased in MNs in the treatment group, and activated caspase-3 and caspase-9 were reduced relative to the control group. By western blot, EAAT2 was elevated with treatment, but there was no significant change in GFAP expression.[187] While gene therapy using HGF has not yet been evaluated, the above studies suggest that it might generate effects similar to those of other neurotrophic factors.

Combination Therapies

With such a broad array of neurotrophic factors to choose from, one might wonder whether combining two or more might lead to increased efficacy. A handful of studies have begun to address this question. Their results suggest that choice of which factors to pair together will be critical in determining whether two factors are better than one.

Dodge et al. combined their IGF-1 and VEGF-165 vectors targeting the choroid plexus and ependymal cell layer. While both had similar effects on alleviating the disease, the combination of the two was no better than the individual factors.[125] It has been hypothesized that the effects of IGF-1 might be mediated by VEGF.[124] This could explain the lack of benefit from the combined therapy.

Haase et al. investigated intramuscular injection of Ad-neurotrophin-3 (NT-3) and/or Ad-CNTF into neonatal *progressive motor neuronopathy (pmn)* mice. This model exhibits muscle paralysis beginning at 2 weeks of age and death at 6–7 weeks. All three treatments (NT-3 alone, CNTF alone, or the combination of the two) increased median life span by about 21 days. Compound muscle action potential amplitudes were higher in treated mice (40% vs 70% of normal). Spike amplitudes increased in treated mice to levels higher than those of wild-type and *pmn* mice, suggesting that these neurotrophic factors could induce peripheral or collateral sprouting. Indeed, examination of muscle showed that Ad-NT-3 and the combination therapy resulted in substantially increased branching of axons that succeeded in reinnervating NMJs that were not innervated in untreated animals. Treatment reduced axonal degeneration in the phrenic nerve fibers. Combining Ad-NT-3 with Ad-CNTF did not further increase life span, but did improve MN counts and the number of myelinated fibers.[95] Although this model is not necessarily a model of ALS, the increased branching driven by the NT-3 vector could be useful for restoring innervation of NMJs in ALS patients, possibly restoring some motor function.

Krakora et al. performed ex vivo gene delivery to express GDNF and/or VEGF in mesenchymal stem cells. Following the induction of muscle damage, these cells were injected into the forelimb triceps brachia, tibialis anterior, and long muscles of the dorsal trunk of SOD1 rats. The individual neurotrophic factors extended life span by 14 days. Combining the two extended life span by 28 days and did a better job preserving NMJ innervation and motor function.[189]

Mitsumoto et al. evaluated the effects of recombinant CNTF and BDNF on the *wobbler* mouse. This model has a defect in a vesicle transport factor and recapitulates many of the abnormalities found in ALS.[190] While the individual factors could merely slow disease progression, the combination of the two halted disease progression for at least 1 month.[191] As noted earlier, there are many factors working against BDNF as a therapy for ALS. One study of BDNF in SOD1 mice failed to find efficacy, albeit with a poor delivery system. Thus, the synergy between CNTF and BDNF is quite remarkable. Neurotrophic factors rarely work in isolation from one another, and this may be a case where one factor is able to sensitize the system to the second, producing a substantially larger effect.

Combining therapies is not limited to neurotrophic factors alone. Kaspar et al. combined IGF-1 therapy with exercise. AAV2-IGF-1 was injected into the intercostal and quadriceps muscles of 90-day-old SOD1 mice. The mice were given exercise wheels at 40 days of age, with different groups being offered the wheel for different lengths of time (2 hours, 6 hours, or *ad libitum*). Exercise alone led to increases in life span (7, 40, and 24 days, respectively). IGF-1 extended life span by 29 days. Combining exercise and gene therapy substantially increased median survival by 83 days.[21]

TETANUS TOXIN HEAVY CHAIN C-FRAGMENT

The tetanus neurotoxin is best known for its ability to induce painful, uncontrolled muscle contractions.[192] The toxin is composed of two proteins: a heavy chain that drives neuron binding, retrograde transport, and translocation of the toxin into the cytoplasm; and a light chain that cleaves proteins necessary for neurotransmitter release. The carboxy-terminal half of the heavy chain TTC mediates neuronal binding and transport.[193] Thus, this portion has been investigated as a targeting platform for delivering therapeutics to MNs, including neurotrophic factors.[100] It was inadvertently discovered that TTC may itself activate neurotrophic signaling and enhance antiapoptotic pathways,[194] activating the PI3K and MAPK signaling pathways.[195]

Moreno-Igoa et al. injected a plasmid expressing TTC into the muscle of 8-week-old SOD1 mice. Ten days later, transgene expression was detectable in both the muscle and spinal cord. Treatment delayed disease onset by 5 days and increased survival by 12 days. Motor function, measured by the rotarod and wire hang tests, also showed similar amounts of improvement. Lumbar MN survival was increased from 43% of wild type to 60% with this treatment. Microgliosis was reduced, but astrogliosis remained unchanged with treatment. A number of changes were observed in the apoptotic network. TTC-treatment reduced caspase-1 and -3 mRNA levels to wild type levels and caspase-3 activation was reduced. Bcl2 mRNA levels remained elevated and unchanged compared to untreated SOD1 mice, but Bax and Bcl2 protein levels were lower than those of untreated SOD1 mice. Akt phosphorylation was also increased twofold, which the authors interpreted as evidence of inhibition of apoptosis. The authors also looked at proteins implicated in calcium-induced cytotoxicity. Neuronal calcium sensor-1 (Ncs1) frequenin protein modulates calcium and calmodulin-dependent enzymes, and these enzymes are abnormal in ALS patients. In SOD1 mice, the Ncs1 mRNA was downregulated. TTC-treatment almost returned Ncs1 expression to normal.[196] These results were confirmed in a subsequent study.[136]

Although TTC might be superior in terms of its ability to selectively target neurons, neurotrophic factors may prove to be more easily translated to the clinic. These factors, being normally expressed, are less likely to induce an immune response to the therapeutic protein. In addition, more than 80% of the population is immunized against the tetanus toxin, including the TTC fragment.[197] Thus, one would expect that the immune system will rapidly destroy the therapeutic protein and any cells expressing it.

ANGIOGENIN

As we saw earlier, hypoxia is a potential contributor to ALS. In addition to VEGF, there are a number of molecules that can contribute to neovascularization, including fibroblast growth factors, platelet-derived growth factor, and angiogenin.[198] Angiogenin, a member of the ribonuclease A superfamily, is of particular interest due to the discovery of mutations in angiogenin in both sporadic ALS and familial ALS patients.[199] In endothelial cells, this molecule is required for signaling of VEGF through the PI3K/Akt pathway.[200] When 1 µg of angiogenin was given systemically once a day to SOD1 mice beginning at 50 days of age (presymptomatically) or at 90 days of age (disease onset), the therapy increased life span by a similar degree, raising median survival from 123 days to 134 days. Motor function, measured by stride length and grip strength, was also improved, and MN survival was increased, even at end stage. Levels of the prosurvival phospho-ERK1/2 and phospho-Akt were also increased.[201] Thus, angiogenin may be a candidate for gene therapy for ALS, possibly in conjunction with neurotrophic factors like VEGF.

SINGLE-CHAIN ANTIBODIES

While only a few percent of ALS cases involve SOD1 mutations, some studies suggest that oxidation of wild-type SOD1 can result in a misfolded protein with similar properties.[202-206] If true, therapies that reduce the misfolded SOD1 burden may find use in a larger fraction of ALS patients than just those that have SOD1 mutations. It had previously been shown that immunization against misfolded SOD1 delayed disease onset and extended life span.[207,208] One antibody, D3H5, has been shown to recognize the misfolded protein generated by several different mutations, suggesting it might be generally useful against misfolded SOD1. The authors engineered a single-chain fragment from this antibody and cloned it into an AAV1 vector. The vector was delivered into SOD1 mice at 45 days of age by intrathecal injection. Life span was increased by 16 days, and disease onset was delayed by a similar amount. Motor function was also improved. Treatment reduced neuronal stress, and MN loss was attenuated. Interestingly, the degree of improvement correlated with the amount of transduction observed.[209] Thus, single-chain antibodies against SOD1 may be another fruitful route toward an ALS therapy.

POSSIBLE FUTURE CANDIDATES

Two other transgenes are worth noting. Although they have not yet been tested in a model of ALS, they seem to have the ability to target specific aspects of the ALS: abnormalities in neurofilaments and nonsense-mediated decay (NMD).

Neurofilaments (NFs) are the main type of intermediate filaments found in myelinated neurons. They are composed of three proteins, designated by size: NF-H, NF-M, and NF-L (heavy, medium, light). It is believed that they form heterotetramers (L and M or L and H). NF-L is essential for NF assembly and is required for NF-M and NF-H transport to the axon, while loss of NF-M and NF-H produce a milder phenotype.[210–213] In a number of neurologic diseases, including ALS, abnormal accumulations of NF are present, often identified as Lewy bodies or spheroids. Interestingly, although there is an accumulation of these proteins, their mRNAs are downregulated in the disease state. In some cases of ALS, mutations have been described in NF-H,[214–217] suggesting a role for the NF system in disease initiation or progression. Furthermore, overexpression of NF-H in mice leads to a disease with perikaryal neurofilamentous accumulations in MNs and axonal transport defects.[218,219] This leads to progressive degeneration of MN axons and muscle degeneration. However, it does not lead to loss of MNs or to a decrease in life span.

Meier et al. investigated the possibility that overexpressing NF-L could suppress the toxicity of NF-H described above, first by creating a double NF-L;NF-H transgenic mouse and then by expressing NF-L from an Ad vector. Body weights were significantly improved in NF-L;NF-H mice, and they did not develop clinical symptoms of motor dysfunction. Electron microscopy analysis of NF-H transgenic mice showed high densities of NFs relative to wild-type mice. Overexpression of NF-L in these mice suppressed this phenotype. They then tested an Ad vector expressing NF-L, injecting it into the tibialis muscle. Transduction of MNs was markedly reduced in NF-H transgenics relative to wild-type controls, likely due to axonal trafficking defects. Still, transduction was of a sufficient level to reduce the number of MNs presenting with swelling of the perikarya. The results of this study suggest that NF subunit stoichiometry is important for normal NF maintenance. Given the existence of mutations in NF-H in some ALS patients, this may point to a role of NF dysfunction in ALS disease progression that could be a target for NF-mediated therapies.[220]

Targeting NMD may be another approach worth considering. TDP-43+ aggregates are found in the majority of ALS patients, and mutations in TDP-43 have been associated with this disease. Abnormal TDP-43 accumulates in the cytoplasm in the form of hyperphosphorylated and ubiquitinated aggregates. In vitro work identified upframeshift protein 1 (UPF1) as a modifier of TDP-43 dysfunction,[221] apparently upregulating NMD.[222] Since SOD1 ALS does not exhibit TDP-43^{+} inclusions, the authors created a TDP-43 model by injecting an AAV9 vector expressing wild-type TDP-43 into neonatal rats. This resulted in an animal that develops paralysis in the hindlimbs. On postnatal day 1, Sprague–Dawley rats received an i.v. injection of AAV9-TDP-43 with or without AAV9-UPF1. Dosing of the TDP-43 was chosen by titration to create a partial lesion. They found that UPF1 therapy improved weight gain and slightly improved motor function relative to rats that just received the TDP-43 transgene.[223] There are two significant limitations to this study. First, while ALS is a disease caused by a combination of defects in neuronal and glial cells, the disease model used here is likely a purely neuronal model. Second, by delivering both the disease-causing and the therapeutic AAV9 vectors simultaneously, the system is, in effect, rigged to treat the cells in which the disease is initiated. This may not be the same as the population of cells that need to be treated in ALS patients. Thus, the results of this study may significantly overestimate how effective this therapy might be.

TIMING OF THERAPY

It would be helpful to discuss the important relationship between the timing and route of therapy. Many of the studies described here initiate treatment before disease onset; in many cases, well before disease onset. Given the lack of diagnostic tools for detecting disease before symptom onset in sporadic ALS and that, even in familial ALS, the majority of people at risk do not have a known mutation for which to screen, it is unlikely for the foreseeable future that therapies will be initiated in the presymptomatic stage in the clinic.

In the studies where the impact of timing was evaluated, initiating treatment presymptomatically almost always outperforms treatment at disease onset. In some cases, particularly muscle directed therapies, a contributing factor for reduced efficacy is the fact that MN axons detach from the NMJ. Once these axons start to die back, it becomes much more difficult to reach the MN from the muscle. Indeed, Azzouz et al. found that retrograde transport of lentivirus was substantially reduced once animals became symptomatic.[124] Thus, treatments that deliver therapeutics to the muscle with the expectation of retrograde transport may be more difficult to successfully translate to the clinic. This problem might be circumvented through the use of AAV9 vectors that, when administered intravenously, can transduce both muscle and the CNS.

CONCLUDING REMARKS

The studies described here attack ALS from a number of different angles. Some target specific abnormalities: increased apoptosis, excitotoxicity, and decreased blood flow. Others, like neurotrophic factors, have the capacity to affect multiple disease pathways simultaneously. The improvements seen in the animal models are generally in line with the effects of Riluzole, the only therapy currently approved for ALS. Thus, these therapies might show similar efficacy in patients as well. All of these therapies do have one similarity: none of them can completely halt the progression of ALS. Combining therapies may produce more substantial gains, as seen with CNTF and BDNF. However, a major limitation is likely the fact that these therapies attack downstream aspects of ALS and not the underlying problem itself.

The nature of ALS is not well understood. While many genes have been associated with ALS, there is no clear biochemical connection between them. Most of the mutations are dominantly inherited, suggesting that they represent gain-of-function, rather than a loss-of-function, mutations. A common thread in ALS is the development of protein aggregates, both soluble and insoluble forms. Some of these aggregates may be the disease initiation factors. Therefore, future therapies may move toward transgenes that can alleviate aggregation, such as heat shock proteins and other chaperones. However, the development of novel transgenes for ALS will likely be hindered until researchers better understand the connection between ALS-associated mutations and the processes that drive disease initiation.

References

1. Fink SL, Cookson BT. Apoptosis, pyroptosis, and necrosis: mechanistic description of dead and dying eukaryotic cells. *Infect Immun* 2005;**73**(4):1907–16.
2. Sathasivam S, Shaw PJ. Apoptosis in amyotrophic lateral sclerosis--what is the evidence? *Lancet Neurol* 2005;**4**(8):500–9.
3. Kerr J, Wyllie A, Currie A. Apoptosis: a basic biological phenomenon with wide ranging implications in tissue kientics. *Br J Cancer* 1972;**26**:239–57.
4. Elmore S. Apoptosis: a review of programmed cell death. *Toxicol Pathol* 2007;**35**(4):495–516.
5. Vukosavic S, Dubois-Dauphin M, Romero N, Przedborski S. Bax and Bcl-2 interaction in a transgenic mouse model of familial amyotrophic lateral sclerosis. *J Neurochem* 1999;**73**(6):2460–8.
6. Gould TW, Buss RR, Vinsant S, et al. Complete dissociation of motor neuron death from motor dysfunction by Bax deletion in a mouse model of ALS. *J Neurosci* 2006;**26**(34):8774–86.
7. Guegan C, Vila M, Rosoklija G, Hays AP, Przedborski S. Recruitment of the mitochondrial-dependent apoptotic pathway in amyotrophic lateral sclerosis. *J Neurosci* 2001;**21**(17):6569–76.
8. Ando Y, Liang Y, Ishigaki S, et al. Caspase-1 and -3 mRNAs are differentially upregulated in motor neurons and glial cells in mutant SOD1 transgenic mouse spinal cord: a study using laser microdissection and real-time RT-PCR. *Neurochem Res* 2003;**28**(6):839–46.

9. Martin L. Neuronal death in amyotrophic lateral sclerosis is apoptosis: possible contribution of a programmed cell death mechanism. *J Neuropathol Exp Neurol* 1999;**58**(5):459–71.

10. Li M, Ona VO, Guegan C, et al. Functional role of caspase-1 and caspase-3 in an ALS transgenic mouse model. *Science* 2000;**288**(5464):335–9.

11. Inoue H, Tsukita K, Iwasato T, et al. The crucial role of caspase-9 in the disease progression of a transgenic ALS mouse model. *Embo J* 2003;**22**(24):6665–74.

12. Wootz H, Hansson I, Korhonen L, Lindholm D. XIAP decreases caspase-12 cleavage and calpain activity in spinal cord of ALS transgenic mice. *Exp Cell Res* 2006;**312**(10): 1890–8.

13. Martinou J, Dubois-Dauphin M, Staple J, et al. Overexpression of BCL-2 in transgenic mice protects neurons from naturally occuring cell death and experimental ischemia. *Neuron* 1994;**13**(4):1017–30.

14. Dubois-Dauphin M, Frankowski H, Tsujimoto Y, Huarte J, Martinou JC. Neonatal motoneurons overexpressing the bcl-2 protooncogene in transgenic mice are protected from axotomy-induced cell death. *Proc Natl Acad Sci U S A* 1994;**91**(8):3309–13.

15. Kostic V, Jackson-Lewis V, de Bilbao F, Dubois-Dauphin M, Przedborski S. Bcl-2: prolonging life in a transgenic mouse model of familial amyotrophic lateral sclerosis. *Science* 1997;**277**:559–62.

16. Sagot Y, Dubois-Dauphin M, Tan S, et al. Bcl-2 overexpression prevents motoneuron cell body loss but not axonal degeneration in a mouse model of a neurodegenerative disease. *J Neoursci* 1995;**15**:7727–33.

17. Garrity-Moses ME, Teng Q, Liu J, Tanase D, Boulis NM. Neuroprotective adeno-associated virus Bcl-(x)L gene transfer in models of motor neuron disease. *Muscle Nerve* 2005;**32**(6):734–44.

18. Schwarze SR, Ho A, Vocero-Akbani A, Dowdy SF. In vivo protein transduction: delivery of a biologically active protein into the mouse. *Science* 1999;**285**(5433):1569–72.

19. Cao G, Pei W, Ge H, et al. In vivo delivery of a Bcl-xL fusion protein containing the TAT Protein Transduction Domain Protects against Ischemic Brain Injury and Neuronal Apoptosis. *J Neurosci* 2002;**22**(13):5423–31.

20. Ohta Y, Kamiya T, Nagai M, et al. Therapeutic benefits of intrathecal protein therapy in a mouse model of amyotrophic lateral sclerosis. *J Neurosci Res* 2008;**86**(13):3028–37.

21. Kaspar BK, Frost LM, Christian L, Umapathi P, Gage FH. Synergy of insulin-like growth factor-1 and exercise in amyotrophic lateral sclerosis. *Ann Neurol* 2005;**57**(5):649–55.

22. Guegan C, Vila M, Teismann P, et al. Instrumental activation of bid by caspase-1 in a transgenic mouse model of ALS. *Mol Cell Neurosci* 2002;**20**(4):553–62.

23. Friedlander R, Brown Jr R, Gagliardini V, Wang J, Yuan J. Inhibition of ICE slows ALS in mice. *Nature* 1997;**388**:31.

24. Azzouz M, Hottinger A, Paterna JC, Zurn AD, Aebischer P, Bueler H. Increased motoneuron survival and improved neuromuscular function in transgenic ALS mice after intraspinal injection of an adeno-associated virus encoding Bcl-2. *Hum Mol Genet* 2000;**9**(5):803–11.

25. Foust KD, Nurre E, Montgomery CL, Hernandez A, Chan CM, Kaspar BK. Intravascular AAV9 preferentially targets neonatal neurons and adult astrocytes. *Nat Biotechnol* 2009;**27**(1):59–65.

26. Lucas DR, Newhouse JP. The toxic effect of sodium L-glutamate on the inner layers of the retina. *AMA Arch Ophthalmol* 1957;**58**(2):193–201.

27. Rothman SM. Synaptic activity mediates death of hypoxic neurons. *Science* 1983;**220**(4596):536–7.

28. Simon RP, Swan JH, Griffiths T, Meldrum BS. Blockade of N-methyl-D-aspartate receptors may protect against ischemic damage in the brain. *Science* 1984;**226**(4676):850–2.

29. Sattler R, Tymianski M. Molecular mechanisms of glutamate receptor-mediated excitotoxic neuronal cell death. *Mol Neurobiol* 2001;**24**(1–3):107–29.

30. Rothstein JD, Dykes-Hoberg M, Pardo CA, et al. Knockout of glutamate transporters reveals a major role for astroglial transport in excitotoxicity and clearance of glutamate. *Neuron* 1996;**16**(3):675–86.
31. Lin C, Bristol L, Jin L. Aberrant RNA processing in a neurodegenerative disease: the cause for absent EAAT2, a glutamate transporter, in amyotrophic lateral sclerosis. *Neuron* 1998;**20**:589–602.
32. Rothstein J, Van Kammen M, Levey A, Martin L, Kuncl R. Selective loss of glial glutamate transporter GLT-1 in amyotrophic lateral sclerosis. *Ann Neurol* 1995;**38**(1):73–84.
33. Rothstein JD, Martin LJ, Kuncl RW. Decreased glutamate transport by the brain and spinal cord in amyotrophic lateral sclerosis. *N Engl J Med* 1992;**326**(22):1464–8.
34. Rothstein J, Tsai G, Kuncl R, et al. Abnormal excitatory amino acid metabolism in amyotrophic lateral sclerosis. *Ann Neurol* 1990;**28**:18–25.
35. Guo H, Lai L, Butchbach ME, et al. Increased expression of the glial glutamate transporter EAAT2 modulates excitotoxicity and delays the onset but not the outcome of ALS in mice. *Hum Mol Genet* 2003;**12**(19):2519–32.
36. Li K, Hala TJ, Seetharam S, Poulsen DJ, Wright MC, Lepore AC. GLT1 overexpression in SOD1(G93A) mouse cervical spinal cord does not preserve diaphragm function or extend disease. *Neurobiol Dis* 2015;**78**:12–23.
37. Rogawski MA. Revisiting AMPA receptors as an antiepileptic drug target. *Epilepsy Curr* 2011;**11**(2):56–63.
38. Seeburg PH. A-to-I editing: new and old sites, functions and speculations. *Neuron* 2002;**35**(1):17–20.
39. Paschen W, Hedreen JC, Ross CA. RNA editing of the glutamate receptor subunits GluR2 and GluR6 in human brain tissue. *J Neurochem* 1994;**63**(5):1596–602.
40. Akbarian S, Smith MA, Jones EG. Editing for an AMPA receptor subunit RNA in prefrontal cortex and striatum in Alzheimer's disease, Huntington's disease and schizophrenia. *Brain Res* 1995;**699**(2):297–304.
41. Kawahara Y, Sun H, Ito K, et al. Underediting of GluR2 mRNA, a neuronal death inducing molecular change in sporadic ALS, does not occur in motor neurons in ALS1 or SBMA. *Neurosci Res* 2006;**54**(1):11–14.
42. Aizawa H, Sawada J, Hideyama T, et al. TDP-43 pathology in sporadic ALS occurs in motor neurons lacking the RNA editing enzyme ADAR2. *Acta Neuropathol* 2010;**120**(1):75–84.
43. Kuner R, Groom AJ, Bresink I, et al. Late-onset motoneuron disease caused by a functionally modified AMPA receptor subunit. *Proc Natl Acad Sci USA* 2005;**102**(16):5826–31.
44. Tateno M, Sadakata H, Tanaka M, et al. Calcium-permeable AMPA receptors promote misfolding of mutant SOD1 protein and development of amyotrophic lateral sclerosis in a transgenic mouse model. *Hum Mol Genet* 2004;**13**(19):2183–96.
45. Hideyama T, Yamashita T, Suzuki T, et al. Induced loss of ADAR2 engenders slow death of motor neurons from Q/R site-unedited GluR2. *J Neurosci* 2010;**30**(36):11917–25.
46. Yamashita T, Chai HL, Teramoto S, et al. Rescue of amyotrophic lateral sclerosis phenotype in a mouse model by intravenous AAV9-ADAR2 delivery to motor neurons. *EMBO Mol Med* 2013;**5**(11):1710–9.
47. Barber SC, Mead RJ, Shaw PJ. Oxidative stress in ALS: a mechanism of neurodegeneration and a therapeutic target. *Biochim Biophys Acta* 2006;**1762**(11–12):1051–67.
48. Dalle-Donne I, Rossi R, Giustarini D, Milzani A, Colombo R. Protein carbonyl groups as biomarkers of oxidative stress. *Clin Chim Acta Int J Clin Chem* 2003;**329**(1–2):23–38.
49. Abe K, Pan LH, Watanabe M, Kato T, Itoyama Y. Induction of nitrotyrosine-like immunoreactivity in the lower motor neuron of amyotrophic lateral sclerosis. *Neurosci Lett* 1995;**199**(2):152–4.
50. Abe K, Pan LH, Watanabe M, Konno H, Kato T, Itoyama Y. Upregulation of protein-tyrosine nitration in the anterior horn cells of amyotrophic lateral sclerosis. *Neurol Res* 1997;**19**(2):124–8.

51. Beal MF, Ferrante RJ, Browne SE, Matthews RT, Kowall NW, Brown Jr. RH. Increased 3-nitrotyrosine in both sporadic and familial amyotrophic lateral sclerosis. *Ann Neurol* 1997;**42**(4):644–54.

52. Shaw PJ, Ince PG, Falkous G, Mantle D. Oxidative damage to protein in sporadic motor neuron disease spinal cord. *Ann Neurol* 1995;**38**(4):691–5.

53. Ferrante RJ, Browne SE, Shinobu LA, et al. Evidence of increased oxidative damage in both sporadic and familial amyotrophic lateral sclerosis. *J Neurochem* 1997;**69**(5):2064–74.

54. Fitzmaurice PS, Shaw IC, Kleiner HE, et al. Evidence for DNA damage in amyotrophic lateral sclerosis. *Muscle Nerve* 1996;**19**(6):797–8.

55. Li B, Xu W, Luo C, Gozal D, Liu R. VEGF-induced activation of the PI3-K/Akt pathway reduces mutant SOD1-mediated motor neuron cell death. *Brain Res Mol Brain Res* 2003;**111**:155–64.

56. Ruttkay-Nedecky B, Nejdl L, Gumulec J, et al. The role of metallothionein in oxidative stress. *Int J Mol Sci* 2013;**14**(3):6044–66.

57. Suzuki KT, Kuroda T. Transfer of copper and zinc from ionic and metallothionein-bound forms to Cu, Zn--superoxide dismutase. *Res Commun Mol Pathol Pharmacol* 1995;**87**(3):287–96.

58. Lyons TJ, Liu H, Goto JJ, et al. Mutations in copper-zinc superoxide dismutase that cause amyotrophic lateral sclerosis alter the zinc binding site and the redox behavior of the protein. *Proc Natl Acad Sci U S A* 1996;**93**(22):12240–4.

59. Crow JP, Sampson JB, Zhuang Y, Thompson JA, Beckman JS. Decreased zinc affinity of amyotrophic lateral sclerosis-associated superoxide dismutase mutants leads to enhanced catalysis of tyrosine nitration by peroxynitrite. *J Neurochem* 1997;**69**(5):1936–44.

60. Goto JJ, Zhu H, Sanchez RJ, et al. Loss of in vitro metal ion binding specificity in mutant copper-zinc superoxide dismutases associated with familial amyotrophic lateral sclerosis. *J Biol Chem* 2000;**275**(2):1007–14.

61. Crow JP, Ye YZ, Strong M, Kirk M, Barnes S, Beckman JS. Superoxide dismutase catalyzes nitration of tyrosines by peroxynitrite in the rod and head domains of neurofilament-L. *J Neurochem* 1997;**69**(5):1945–53.

62. Puttaparthi K, Gitomer WL, Krishnan U, Son M, Rajendran B, Elliott JL. Disease progression in a transgenic model of familial amyotrophic lateral sclerosis is dependent on both neuronal and non-neuronal zinc binding proteins. *J Neurosci* 2002;**22**(20):8790–6.

63. Hashimoto K, Hayashi Y, Watabe K, Inuzuka T, Hozumi I. Metallothionein-III prevents neuronal death and prolongs life span in amyotrophic lateral sclerosis model mice. *Neuroscience* 2011;**189**:293–8.

64. Sarlette A, Krampfl K, Grothe C, Neuhoff N, Dengler R, Petri S. Nuclear erythroid 2-related factor 2-antioxidative response element signaling pathway in motor cortex and spinal cord in amyotrophic lateral sclerosis. *J Neuropathol Exp Neurol* 2008;**67**(11):1055–62.

65. Nanou A, Higginbottom A, Valori CF, et al. Viral delivery of antioxidant genes as a therapeutic strategy in experimental models of amyotrophic lateral sclerosis. *Mol Ther* 2013;**21**(8):1486–96.

66. Russo VC, Gluckman PD, Feldman EL, Werther GA. The insulin-like growth factor system and its pleiotropic functions in brain. *Endocr Rev* 2005;**26**(7):916–43.

67. Otrock ZK, Makarem JA, Shamseddine AI. Vascular endothelial growth factor family of ligands and receptors: review. *Blood Cells Mol Dis* 2007;**38**(3):258–68.

68. Lu B, Nagappan G, Guan X, Nathan PJ, Wren P. BDNF-based synaptic repair as a disease-modifying strategy for neurodegenerative diseases. *Nat Rev Neurosci* 2013;**14**(6):401–16.

69. Lapchak PA, Jiao S, Miller PJ, et al. Pharmacological characterization of glial cell line-derived neurotrophic factor (GDNF): implications for GDNF as a therapeutic molecule for treating neurodegenerative diseases. *Cell Tissue Res* 1996;**286**(2):179–89.

70. Schneider A, Kuhn HG, Schabitz WR. A role for G-CSF (granulocyte-colony stimulating factor) in the central nervous system. *Cell Cycle* 2005;**4**(12):1753–7.

71. Garbers C, Hermanns HM, Schaper F, et al. Plasticity and cross-talk of interleukin 6-type cytokines. *Cytokine Growth Factor Rev* 2012;**23**(3):85–97.
72. Zhou FQ, Snider WD. Intracellular control of developmental and regenerative axon growth. *Philos Trans R Soc Lond B Biol Sci* 2006;**361**(1473):1575–92.
73. Moore DL, Goldberg JL. Multiple transcription factor families regulate axon growth and regeneration. *Dev Neurobiol* 2011;**71**(12):1186–211.
74. Zhang X, Tang N, Hadden TJ, Rishi AK. Akt, FoxO and regulation of apoptosis. *Biochim Biophys Acta* 2011;**1813**(11):1978–86.
75. Datta SR, Brunet A, Greenberg ME. Cellular survival: a play in three Akts. *Genes Dev* 1999;**13**(22):2905–27.
76. McCubrey JA, Steelman LS, Chappell WH, et al. Roles of the Raf/MEK/ERK pathway in cell growth, malignant transformation and drug resistance. *Biochim Biophys Acta* 2007;**1773**(8):1263–84.
77. Battle TE, Frank DA. The role of STATs in apoptosis. *Curr Mol Med* 2002;**2**(4):381–92.
78. Song G, Ouyang G, Bao S. The activation of Akt/PKB signaling pathway and cell survival. *J Cell Mol Med* 2005;**9**(1):59–71.
79. Kurland JF, Voehringer DW, Meyn RE. The MEK/ERK pathway acts upstream of NF kappa B1 (p50) homodimer activity and Bcl-2 expression in a murine B-cell lymphoma cell line. MEK inhibition restores radiation-induced apoptosis. *J Biol Chem* 2003;**278**(34):32465–70.
80. Madamanchi NR, Li S, Patterson C, Runge MS. Reactive oxygen species regulate heat-shock protein 70 via the JAK/STAT pathway. *Arterioscler Thromb Vasc Biol* 2001;**21**(3):321–6.
81. Cntf A. A double-blind placebo-controlled clinical trial of subcutaneous recombinant human ciliary neurotrophic factor (rHCNTF) in amyotrophic lateral sclerosis. *Neurology* 1996;**46**:1244–9.
82. BDNF sg. (Phase III) A controlled trial of recombinant methionyl human BDNF in ALS. *Neurology* 1999;**52**:1427–33.
83. Lai EC, Felice KJ, Festoff BW, et al. Effect of recombinant human insulin-like growth factor-I on progression of ALS. A placebo-controlled study. The North America ALS/IGF-I Study Group. *Neurology* 1997;**49**(6):1621–30.
84. Borasio GD, Robberecht W, Leigh PN, et al. A placebo-controlled trial of insulin-like growth factor-I in amyotrophic lateral sclerosis. European ALS/IGF-I Study Group. *Neurology* 1998;**51**(2):583–6.
85. Nagano I, Shiote M, Murakami T, et al. Beneficial effects of intrathecal IGF-1 administration in patients with amyotrophic lateral sclerosis. *Neurol Res* 2005;**27**(7):768–72.
86. Arakawa Y, Sendtner M, Thoenen H. Survival effect of ciliary neurotrophic factor (CNTF) on chick embryonic motoneurons in culture: comparison with other neurotrophic factors and cytokines. *J Neurosci* 1990;**10**(11):3507–15.
87. Martinou JC, Martinou I, Kato AC. Cholinergic differentiation factor (CDF/LIF) promotes survival of isolated rat embryonic motoneurons in vitro. *Neuron* 1992;**8**(4):737–44.
88. Pennica D, Arce V, Swanson TA, et al. Cardiotrophin-1, a cytokine present in embryonic muscle, supports long-term survival of spinal motoneurons. *Neuron* 1996;**17**(1):63–74.
89. Sendtner M, Kreutzberg GW, Thoenen H. Ciliary neurotrophic factor prevents the degeneration of motor neurons after axotomy. *Nature* 1990;**345**(6274):440–1.
90. Mitsumoto H, Ikeda K, Holmlund T, et al. The effects of ciliary neurotrophic factor on motor dysfunction in wobbler mouse motor neuron disease. *Ann Neurol* 1994;**36**(2):142–8.
91. Ikeda K, Iwasaki Y, Tagaya N, Shiojima T, Kinoshita M. Neuroprotective effect of cholinergic differentiation factor/leukemia inhibitory factor on wobbler murine motor neuron disease. *Muscle Nerve* 1995;**18**(11):1344–7.
92. Sendtner M, Schmalbruch H, Stockli KA, Carroll P, Kreutzberg GW, Thoenen H. Ciliary neurotrophic factor prevents degeneration of motor neurons in mouse mutant progressive motor neuronopathy. *Nature* 1992;**358**(6386):502–4.

93. Sagot Y, Tan SA, Baetge E, Schmalbruch H, Kato AC, Aebischer P. Polymer encapsulated cell lines genetically engineered to release ciliary neurotrophic factor can slow down progressive motor neuronopathy in the mouse. *Eur J Neurosci* 1995;**7**(6):1313–22.

94. Bordet T, Schmalbruch H, Pettmann B, et al. Adenoviral cardiotrophin-1 gene transfer protects pmn mice from progressive motor neuronopathy. *J Clin Invest* 1999;**104**(8):1077–85.

95. Haase G, Pettmann B, Bordet T, et al. Therapeutic benefit of ciliary neurotrophic factor in progressive motor neuronopathy depends on the route of delivery. *Ann Neurol* 1999;**45**(3):296–304.

96. Bordet T, Lesbordes JC, Rouhani S, et al. Protective effects of cardiotrophin-1 adenoviral gene transfer on neuromuscular degeneration in transgenic ALS mice. *Hum Mol Genet* 2001;**10**(18):1925–33.

97. The pharmacokinetics of subcutaneously administered recombinant human ciliary neurotrophic factor (rHCNTF) in patients with amyotrophic lateral sclerosis: relation to parameters of the acute-phase response. The ALS CNTF Treatment Study (ACTS) Phase I-II Study Group. *Clin Neuropharmacol* 1995;**18**(6):500–14.

98. Miller RG, Bryan WW, Dietz MA, et al. Toxicity and tolerability of recombinant human ciliary neurotrophic factor in patients with amyotrophic lateral sclerosis. *Neurology* 1996;**47**(5):1329–31.

99. Ikeda K, Wong V, Holmlund TH, et al. Histometric effects of ciliary neurotrophic factor in wobbler mouse motor neuron disease. *Ann Neurol* 1995;**37**(1):47–54.

100. Bordet T, Castelnau-Ptakhine L, Fauchereau F, Friocourt G, Kahn A, Haase G. Neuronal targeting of cardiotrophin-1 by coupling with tetanus toxin C fragment. *Mol Cell Neurosci* 2001;**17**(5):842–54.

101. Burgess AW, Metcalf D. Characterization of a serum factor stimulating the differentiation of myelomonocytic leukemic cells. *Int J Cancer* 1980;**26**(5):647–54.

102. Nagata S, Tsuchiya M, Asano S, et al. Molecular cloning and expression of cDNA for human granulocyte colony-stimulating factor. *Nature* 1986;**319**(6052):415–8.

103. Schabitz WR, Kollmar R, Schwaninger M, et al. Neuroprotective effect of granulocyte colony-stimulating factor after focal cerebral ischemia. *Stroke* 2003;**34**(3):745–51.

104. Schneider A, Kruger C, Steigleder T, et al. The hematopoietic factor G-CSF is a neuronal ligand that counteracts programmed cell death and drives neurogenesis. *J Clin Invest* 2005;**115**(8):2083–98.

105. Yamasaki R, Tanaka M, Fukunaga M, et al. Restoration of microglial function by granulocyte-colony stimulating factor in ALS model mice. *J Neuroimmunol* 2010;**229**(1–2):51–62.

106. Henriques A, Pitzer C, Dittgen T, Klugmann M, Dupuis L, Schneider A. CNS-targeted viral delivery of G-CSF in an animal model for ALS: improved efficacy and preservation of the neuromuscular unit. *Mol Ther J Am Soc Gene Ther* 2011;**19**(2):284–92.

107. Leung DW, Cachianes G, Kuang WJ, Goeddel DV, Ferrara N. Vascular endothelial growth factor is a secreted angiogenic mitogen. *Science* 1989;**246**(4935):1306–9.

108. Keck PJ, Hauser SD, Krivi G, et al. Vascular permeability factor, an endothelial cell mitogen related to PDGF. *Science (New York, N.Y.)* 1989;**246**:1309–12.

109. Tischer E, Mitchell R, Hartman T, et al. The human gene for vascular endothelial growth factor. Multiple protein forms are encoded through alternative exon splicing. *J Biol Chem* 1991;**266**(18):11947–54.

110. Neufeld G, Cohen T, Gengrinovitch S, Poltorak Z. Vascular endothelial growth factor (VEGF) and its receptors. *FASEB J* 1999;**13**(1):9–22.

111. Sheen IS, Jeng KS, Shih SC, et al. Clinical significance of the expression of isoform 165 vascular endothelial growth factor mRNA in noncancerous liver remnants of patients with hepatocellular carcinoma. *World J Gastroenterol* 2005;**11**(2):187–92.

112. Jin K, Zhu Y, Sun Y, Mao XO, Xie L, Greenberg DA. Vascular endothelial growth factor (VEGF) stimulates neurogenesis in vitro and in vivo. *Proc Natl Acad Sci USA* 2002;**99**:11946–50.

113. Schänzer A, Wachs F-P, Wilhelm D, et al. Direct stimulation of adult neural stem cells in vitro and neurogenesis in vivo by vascular endothelial growth factor. *Brain Pathol (Zurich, Switzerland)* 2004;**14**:237–48.

114. Hashimoto T, Zhang X-M, Chen BY-k, Yang X-J. VEGF activates divergent intracellular signaling components to regulate retinal progenitor cell proliferation and neuronal differentiation. *Dev (Cambridge, England)* 2006;**133**:2201–10.

115. Erskine L, Reijntjes S, Pratt T, et al. VEGF signaling through neuropilin 1 guides commissural axon crossing at the optic chiasm. *Neuron* 2011;**70**:951–65.

116. Ruiz de Almodovar C, Fabre PJ, Knevels E, et al. VEGF mediates commissural axon chemoattraction through its receptor Flk1. *Neuron* 2011;**70**:966–78.

117. Matsuzaki H, Tamatani M, Yamaguchi A, et al. Vascular endothelial growth factor rescues hippocampal neurons from glutamate-induced toxicity: signal transduction cascades. *FASEB J* 2001;**15**:1218–20.

118. Svensson B, Peters M, König H-G, et al. Vascular endothelial growth factor protects cultured rat hippocampal neurons against hypoxic injury via an antiexcitotoxic, caspase-independent mechanism. *J Cereb Blood Flow Metab* 2002;**22**:1170–5.

119. Jin KL, Mao XO, Greenberg DA. Vascular endothelial growth factor: direct neuroprotective effect in in vitro ischemia. *Proc Natl Acad Sci USA* 2000;**97**(18):10242–7.

120. Oosthuyse B, Moons L, Storkebaum E, et al. Deletion of the hypoxia-response element in the vascular endothelial growth factor promoter causes motor neuron degeneration. *Nat Genet* 2001;**28**(2):131–8.

121. Wang Y, Mao XO, Xie L, et al. Vascular endothelial growth factor overexpression delays neurodegeneration and prolongs survival in amyotrophic lateral sclerosis mice. *J Neurosci* 2007;**27**(2):304–7.

122. Zheng C, Nennesmo I, Fadeel B, Henter J-I. Vascular endothelial growth factor prolongs survival in a transgenic mouse model of ALS. *Ann Neurol* 2004;**56**:564–7.

123. Storkebaum E, Lambrechts D, Dewerchin M, et al. Treatment of motoneuron degeneration by intracerebroventricular delivery of VEGF in a rat model of ALS. *Nat Neurosci* 2005;**8**:85–92.

124. Azzouz M, Ralph GS, Storkebaum E, et al. VEGF delivery with retrogradely transported lentivector prolongs survival in a mouse ALS model. *Nature* 2004;**429**:413–7.

125. Dodge JC, Treleaven CM, Fidler JA, et al. AAV4-mediated expression of IGF-1 and VEGF within cellular components of the ventricular system improves survival outcome in familial ALS mice. *Mol Ther* 2010;**18**(12):2075–84.

126. Carmeliet P. Blood vessels and nerves: common signals, pathways and diseases. *Nat Rev Genet* 2003;**4**(9):710–20.

127. Amano H, Hackett NR, Kaner RJ, Whitlock P, Rosengart TK, Crystal RG. Alteration of splicing signals in a genomic/cDNA hybrid VEGF gene to modify the ratio of expressed VEGF isoforms enhances safety of angiogenic gene therapy. *Mol Ther* 2005;**12**(4):716–24.

128. Whitlock PR, Hackett NR, Leopold PL, Rosengart TK, Crystal RG. Adenovirus-mediated transfer of a minigene expressing multiple isoforms of VEGF is more effective at inducing angiogenesis than comparable vectors expressing individual VEGF cDNAs. *Mol Ther* 2004;**9**(1):67–75.

129. Rebar EJ, Huang Y, Hickey R, et al. Induction of angiogenesis in a mouse model using engineered transcription factors. *Nat Med* 2002;**8**(12):1427–32.

130. Sakowski SA, Heavener SB, Lunn JS, et al. Neuroprotection using gene therapy to induce vascular endothelial growth factor-A expression. *Gene Ther* 2009;**16**(11):1292–9.

131. Kliem MA, Heeke BL, Franz CK, et al. Intramuscular administration of a VEGF zinc finger transcription factor activator (VEGF-ZFP-TF) improves functional outcomes in SOD1 rats. *Amyotroph Lateral Scler Off Publ World Fed Neurol Res Group Motor Neuron Dis* 2011;**12**(5):331–9.

132. Lindholm D, Dechant G, Heisenberg CP, Thoenen H. Brain-derived neurotrophic factor is a survival factor for cultured rat cerebellar granule neurons and protects them against glutamate-induced neurotoxicity. *Eur J Neurosci* 1993;**5**(11):1455–64.
133. Arancibia S, Silhol M, Mouliere F, et al. Protective effect of BDNF against beta-amyloid induced neurotoxicity in vitro and in vivo in rats. *Neurobiol Dis* 2008;**31**(3):316–26.
134. Ferenz KB, Gast RE, Rose K, Finger IE, Hasche A, Krieglstein J. Nerve growth factor and brain-derived neurotrophic factor but not granulocyte colony-stimulating factor, nimodipine and dizocilpine, require ATP for neuroprotective activity after oxygen-glucose deprivation of primary neurons. *Brain Res* 2012;**1448**:20–6.
135. Cohen-Cory S, Fraser S. Effects of brain-derived neurotrophic factor on optic axons branching and remodelling in vivo. *Nature* 1995;**378**(9):192–6.
136. Calvo AC, Moreno-Igoa M, Mancuso R, et al. Lack of a synergistic effect of a non-viral ALS gene therapy based on BDNF and a TTC fusion molecule. *Orphanet J Rare Dis* 2011;**6**:10.
137. L.B. Tovar-Y-Romo, Ramirez-Jarquin U.N., Lazo-Gomez R., Tapia R. Trophic factors as modulators of motor neuron physiology and survival: implications for ALS therapy. Front Cell Neurosci. 2014;8:61.
138. Fryer HJ, Wolf DH, Knox RJ, et al. Brain-derived neurotrophic factor induces excitotoxic sensitivity in cultured embryonic rat spinal motor neurons through activation of the phosphatidylinositol 3-kinase pathway. *J Neurochem* 2000;**74**(2):582–95.
139. Tovar-y-Romo LB, Tapia R. Delayed administration of VEGF rescues spinal motor neurons from death with a short effective time frame in excitotoxic experimental models in vivo. *ASN Neuro* 2012;**4**:2.
140. Yanpallewar SU, Barrick CA, Buckley H, Becker J, Tessarollo L. Deletion of the BDNF truncated receptor TrkB.T1 delays disease onset in a mouse model of amyotrophic lateral sclerosis. *PLoS One* 2012;**7**(6):e39946.
141. Suter-Crazzolara C, Unsicker K. GDNF is expressed in two forms in many tissues outside the CNS. *Neuroreport* 1994;**5**(18):2486–8.
142. Trupp M, Ryden M, Jornvall H, et al. Peripheral expression and biological activities of GDNF, a new neurotrophic factor for avian and mammalian peripheral neurons. *J Cell Biol* 1995;**130**(1):137–48.
143. Springer JE, Seeburger JL, He J, Gabrea A, Blankenhorn EP, Bergman LW. cDNA sequence and differential mRNA regulation of two forms of glial cell line-derived neurotrophic factor in Schwann cells and rat skeletal muscle. *Exp Neurol* 1995;**131**(1):47–52.
144. Lin LF, Doherty DH, Lile JD, Bektesh S, Collins F. GDNF: a glial cell line-derived neurotrophic factor for midbrain dopaminergic neurons. *Science* 1993;**260**(5111):1130–2.
145. Ryu H, Jeon GS, Cashman NR, Kowall NW, Lee J. Differential expression of c-Ret in motor neurons versus non-neuronal cells is linked to the pathogenesis of ALS. *Lab Invest* 2011;**91**(3):342–52.
146. Henderson CE, Phillips HS, Pollock RA, et al. GDNF: a potent survival factor for motoneurons present in peripheral nerve and muscle. *Science* 1994;**266**(5187):1062–4.
147. Keir S, Xiao X, Li J, Kennedy P. Adeno-associated virus-mediated delivery of glial cell line-derived neurotrophic factor protects motor neuron-like cells from apoptosis. *J Neurovirol* 2001;**7**(5):437–46.
148. Oppenheim RW, Houenou LJ, Johnson JE, et al. Developing motor neurons rescued from programmed and axotomy-induced cell death by GDNF. *Nature* 1995;**373**(6512):344–6.
149. Corse AM, Bilak MM, Bilak SR, Lehar M, Rothstein JD, Kuncl RW. Preclinical testing of neuroprotective neurotrophic factors in a model of chronic motor neuron degeneration. *Neurobiol Dis* 1999;**6**(5):335–46.
150. Manabe Y, Nagano I, Gazi MS, et al. Glial cell line-derived neurotrophic factor protein prevents motor neuron loss of transgenic model mice for amyotrophic lateral sclerosis. *Neurol Res* 2003;**25**(2):195–200.

151. Ciriza J, Moreno-Igoa M, Calvo AC, et al. A genetic fusion GDNF-C fragment of tetanus toxin prolongs survival in a symptomatic mouse ALS model. *Restor Neurol Neurosci* 2008;**26**(6):459–65.
152. Mohajeri M, Figlewicz D, Bohn M. Intramuscular grafts of myoblasts genetically modified to secrete glial cell line-derived neurotrophic factor prevent motoneuron loss and disease progression in a mouse model of familial amyotrophic lateral sclerosis. *Hum Gene Ther* 1999;**10**(11):1853–66.
153. Lu YY, Wang LJ, Muramatsu S, et al. Intramuscular injection of AAV-GDNF results in sustained expression of transgenic GDNF, and its delivery to spinal motoneurons by retrograde transport. *Neurosci Res* 2003;**45**(1):33–40.
154. Wang LJ, Lu YY, Muramatsu S, et al. Neuroprotective effects of glial cell line-derived neurotrophic factor mediated by an adeno-associated virus vector in a transgenic animal model of amyotrophic lateral sclerosis. *J Neurosci* 2002;**22**(16):6920–8.
155. Yamamoto M, Kobayashi Y, Li M, et al. In vivo gene electroporation of glial cell line-derived neurotrophic factor (GDNF) into the skeletal muscle of SOD1 mutant mice. *Neurochem Res* 2001;**26**(11):1201–7.
156. Moreno-Igoa M, Calvo AC, Ciriza J, Munoz MJ, Zaragoza P, Osta R. Non-viral gene delivery of the GDNF, either alone or fused to the C-fragment of tetanus toxin protein, prolongs survival in a mouse ALS model. *Restor Neurol Neurosci* 2012;**30**(1):69–80.
157. Foust KD, Flotte TR, Reier PJ, Mandel RJ. Recombinant adeno-associated virus-mediated global anterograde delivery of glial cell line-derived neurotrophic factor to the spinal cord: comparison of rubrospinal and corticospinal tracts in the rat. *Hum Gene Ther* 2008;**19**(1):71–82.
158. Guillot S, Azzouz M, Deglon N, Zurn A, Aebischer P. Local GDNF expression mediated by lentiviral vector protects facial nerve motoneurons but not spinal motoneurons in SOD1(G93A) transgenic mice. *Neurobiol Dis* 2004;**16**(1):139–49.
159. Klein SM, Behrstock S, McHugh J, et al. GDNF delivery using human neural progenitor cells in a rat model of ALS. *Hum Gene Ther* 2005;**16**(4):509–21.
160. Suzuki M, McHugh J, Tork C, et al. GDNF secreting human neural progenitor cells protect dying motor neurons, but not their projection to muscle, in a rat model of familial ALS. *PLoS One* 2007;**2**(1):e689.
161. Acsadi G, Anguelov RA, Yang H, et al. Increased survival and function of SOD1 mice after glial cell-derived neurotrophic factor gene therapy. *Hum Gene Ther* 2002;**13**(9):1047–59.
162. Manabe Y, Nagano I, Gazi MS, et al. Adenovirus-mediated gene transfer of glial cell line-derived neurotrophic factor prevents motor neuron loss of transgenic model mice for amyotrophic lateral sclerosis. *Apoptosis* 2002;**7**(4):329–34.
163. Salmon Jr. WD, Daughaday WH. A hormonally controlled serum factor which stimulates sulfate incorporation by cartilage in vitro. *J Lab Clin Med* 1957;**49**(6):825–36.
164. Rinderknecht E, Humbel RE. Polypeptides with nonsuppressible insulin-like and cell-growth promoting activities in human serum: isolation, chemical characterization, and some biological properties of forms I and II. *Proc Natl Acad Sci U S A* 1976;**73**(7):2365–9.
165. Daughaday WH, Hall K, Raben MS, Salmon Jr. WD, van den Brande JL, van Wyk JJ. Somatomedin: proposed designation for sulphation factor. *Nature* 1972;**235**(5333):107.
166. Rabinovsky ED, Gelir E, Gelir S, et al. Targeted expression of IGF-1 transgene to skeletal muscle accelerates muscle and motor neuron regeneration. *FASEB J* 2003;**17**(1):53–5.
167. Neff NT, Prevette D, Houenou LJ, et al. Insulin-like growth factors: putative muscle-derived trophic agents that promote motoneuron survival. *J Neurobiol* 1993;**24**(12):1578–88.
168. Kaspar BK, Llado J, Sherkat N, Rothstein JD, Gage FH. Retrograde viral delivery of IGF-1 prolongs survival in a mouse ALS model. *Science* 2003;**301**(5634):839–42.

169. Lepore AC, Haenggeli C, Gasmi M, et al. Intraparenchymal spinal cord delivery of adeno-associated virus IGF-1 is protective in the SOD1G93A model of ALS. *Brain Res* 2007;**1185**:256–65.

170. Franz CK, Federici T, Yang J, et al. Intraspinal cord delivery of IGF-I mediated by adeno-associated virus 2 is neuroprotective in a rat model of familial ALS. *Neurobiol Dis* 2009;**33**(3):473–81.

171. Dodge JC, Haidet AM, Yang W, et al. Delivery of AAV-IGF-1 to the CNS extends survival in ALS mice through modification of aberrant glial cell activity. *Mol Ther J Am Soc Gene Ther* 2008;**16**(6):1056–64.

172. Dobrowolny G, Giacinti C, Pelosi L, et al. Muscle expression of a local Igf-1 isoform protects motor neurons in an ALS mouse model. *J Cell Biol* 2005;**168**(2):193–9.

173. Nakamura T, Nawa K, Ichihara A. Partial purification and characterization of hepatocyte growth factor from serum of hepatectomized rats. *Biochem Biophys Res Commun* 1984;**122**(3):1450–9.

174. Nakamura T, Nishizawa T, Hagiya M, et al. Molecular cloning and expression of human hepatocyte growth factor. *Nature* 1989;**342**(6248):440–3.

175. Maina F, Klein R. Hepatocyte growth factor, a versatile signal for developing neurons. *Nat Neurosci* 1999;**2**(3):213–7.

176. Sonnenberg E, Meyer D, Weidner KM, Birchmeier C. Scatter factor/hepatocyte growth factor and its receptor, the c-met tyrosine kinase, can mediate a signal exchange between mesenchyme and epithelia during mouse development. *J Cell Biol* 1993;**123**(1):223–35.

177. Jung W, Castren E, Odenthal M, et al. Expression and functional interaction of hepatocyte growth factor-scatter factor and its receptor c-met in mammalian brain. *J Cell Biol* 1994;**126**(2):485–94.

178. Achim CL, Katyal S, Wiley CA, et al. Expression of HGF and cMet in the developing and adult brain. *Brain Res Dev Brain Res* 1997;**102**(2):299–303.

179. Thewke DP, Seeds NW. Expression of hepatocyte growth factor/scatter factor, its receptor, c-met, and tissue-type plasminogen activator during development of the murine olfactory system. *J Neurosci* 1996;**16**(21):6933–44.

180. Di Renzo MF, Bertolotto A, Olivero M, et al. Selective expression of the Met/HGF receptor in human central nervous system microglia. *Oncogene* 1993;**8**(1):219–22.

181. Krasnoselsky A, Massay MJ, DeFrances MC, Michalopoulos G, Zarnegar R, Ratner N. Hepatocyte growth factor is a mitogen for Schwann cells and is present in neurofibromas. *J Neurosci* 1994;**14**(12):7284–90.

182. Ebens A, Brose K, Leonardo ED, et al. Hepatocyte growth factor/scatter factor is an axonal chemoattractant and a neurotrophic factor for spinal motor neurons. *Neuron* 1996;**17**(6):1157–72.

183. Wong V, Glass DJ, Arriaga R, Yancopoulos GD, Lindsay RM, Conn G. Hepatocyte growth factor promotes motor neuron survival and synergizes with ciliary neurotrophic factor. *J Biol Chem* 1997;**272**(8):5187–91.

184. Yamamoto Y, Livet J, Pollock RA, et al. Hepatocyte growth factor (HGF/SF) is a muscle-derived survival factor for a subpopulation of embryonic motoneurons. *Development* 1997;**124**(15):2903–13.

185. Okura Y, Arimoto H, Tanuma N, et al. Analysis of neurotrophic effects of hepatocyte growth factor in the adult hypoglossal nerve axotomy model. *Eur J Neurosci* 1999;**11**(11):4139–44.

186. Novak KD, Prevette D, Wang S, Gould TW, Oppenheim RW. Hepatocyte growth factor/scatter factor is a neurotrophic survival factor for lumbar but not for other somatic motoneurons in the chick embryo. *J Neurosci* 2000;**20**(1):326–37.

187. Ishigaki A, Aoki M, Nagai M, et al. Intrathecal delivery of hepatocyte growth factor from amyotrophic lateral sclerosis onset suppresses disease progression in rat amyotrophic lateral sclerosis model. *J Neuropathol Exp Neurol* 2007;**66**(11):1037–44.
188. Sun W, Funakoshi H, Nakamura T. Overexpression of HGF retards disease progression and prolongs life span in a transgenic mouse model of ALS. *J Neurosci* 2002;**22**(15):6537–48.
189. Krakora D, Mulcrone P, Meyer M, et al. Synergistic effects of GDNF and VEGF on lifespan and disease progression in a familial ALS rat model. *Mol Ther* 2013;**21**(8):1602–10.
190. Moser JM, Bigini P, Schmitt-John T. The wobbler mouse, an ALS animal model. *Mol Genet Genom MGG* 2013;**288**(5–6):207–29.
191. Mitsumoto H, Ikeda K, Klinkosz B, Cedarbaum J, Wong V, Lindsay R. Arrest of motor neuron disease in wobbler mice cotreated with CNTF and BDNF. *Science* 1994;**265**:1107–10.
192. Farrar JJ, Yen LM, Cook T, et al. Tetanus. *J Neurol Neurosurg Psychiatry* 2000;**69**(3):292–301.
193. Toivonen JM, Olivan S, Osta R. Tetanus toxin C-fragment: the courier and the cure? *Toxins (Basel)* 2010;**2**(11):2622–44.
194. Chaib-Oukadour I, Gil C, Aguilera J. The C-terminal domain of the heavy chain of tetanus toxin rescues cerebellar granule neurones from apoptotic death: involvement of phosphatidylinositol 3-kinase and mitogen-activated protein kinase pathways. *J Neurochem* 2004;**90**(5):1227–36.
195. Ciriza J, García-Ojeda M, Martín-Burriel I, Agulhon C, Miana-Mena FJ, Muñoz MJ, et al. Antiapoptotic activity maintenance of brain derived neurotrophic factor and the C fragment of the tetanus toxin genetic fusion protein. *Cent Eur J Biol* 2008;**3**:105–12.
196. Moreno-Igoa M, Calvo AC, Penas C, et al. Fragment C of tetanus toxin, more than a carrier. Novel perspectives in non-viral ALS gene therapy. *J Mol Med (Berl)* 2010;**88**(3):297–308.
197. Walker AT, Smith PJ, Kolasa M. Reduction of racial/ethnic disparities in vaccination coverage, 1995–2011. *MMWR Surveill Summ* 2014(63 Suppl. 1):7–12.
198. Distler JH, Hirth A, Kurowska-Stolarska M, Gay RE, Gay S, Distler O. Angiogenic and angiostatic factors in the molecular control of angiogenesis. *Q J Nucl Med* 2003;**47**(3):149–61.
199. Greenway MJ, Andersen PM, Russ C, et al. ANG mutations segregate with familial and 'sporadic' amyotrophic lateral sclerosis. *Nat Genet* 2006;**38**(4):411–3.
200. Kishimoto K, Liu S, Tsuji T, Olson KA, Hu GF. Endogenous angiogenin in endothelial cells is a general requirement for cell proliferation and angiogenesis. *Oncogene* 2005;**24**(3):445–56.
201. Kieran D, Sebastia J, Greenway MJ, et al. Control of motoneuron survival by angiogenin. *J Neurosci* 2008;**28**(52):14056–61.
202. Ezzi SA, Urushitani M, Julien JP. Wild-type superoxide dismutase acquires binding and toxic properties of ALS-linked mutant forms through oxidation. *J Neurochem* 2007;**102**(1):170–8.
203. Gruzman A, Wood WL, Alpert E, et al. Common molecular signature in SOD1 for both sporadic and familial amyotrophic lateral sclerosis. *Proc Natl Acad Sci USA* 2007;**104**(30):12524–9.
204. Kabashi E, Valdmanis PN, Dion P, Rouleau GA. Oxidized/misfolded superoxide dismutase-1: the cause of all amyotrophic lateral sclerosis? *Ann Neurol* 2007;**62**(6):553–9.
205. Bosco DA, Morfini G, Karabacak NM, et al. Wild-type and mutant SOD1 share an aberrant conformation and a common pathogenic pathway in ALS. *Nat Neurosci* 2010;**13**(11):1396–403.
206. Pokrishevsky E, Grad LI, Yousefi M, Wang J, Mackenzie IR, Cashman NR. Aberrant localization of FUS and TDP43 is associated with misfolding of SOD1 in amyotrophic lateral sclerosis. *PLoS One* 2012;**7**(4):e35050.

207. Urushitani M, Ezzi SA, Julien JP. Therapeutic effects of immunization with mutant superoxide dismutase in mice models of amyotrophic lateral sclerosis. *Proc Natl Acad Sci U S A* 2007;**104**(7):2495–500.

208. Takeuchi S, Fujiwara N, Ido A, et al. Induction of protective immunity by vaccination with wild-type apo superoxide dismutase 1 in mutant SOD1 transgenic mice. *J Neuropathol Exp Neurol* 2010;**69**(10):1044–56.

209. Patel P, Kriz J, Gravel J, et al. Adeno-associated virus-mediated delivery of a recombinant single-chain antibody against misfolded superoxide dismutase for treatment of amyotrophic lateral sclerosis. *Mol Ther J Am Soc Gene Ther* 2014;**22**(3):498–510.

210. Elder GA, Friedrich Jr. VL, Bosco P, et al. Absence of the mid-sized neurofilament subunit decreases axonal calibers, levels of light neurofilament (NF-L), and neurofilament content. *J Cell Biol* 1998;**141**(3):727–39.

211. Elder GA, Friedrich Jr. VL, Kang C, et al. Requirement of heavy neurofilament subunit in the development of axons with large calibers. *J Cell Biol* 1998;**143**(1):195–205.

212. Rao MV, Houseweart MK, Williamson TL, Crawford TO, Folmer J, Cleveland DW. Neurofilament-dependent radial growth of motor axons and axonal organization of neurofilaments does not require the neurofilament heavy subunit (NF-H) or its phosphorylation. *J Cell Biol* 1998;**143**(1):171–81.

213. Zhu Q, Lindenbaum M, Levavasseur F, Jacomy H, Julien JP. Disruption of the NF-H gene increases axonal microtubule content and velocity of neurofilament transport: relief of axonopathy resulting from the toxin beta,beta'-iminodipropionitrile. *J Cell Biol* 1998;**143**(1):183–93.

214. Julien JP. Neurofilaments and motor neuron disease. *Trends Cell Biol* 1997;**7**(6):243–9.

215. Figlewicz DA, Krizus A, Martinoli MG, et al. Variants of the heavy neurofilament subunit are associated with the development of amyotrophic lateral sclerosis. *Hum Mol Genet* 1994;**3**(10):1757–61.

216. Tomkins J, Usher P, Slade JY, et al. Novel insertion in the KSP region of the neurofilament heavy gene in amyotrophic lateral sclerosis (ALS). *Neuroreport* 1998;**9**(17):3967–70.

217. Al-Chalabi A, Andersen PM, Nilsson P, et al. Deletions of the heavy neurofilament subunit tail in amyotrophic lateral sclerosis. *Hum Mol Genet* 1999;**8**(2):157–64.

218. Cote F, Collard J, Julien J. Progressive neuronopathy in transgenic mice expressing the human neurofilament heavy gene: a mouse model of amyotrophic lateral sclerosis. *Cell* 1993;**73**(1):35–46.

219. Collard JF, Cote F, Julien JP. Defective axonal transport in a transgenic mouse model of amyotrophic lateral sclerosis. *Nature* 1995;**375**(6526):61–4.

220. Meier J, Couillard-Despres S, Jacomy H, Gravel C, Julien JP. Extra neurofilament NF-L subunits rescue motor neuron disease caused by overexpression of the human NF-H gene in mice. *J Neuropathol Exp Neurol* 1999;**58**(10):1099–110.

221. Ju S, Tardiff DF, Han H, et al. A yeast model of FUS/TLS-dependent cytotoxicity. *PLoS Biol* 2011;**9**(4):e1001052.

222. Barmada SJ, Ju S, Arjun A, et al. Amelioration of toxicity in neuronal models of amyotrophic lateral sclerosis by hUPF1. *Proc Natl Acad Sci USA* 2015;**112**(25):7821–6.

223. Jackson KL, Dayton RD, Orchard EA, et al. Preservation of forelimb function by UPF1 gene therapy in a rat model of TDP-43-induced motor paralysis. *Gene Ther* 2015;**22**(1):20–8.

Stem Cell Therapy for Amyotrophic Lateral Sclerosis

K.S. Chen and E.L. Feldman

University of Michigan, Ann Arbor, MI, United States

OUTLINE

Molecular and Cellular Therapies for Motor Neuron Diseases.
DOI: http://dx.doi.org/10.1016/B978-0-12-802257-3.00009-2

207

INTRODUCTION

Amyotrophic lateral sclerosis (ALS) is a terrifying diagnosis that strikes at the core of an individual's function and productivity. The disease is characterized by progressive and selective degeneration of upper and lower motor neurons.[1] This manifests as an insidious, inexorable decline in motor function, with progressively compromised strength, coordination, gait, swallowing, speech, and respiratory function. For nearly all those afflicted, this means first relying on canes or walkers, then wheelchair-dependence, progressing to ventilator-dependence and a bedbound status, all the while with sensation and cognition preserved. Complications arising from this decline lead to death in an average of 3–5 years from time of diagnosis. The estimated cost of this illness to society ranges from $256–433 million to over $1 billion,[2] yet few treatment options exist. The only US Food and Drug Administration (FDA)-approved pharmacotherapy, riluzole, extends life span by a matter of mere months.[3] In the more than two decades since approval of riluzole, no disease-modifying therapies have come to fruition.

The reasons behind this lack of progress lie in the fact that mechanisms of motor neuron loss in ALS remain a mystery. Many models exist to explain how motor neurons are selectively killed including excitotoxicity,[4–8] loss of neurotrophic factors,[9–11] impaired RNA metabolism,[12,13] protein aggregation,[14,15] inflammatory signaling,[16] mitochondrial pathology,[17,18] endoplasmic reticulum dysfunction leading to misfolded proteins,[19] and many others. Cases with genetic underpinnings offer some clues, with approximately 15% of cases associated with mutations in genes such as Cu^{2+}-Zn^{2+} superoxide dismutase-1 (SOD1), transactive response DNA-binding protein 43 (TDP43), fused in sarcoma, and the recently identified hexanucleotide repeat in chromosome 9 open reading frame 72 (c9orf72).[20] Indeed, various animal models expressing these mutations, particularly the mutant SOD1 protein, have been used in most of the preclinical studies described. However, the large majority of ALS cases appear sporadic in nature and many treatments targeted at the above-mentioned pathways have failed in large-scale clinical trials.[21–28]

In this stark landscape of ALS therapeutics, stem cells have emerged as a promising new strategy to combat the multifaceted pathology of this disease. In the short time since stem cells were first described, there has been exponential growth in scientific understanding of stem cell biology as well as potential applications in disease.[29] In this chapter, we will begin by describing the prevailing rationale underlying stem cell therapy, then touch on current experience using different types of stem cells in ALS.

STEM CELLS IN AMYOTROPHIC LATERAL SCLEROSIS: MICROENVIRONMENT MODULATION

Stem cells are defined by the fundamental property of being able to undergo asymmetric cell division, with one daughter cell able to differentiate into one or more different specific cell types while the other daughter cell retains the capacity for self-renewal.[30,31] There are a number of different cell types that fall under the category of "stem cell," each with unique properties and varying degrees of differentiation potential (Fig. 9.1).

Early interest in using stem cell technology in ALS was focused on the possibility of regenerating the lost motor neuron pool. Experiments using murine stem cells showed that, after in vitro exposure to retinoic acid and the hedgehog agonist Hh-Ag1.3, these cells had the ability to differentiate into motor neurons and form neuromuscular junctions after transplantation into chick embryos.[32] However, this enthusiasm was quickly tempered, as this initial success was not reproducible in rodent models of ALS.[33] While stem cells can indeed be induced to form motor neurons in the spinal cord, these cells must integrate into local circuitry and receive input from interneurons and descending axons, grow new axons through an impermissible central nervous system (CNS), travel significant lengths in the periphery to connect with skeletal muscle, achieve remyelination, and finally form mature neuromuscular contacts. And while these are the minimum criteria for the creation of new motor units, successful functional improvements depend on formation of sufficient numbers of motor units on proper agonist/antagonist muscles within a time window prior to irreversible atrophy of muscle tissue. With our current understanding and abilities, this goal of motor system reconstruction is currently out of reach.

Furthermore, motor neuron replacement strategies face another challenge regarding inherent ALS pathophysiology. It is now understood that while motor neuron loss is the most obvious marker of ALS, factors within the motor neuron microenvironment contribute to the pathology. In chimeric mice expressing a mutant SOD1 protein in either motor neurons or astrocytes, motor neuron survival was preserved, despite mutant SOD1 expression, if surrounding astrocytes expressed the normal gene.[34] In contrast, motor neurons expressing the wild-type SOD1 protein still exhibited degeneration if neighboring astrocytes expressed mutant SOD1.[35-38] This has led to the widely held notion that motor neuron death in ALS is not a cell autonomous process. Accumulating evidence suggests interactions with interneurons, microglia, inflammatory signaling, vasculature, Schwann cells, and skeletal muscle play a role in ALS pathology.[39-47] Transplanted motor neurons would still need to survive in this hostile environment in order for cell replacement to succeed.

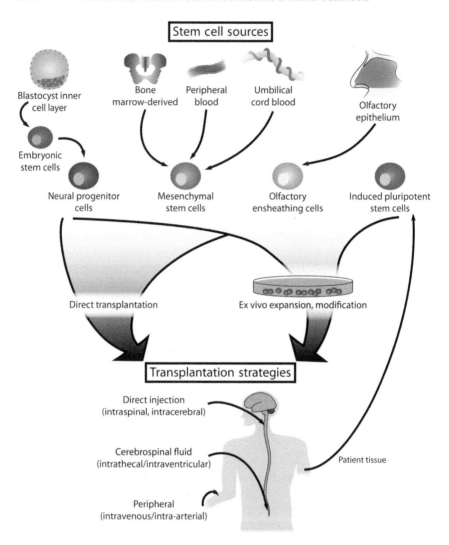

FIGURE 9.1   Stem cell-based strategies to treat amyotrophic lateral sclerosis (ALS) are diverse in terms of cell types and treatment techniques used. Many sources of stem cells have been explored for treatment of ALS. These stem cells can be transplanted directly after isolation, or can be expanded and even modified ex vivo prior to implantation. The strategies for stem cell implantation are also varied, ranging from benign peripheral infusion to more invasive intraspinal/intracerebral injections. A new frontier involves utilizing a patient's own tissue to generate inducible pluripotent stem cells, which can be used to study ALS pathophysiology, and eventually be modified ex vivo prior to reimplantation.

Given this more nuanced and global understanding of ALS pathology, stem cells still rise to the forefront as a promising therapeutic option for ALS. Stem cells can succeed where pharmacotherapy has failed by having simultaneous impact on the multifactorial pathways leading to motor neuron death. Pharmacologic therapies often achieve their effects by targeting one specific signaling pathway, and pharmacologic action in nonaffected tissues may give rise to side effects. Stem cells, on the other hand, can act locally via intercellular and paracrine mechanisms, while retaining the ability to simultaneously influence many signaling cascades that may be deranged. Thus, the majority of current stem cell therapy investigations are aimed at modulating the local microenvironment and preventing the further loss of motor neurons.

Nearly every type of stem cell has been applied in ALS animal models (Fig. 9.1), including embryonic stem cells (ESCs), olfactory ensheathing cells (OECs), bone marrow-derived mesenchymal stem cells (MSCs), peripheral blood stem cells (PBSCs), umbilical cord stem cells (UBSCs), and neural progenitor cells (NPCs). While results and progress in translation to humans have been variable, the fervor with which stem cell therapy in ALS is being pursued conveys a general sense of excitement and optimism.

EMBRYONIC STEM CELLS

The cell with the most diverse differentiation capability is the ESC. This cell, derived from the embryonic blastocyst, is considered totipotent, capable of forming all tissue types if given the correct environment. While the original interest in ESCs was for motor neuron replacement, it was found that human ESCs inefficiently formed mature motor neurons. Rather, these implanted cells were locally neuroprotective, perhaps through local secretion of transforming growth factor α and brain-derived neurotrophic factors.[48]

The robust replicative potential of ESCs is in some ways a double-edged sword. Early use of ESCs in the mutant SOD1 rat model demonstrated tumor formation at sites of cell implantation (Fig. 9.2).[49] Later explorations of ESCs in this and other models of motor neuron degeneration would utilize in vitro differentiation of murine ESCs toward motor neuron fates prior to implantation.[33,50,51]

Part of the difficulty with using ESCs also involves the social, ethical, and political ramifications of establishing and utilizing new cell lines derived from human embryos.[52] Therefore, while ESCs possess extraordinary potential, the attendant caveats have made ESCs a less attractive option for clinical translation. To date, no clinical trials in humans have been conducted using human ESCs.

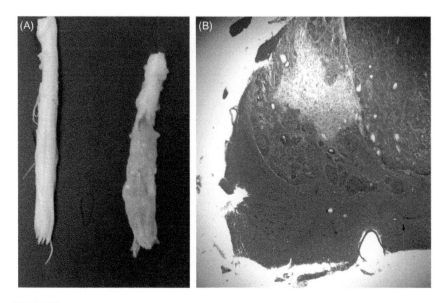

FIGURE 9.2   Embryonic stem cell (ESC) injection carries a risk of tumor formation. (A) Rat spinal cord on the left demonstrates normal gross anatomy after vehicle control injection, compared to rat spinal cord implanted with mouse ESCs on the right showing a bulky tumor which led to rapid neurologic decline. (B) Histological studies confirmed large tumor formation at the site of ESC implantation in the ventral horn of the spinal cord. Subsequent studies using more lineage-committed neural progenitor cells have not shown the same degree of tumorigenicity.

BONE MARROW-DERIVED MESENCHYMAL STEM CELLS

Another well-described pool of stem cells are MSCs that are enriched in bone marrow. These cells normally support hematopoietic stem cells in the bone marrow, contribute to formation of mesenchymal tissues (adipocytes, chondrocytes, osteocytes, etc.),[53] and can even transdifferentiate to cell fates of nonmesenchymal origin.[54,55] Despite the capacity to form new neural and glial tissue,[56,57] the usefulness of MSCs likely lies in their immunomodulatory characteristics and their role as paracrine "feeder" cells. Experiments with transplantation of bone marrow MSCs in ALS models demonstrated immune modulation and microglial formation,[58–60] as well as neurotrophic support by glial derived neurotrophic factor, vascular endothelial growth factor, and other growth factors.[61–65] These studies have demonstrated increased motor neuron survival, as well as functional improvements by a variety of physiological measures, and have been repeated using rat or human MSCs in mutant SOD1 animal models with similar findings.[66–73]

An advantage of using bone marrow-derived MSCs is the presence of well-established experience in harvesting and expanding clinical-grade MSCs from patient bone marrow. Pilot studies using infusion of MSCs have demonstrated tolerability and safety.[74–77] In one study of 10 patients receiving intrathecal MSCs harvested by bone marrow aspiration, no serious adverse events (AEs) were reported, and some showed stabilization of disease by the ALS Functional Rating Score-Revised (ALSFRS-R).[76] Another study enrolled 19 patients who received intrathecal MSCs harvested by bone marrow aspiration. No significant AEs were reported and a 6-month period of disease stability was seen after the procedure.[75] Building on the idea of immune-mediated neurotoxicity and immunomodulation, T-cell vaccination was used in conjunction with MSCs differentiated along a neural precursor path. The T-cell vaccination was inspired by experiments in multiple sclerosis showing induction of immune tolerance, vis-à-vis the plethora of data supporting the role of neuro-inflammation in ALS. This protocol involves obtaining peripheral blood mononuclear cells by apheresis, immune-negative selecting for T cells, inactivating the T cells by radiation, then administering the T cells to patients by IV infusion every 28 days. Subsequently, seven patients completed a protocol involving intra-arterial bone marrow infusion, intravenous administration of autologous in vitro-expanded effector T cells, followed by intra-arterial administration of autologous MSC-derived Neural stem cells. No significant AEs were seen and some patients experienced a transient improvement in symptomatology.[77] Another ongoing paradigm evaluating both intrathecal and intramuscular injection of MSCs modified to secrete neurotrophic factors is currently recruiting patients (BrainStorm Cell Therapeutics, NCT02017912).

More invasive intraspinal injection of bone marrow-derived MSCs has also been performed. Consecutive phase I studies in Italy studied 9 patients initially,[78,79] followed by an additional 10 patients,[80] for a total of 19 patients.[81] In these studies, bone marrow obtained from the iliac crest was expanded in vitro prior to direct surgical implantation of the cells into the dorsal spinal cord. The procedure was well tolerated and led to a slowed disease course in some patients. Another group evaluated 11 patients following injection of autologous bone marrow MSCs into the spinal cord.[82,83] No serious AEs were reported, nor were there any reported changes in the disease course; however, MNs near the areas of grafting showed fewer degenerative signs on histopathology. Finally, a study assessing 13 patients receiving bone marrow-derived MSCs by various routes indicated that the procedure was safe.[84] In some cases motor improvement was also reported. The translation of MSCs to modulate motor neuron microenvironments appears promising and we await results of ongoing clinical trials with excitement.

PERIPHERAL BLOOD STEM CELLS

MSCs have also been isolated from adult peripheral blood, thus allowing clinicians to forego bone marrow aspiration. Typically granulocyte-colony stimulating factor (G-CSF) is administered to promote the mobilization of CD34+ stem cells from the bone marrow into the peripheral blood. These can be collected by apheresis and expanded for later re-infusion. These CD34+ PBSCs were initially found in cancer patients receiving chemotherapy,[85] and it was later shown that these circulating hematopoietic progenitor cells could migrate into the CNS and provide support for diseased MNs.[86] G-CSF itself has also been suggested to have a neuroprotective effect.[87] Thus, a number of strategies have been implemented to collect and redistribute PBSCs to the CNS in ALS.

Based on the presumption that PBSCs can migrate into the CNS, a study using subcutaneous G-CSF given to 13 patients showed a slowing of disease progression by ALSFRS-R and compound muscle action potentials.[88] Another study in 17 patients comparing subcutaneous G-CSF to a placebo showed elevated CD34+ PBSCs but no difference in disease progression.[89] Similarly, the STEMALS trial utilized subcutaneous G-CSF along with a 5-day course of mannitol with the hopes of increasing permeability across the blood–brain barrier in 26 patients.[90,91] Again, there was an increase in circulating CD34+ PBSCs, and decreases in the proinflammatory cytokines monocyte chemotactic protein-1 (MCP-1) and interleukin-17 (IL-17) were also noted, but no change in ALSFRS-R was detected.

Some groups have tried to enrich the number of circulating PBSCs first by collection of CD34+ cells induced by G-CSF, followed by re-administration. One study utilized subcutaneous G-CSF in eight patients prior to PBSC isolation and infusion peripherally.[92] While no AEs were reported, clinical and imaging measures did not seem to be significantly impacted. Finally, an aggressive strategy was trialed in six ALS patients who received total body radiation followed by peripheral infusion of G-CSF-primed PBSCs from human leukocyte antigen (HLA) matched siblings.[93] Patients were immunosuppressed with methotrexate and tacrolimus, and graft versus host disease was seen in half of the patients. Although donor hematopoietic stem cells entered the CNS at sites of MN degeneration and engrafted as immunomodulatory cells, no clinical benefit was detected and the study was halted due to a lack of benefit and impaired quality of life.

In contrast to the above G-CSF studies, others have examined more invasive procedures for PBSC delivery in order to bypass the blood–brain barrier. An early study tested three patients using a protocol of subcutaneous G-CSF therapy and isolation of CD34+ stem cells, with subsequent intrathecal administration of collected stem cells showing minimal AEs.[94] A separate group has focused on CD133+ cells mobilized by G-CSF with direct injection into cortical motor areas of the brain. In the initial study,

10 patients were enrolled and compared to 10 control patients not accepting treatment or who applied after the study period.[95] The treatment group showed an improvement in baseline ALSFRS-R scores; however, this control group did have a higher ALSFRS-R score at baseline. An additional 67 patients were then evaluated after having undergone the same procedure.[96] Two postoperative deaths were reported: one attributable to a postoperative myocardial infarction along with an acute subdural hematoma requiring operative evacuation, whereas the other was attributed to progression of disease. No long-term outcome data from this larger cohort have yet been reported.

Overall, studies with PBSCs appear to demonstrate clinical safety, with a suggestion of clinical efficacy in some cases. Based on preclinical models, this is an area that does demonstrate therapeutic potential; however, it is clear that these trials struggle with the technique of stem cell delivery, balancing widespread yet inefficient distribution of blood-borne PBSCs against expedient yet risky surgical methods. Hence, moving forward, this method of therapy will benefit from some agreement on G-CSF delivery strategies as well as with good clinical trial design utilizing large numbers of well-defined patient populations and standardized outcome measures.

UMBILICAL CORD BLOOD STEM CELLS

MSCs are also found in umbilical cord blood. Soon after an infant's delivery, blood from the placenta and umbilical cord is extracted and UBSCs can be isolated. Previously, UBSCs have found application in diverse conditions, reviewed elsewhere.[97–99] They hold particular promise in neurodegenerative disease due to their immunomodulatory effects as well as their ability to differentiate into cells bearing neuronal and glial markers.[100,101]

Initial work utilized intravenous administration of human UBSCs in murine models of ALS. This method demonstrated that human UBSCs could enter the CNS diffusely, and formed cells with astrocytic and neuronal characteristics.[102,103] This paradigm did also appear to modestly delay disease onset[104,105] along with improvements in neuromuscular function by electrodiagnostic measurement.[106] More invasive intraventricular introduction of UCSBs also slowed motor decline and improved survival in the mutant SOD1 mouse.[107] Initial trials with intraspinal injection were disappointing,[108] but later injection of a population purified for CD34+ UCSBs showed clinical efficacy and improved survival.[109]

An attempt at using UCBs in humans has been reported in one German patient.[110] When this patient's granddaughter was born 6 months after his diagnosis of ALS, CD34+ UCBs from this infant were cryopreserved and banked. Later, these CD34+ UCBs were thawed and the patient received 1.34×10^7 cells divided between six total injections at the T8 spinal level. For a period of about 24 months after UCB injection, the patient's ALSFRS-R score plateaued (interestingly, effects were seen in lower extremity and bulbar function but not upper extremity function). The patient's disease later began to progress, but no AEs were attributable to the surgical procedure.

OLFACTORY ENSHEATHING CELLS

While many of the MSCs described above are more easily accessible, true CNS stem cells are less accessible and are produced in far fewer numbers during adulthood. While stem cell generation has been observed in the dentate gyrus and subventricular zone,[111] the degree of stem cell proliferation and the difficulty in harvesting these cells make the use of these cells in human adults impractical at this time. The other more accessible site of stem cell generation in adults is the olfactory epithelium. Here, OECs regenerate olfactory neurons and supporting glial/sustentacular cells. OECs have been applied to models of spinal cord injury,[112–114] stroke,[115,116] Parkinson's disease,[117] and have been shown to modulate immune responses, provide a scaffold for regenerating axons, promote remyelination, and even differentiate into neuronal and glial tissues.[118] Harnessing this capability has formed the rationale for using OECs in neurodegenerative diseases like ALS.

Initial studies isolated neurosphere-forming cells from mouse olfactory bulb. After transplantation in mutant SOD1 mouse spinal cords, formation of neurons expressing choline acetyltransferase, interneurons, and various glia was observed.[119] Native motor neuron survival was increased, and mice receiving cell transplants demonstrated a delay in disease onset as well as prolonged lifespan. Remyelination may also have contributed to functional improvement in a study with mutant SOD1 rats.[120] Implantation of OECs in the ventricular system also demonstrated dissemination of stem cells; however, in this case mutant SOD1 mice receiving OECs showed no improvement in motor performance (in contrast, this study showed an improvement in function in females that received ventricular implantation of MSCs).[121]

Studies with implantation of OECs into the corona radiata of mutant SOD1 rats demonstrated downstream rescue of distant spinal cord motor neurons.[122] On the basis of these data, a treatment paradigm has been implemented utilizing surgical implantation of fetal OECs into the bilateral corona radiata.[123,124] Early results suggested reduction in disease progression by patient or caregiver self-report.

A larger study of 507 patients was then pursued, showing an increase in ALSFRS-R scores.[125] In this larger study, no control groups or dose-responsive effects were reported. Some concerns have been raised regarding patients enrolled in these trials given the undefined trial protocols. One report on seven Dutch patients evaluated at a local institution demonstrated no change in disease progression.[126] Another Spanish study on three patients undergoing this procedure again did not show a change in disease course.[127] A postmortem analysis of two patients who underwent the procedure showed graft encasement with no evidence to suggest alteration of ALS neuropathology.[128] Finally, one patient from the United States was reported to show disease progression and possible trauma with hemorrhage and vasogenic edema at the injection site.[129] With these concerns raised in the literature, ongoing implementation of OECs in ALS should proceed with caution, close scrutiny, and ethical and robust clinical trial design.

NEURAL PROGENITOR CELLS

NPCs are pluripotent stem cells, having a differentiation profile restricted to neuronal and glial cell types. Thus, they carry the potential for generation of CNS-relevant tissues without the tumorigenic potential of more primitive cells such as ESCs.

One such cell line, NSI-566RSC, has been extensively studied in preclinical experiments.[130–134] An immortalized cell line derived from donated human fetal spinal cord, NSI-566RSC cells implanted into mutant SOD1 rats rescued motor neurons, correlating with improved motor function and prolonged lifespan.[132,134–136] The concept of motor neuron rescue by local effects was reinforced in these studies. Neurotrophic signaling has been demonstrated, as well as formation of inhibitory synapses[132,134] favoring motor neuron survival in the region of cell implantation but not distantly.[133]

With this promising preclinical data in hand, development of techniques for intraspinal injection of stem cells was also pursued. Given the spatially restricted effect of NPCs on local motor neurons, any technique for spinal injection would require accurate and reproducible placement of cells in the anterior horn of spinal cord gray matter, while minimizing trauma to the exquisitely sensitive surrounding spinal cord. A specialized spinal cord injection platform was devised (Fig. 9.3)[137] and tested in the Gottingen minipig.[138–140] Features aimed at maximizing safety for use in humans were incorporated into the design of the injection platform. These include a frame that is anchored to the bony spine, facilitating platform stability if the operative table requires adjustment or the patient exhibits any intra-operative movements. A traveling gondola carries a multi-axial "Z-drive" allowing precise adjustments in Cartesian as well as polar coordinate planes. The injection needle is configured with a hub that, when

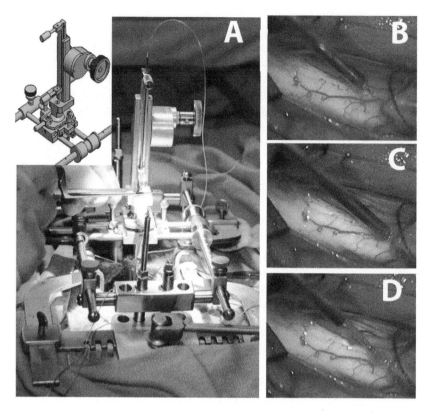

FIGURE 9.3   Spinal cord injection system for intraspinal stem cell transplantation. The spinal injection platform consists of two bridge rails (blue) and is anchored to the spine. (A) A gondola (green) travels in a cranial–caudal dimension and carries a mechanical Z drive (orange) which allows precise manipulation of a floating cannula. (B) The cannula tip is positioned over the dorsal surface over the ventral horn of spinal cord. (C) Needle penetrates into spinal cord. The hub ensures needle penetration is precisely to the level of the ventral horn. (D) Once the needle tip is at the target, the metal outer sleeve is pulled up, allowing flexible tubing to accommodate cardiorespiratory pulsations of spinal cord. *Reproduced with permission from Boulis NM, Federici T, Glass JD, et al. Translational stem cell therapy for amyotrophic lateral sclerosis. Nat Rev Neurol 2011;8(3):172–76.*

flush with the dorsal surface of the spinal cord, places the tip of the needle within the anterior horn (3–5 mm from dorsal surface). This needle is supported by flexible tubing that allows the needle to "float" with the cardiorespiratory pulsations of the spinal cord, and thus minimize trauma.

In the minipig, experiments showed that these animals tolerated 5–10 unilateral cervical spinal cord injections with all recovering to preoperative baseline neurologic function by postoperative day 14.[138] Immunosuppressive regimens were also optimized in these animals to maximize cell survivability.[140] This work paved the way for a pivotal Phase I clinical trial using NPCs in patients with ALS.

With supportive preclinical data for NSI-566RSC in the mutant SOD1 rat model and reassuring safety data with the spinal injection device in the Gottingen minipig, the US FDA gave approval for "A Phase l, Open-label, First in Human, Feasibility and Safety Study of Human Spinal Cord Derived Neural Stem Cell Transplantation for the Treatment of Amyotrophic Lateral Sclerosis" (NCT01348451). This study employed a "risk escalation" design to establish safety. First enrolled were nonambulatory ALS patients receiving five unilateral lumbar stem cell injections (100,000 cells in 10 µL per injection) or five bilateral (10 total) lumbar stem cell injections at the same cell dose. Then, the same implantation paradigm was used in ambulatory ALS patients. Next, five unilateral cervical injections (at the same dose) were made in ambulatory ALS patients. Finally, the cohort receiving five bilateral lumbar injections then underwent five unilateral cervical injections as the final "highest-risk" group.[141,142] Safety and tolerability of the NPC implantation protocol were firmly established, with most AEs attributable to the immunosuppressive regimen required postoperatively and no complications related to the procedure itself.[142]

In a number of patients, the slope of disease progression was noted to improve after surgery.[141] Using a sliding-window analysis of ALSFRS-R scores, many patients demonstrated stability or even improvement in the ALSFRS-R, particularly in patients who received both lumbar and cervical injections (Fig. 9.4). A majority of patients demonstrated improvement in grip strength testing, hand-held dynamometry, and electrical impedance myography.[141] These improvements can be correlated to the implanted stem cells that were shown to survive over the long term.[143] To date, seven deaths have been recorded among the patients enrolled in the Phase I trial (six due to ALS disease progression and one due to a previously undiagnosed congenital cardiac defect). Autopsy data demonstrated no tumor formation in any subject. Intriguingly, analysis of these spinal cords demonstrated live donor cells in recipient tissue (Fig. 9.5) by histologic staining.[143] In female subjects, fluorescence in situ hybridization for donor male XY chromosomes showed transplanted cells surviving amid recipient XX cells. Quantitative polymerase chain reaction also showed persistence of donor DNA to some degree in all postmortem specimens.

These tantalizing data allowed the FDA to give approval for "A Phase IIa, Open-label, Dose Escalation and Safety Study of Human Spinal Cord Derived Neural Stem Cell Transplantation for the Treatment of Amyotrophic Lateral Sclerosis" (NCT01730716). Again, the design of this trial incorporated a stepwise escalation of risk to determine a maximal dose of NPCs. With a long-term goal in mind of improving respiratory function, cervical spinal cord was targeted in most patients. Group A (n = 3) received five bilateral (10 total) injections in the cervical spinal cord (200,000

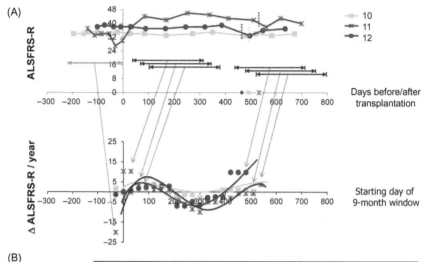

	Presurgery	Postlumbar surgery		Postcervical surgery	
Patient ID	Δ ALSFRS-R (per year)	Active period	Δ ALSFRS-R (per year)	Active period	Δ ALSFRS-R (per year)
10	1.8	4–13 mos	4	0–9 mos	4.1
11	−20.1	0–9 mos	10.2	0–9 mos	3.9
12	−1.3	4–13 mos	3	0–9 mos	9.8

FIGURE 9.4   Analysis of potential biological activity windows in patients receiving intraspinal injection of neural progenitor cells (NPCs). Postsurgery data points for patients receiving both lumbar and cervical NPC injections (subjects 10, 11, and 12) were divided into a series of sliding 9-month windows beginning each month postimplantation. Slopes of disease progression were calculated across each window using ALSFRS-R and compared to presurgical windows. (A) The top panel demonstrates ALSFRS-R scores for these subjects during the presurgical period (green) and representative ranges associated with the various sliding postsurgical 9-month windows (dark blue). The bottom panel demonstrates the slopes obtained for each sliding window, with the x-axis corresponding to the first month for each 9-month window. The first plotted slope for each subject corresponds to their presurgical progression rate. Slope values higher than the presurgical slope at baseline represent improved or attenuated progression rates during the designated window. Note that the starting month of the final sliding window for each patient coincides with the dates of the second surgery (cervical), which occurred at 17.5, 19, and 16.6 months after the initial lumbar surgery (time 0) for subjects 10, 11, and 12, respectively. (B) The presurgical slope and postsurgical slopes associated with the window correlating to the peak benefit windows for both the lumbar and cervical postsurgery time frames are summarized. ALSFRS-R, Amyotrophic Lateral Sclerosis Functional Rating Scale-Revised. *Reproduced with permission from Feldman EL, Boulis NM, Hur J, et al. Intraspinal neural stem cell transplantation in amyotrophic lateral sclerosis: phase 1 trial outcomes. Ann Neurol 2014;75(3):363–73.*

cells per injection for a total of 2 million cells). Subsequent patients in Groups B, C, and D (n = 3 each) then received 10 bilateral (20 total) cervical injections of escalating doses for a total of 4, 6, and 8 million cells, respectively. Finally, Group E patients (n = 3) first received 8 million cells in 10

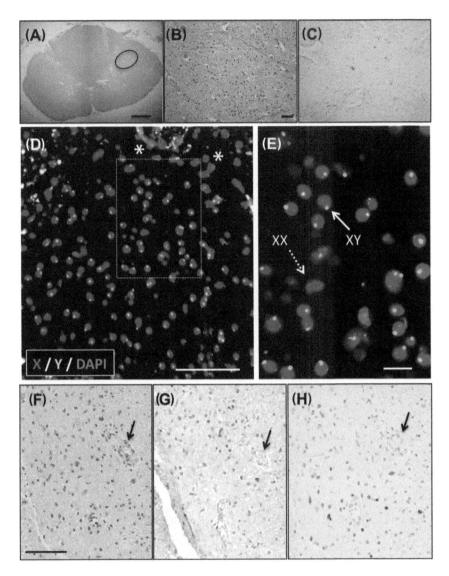

FIGURE 9.5 Donor neural progenitor cell (NPC) localization and characterization in a female amyotrophic lateral sclerosis (ALS) recipient. (A, B) Hemotoxylin and Eosin (H&E) staining shows nests of cells in the spinal cord (circle). (C) Sections stained with glial fibrillary acidic protein (GFAP) show lack of labeling of in nest of cells. (D, E) Fluorescence in situ hybridization (FISH) labeling shows numerous X (red) Y (green)-positive donor cells counterstained with 4′,6-diamidino-2-phenylindole (DAPI) nuclear stain (blue). This is seen in the context of XX-positive recipient cells in the surrounding regions. (F–H) H&E labeling of NPCs graft (F), with IHC labeling of stem cell marker SOX2 (G) and NeuN (H). Scale bars: 1 mm (A), 50 μm (B–D), 10 μm (E), 100 μm (F–H). IHC, immunohistochemistry; NeuN, neuronal nuclear antigen; SOX2, sex-determining region Y-box 2. *Reproduced with permission from Tadesse T, Gering M, Senitzer D, et al. Analysis of graft survival in a trial of stem cell transplant in ALS. Ann Clin Transl Neurol 2014;**1**(11):900–8.*

bilateral (20 total) cervical injections, followed a month later by 8 million cells in 10 bilateral (20 total) lumbar cord injections, giving a total of 16 million cells for this cohort. The final surgery was performed in July 2014 and results of data analysis are pending as of the writing of this article.

THE FUTURE OF STEM CELLS IN AMYOTROPHIC LATERAL SCLEROSIS

While the above discussion has described a cornucopia of methods and strategies for using stem cells in ALS, perhaps even more exciting are new technologies that are rapidly developing and can complement stem cell therapies. One such technique is in the creation of inducible pluripotent stem cells (iPS). Described by Yamanaka and colleagues, the expression of Oct3/4, Sox2, c-myc, and Klf4 gives researchers the ability to "de-differentiate" cells such as fibroblasts into pluripotent stem cells.[144,145] These fibroblasts can be easily obtained from ALS patients and subsequently induced to form motor neurons for in vitro studies.[146,147] Suddenly even sporadic forms of ALS can be studied in the laboratory[148] and high-throughput screening of potential therapeutics can be applied.[149] Furthermore, lessons learned from previous trials with stem cells could be applied to iPS cells, and patient-derived cells could be autologously retransplanted after in vitro manipulation, thus eliminating concerns for rejection and immunosuppression.

Cell-based therapy also carries the unique possibility of bridging the gap from the laboratory to the clinical realm. While many of the stem cells described above are available only in a limited number and require invasive methods to obtain them, even a small number of cells can be expanded to usable quantities in the laboratory. Additionally, the laboratory setting allows manipulation of cells prior to reimplantation. Some have taken ESCs or NPCs and directed differentiation toward certain cell types (oligodendrocytes or astrocytes, for example)[33,108,150–153] to enrich neuroprotective processes while minimizing tumorigenicity. Others have genetically modified stem cells to secrete various growth factors, creating cellular "factories" that can elaborate supportive neurotrophic factors into a diseased local microenvironment.[71,73,154–157] With this cycle between the laboratory and the clinic ongoing, stem cell therapy holds promise not only in ALS, but indeed can be implemented in other neurologic conditions in dire need of treatment options.

Perhaps the most important lesson to be learned from recent experience is that translation of stem cell therapy for ALS must still be approached carefully. While many of the above-mentioned studies with various stem cell types have shown promise, the allure of stem cell therapy has also resulted in the proliferation of less well-regulated trials.[158,159] As the ALS community ventures into the unexplored frontier of stem cell research, we must remain steadfast in our commitment to stringent preclinical testing, rational clinical trial design, conflict of interest prevention, transparency in data reporting, and prioritizing patient safety. Only in this way will the benefits of stem cell therapy be maximally realized.

CONCLUSIONS

We have outlined here the basic considerations in stem cell therapy for ALS as well as nascent trials in humans. While ALS stem cell therapy is still a young discipline, stem cells have the potential to drastically impact the course of disease by affecting the environment around diseased motor neurons and promote survival and thus function. And while many types of stem cells have been trialed as therapeutic options (and other techniques are just over the horizon), a few fundamental tenets should be considered with every potential therapy: (1) stem cells should be readily obtainable in numbers sufficient for clinical use; (2) cells should minimize the potential for tumor formation yet survive in sufficient numbers to have clinical impact on remaining motor neurons; (3) the paradigm for cell delivery should balance CNS penetration and therapeutic potential with procedural reproducibility and safety; (4) clinical trial design should be rigorous with sufficient subjects to glean meaningful data, yet be sensitive to the variable course of disease and ethically considerate of the lack of other treatment options; (5) trial outcomes should remain objective with measurements such as the ALSFRS-R and pulmonary function testing; and (6) subsequent analysis should describe the survival, function, and potential mechanism of transplanted cells. Keeping these principles in mind, stem cell technology has now completed its first steps and we may remain optimistic that further trials will make great strides toward a real treatment for ALS.

Acknowledgments

Funding support during the preparation of this manuscript was received from the University of Michigan Program for Neurology Research & Discovery, the A. Alfred Taubman Medical Research Institute, and the National Institutes of Health (University of Michigan Clinician Scientist Training Program, NINDS R25NS089450 to K.S.C.).

References

1. Gordon PH. Amyotrophic lateral sclerosis: an update for 2013 clinical features, pathophysiology, management and therapeutic trials. *Aging Dis* 2013;4(5):295–310.
2. Larkindale J, Yang W, Hogan PF, et al. Cost of illness for neuromuscular diseases in the United States. *Muscle Nerve* 2014;49(3):431–8.

3. Bensimon G, Lacomblez L, Meininger V. A controlled trial of riluzole in amyotrophic lateral sclerosis. ALS/Riluzole Study Group. *New Engl J Med* 1994;**330**(9):585–91.
4. Bellingham MC. A review of the neural mechanisms of action and clinical efficiency of riluzole in treating amyotrophic lateral sclerosis: what have we learned in the last decade? *CNS Neurosci Therap* 2011;**17**(1):4–31.
5. Grosskreutz J, Van Den Bosch L, Keller BU. Calcium dysregulation in amyotrophic lateral sclerosis. *Cell Calcium* 2010;**47**(2):165–74.
6. Martin LJ, Chang Q. Inhibitory synaptic regulation of motoneurons: a new target of disease mechanisms in amyotrophic lateral sclerosis. *Mol Neurobiol* 2012;**45**(1):30–42.
7. Redler RL, Dokholyan NV. The complex molecular biology of amyotrophic lateral sclerosis (ALS). *Prog Mol Biol Transl Sci* 2012;**107**:215–62.
8. Wainger BJ, Kiskinis E, Mellin C, et al. Intrinsic membrane hyperexcitability of amyotrophic lateral sclerosis patient-derived motor neurons. *Cell Rep* 2014;**7**(1):1–11.
9. Lunn JS, Sakowski SA, Kim B, Rosenberg AA, Feldman EL. Vascular endothelial growth factor prevents G93A-SOD1-induced motor neuron degeneration. *Dev Neurobiol* 2009;**69**(13):871–84.
10. Tovar YRLB, Ramirez-Jarquin UN, Lazo-Gomez R, Tapia R. Trophic factors as modulators of motor neuron physiology and survival: implications for ALS therapy. *Front Cell Neurosci* 2014;**8**:61.
11. Vincent AM, Mobley BC, Hiller A, Feldman EL. IGF-I prevents glutamate-induced motor neuron programmed cell death. *Neurobiol Dis* 2004;**16**(2):407–16.
12. Heutink P, Jansen IE, Lynes EM. C9orf72; abnormal RNA expression is the key. *Exp Neurol* 2014;**262**(Pt B):102–10.
13. Paez-Colasante X, Figueroa-Romero C, Sakowski SA, Goutman SA, Feldman EL. Amyotrophic lateral sclerosis: mechanisms and therapeutics in the epigenomic era. *Nat Rev Neurol* 2015;**11**(5):266–79.
14. Lee S, Kim HJ. Prion-like mechanism in amyotrophic lateral sclerosis: are protein aggregates the key? *Exp Neurobiol* 2015;**24**(1):1–7.
15. Ogawa M, Furukawa Y. A seeded propagation of Cu, Zn-superoxide dismutase aggregates in amyotrophic lateral sclerosis. *Front Cell Neurosci* 2014;**8**:83.
16. Rizzo F, Riboldi G, Salani S, et al. Cellular therapy to target neuroinflammation in amyotrophic lateral sclerosis. *Cell Mol Life Sci CMLS* 2014;**71**(6):999–1015.
17. Cozzolino M, Ferri A, Valle C, Carri MT. Mitochondria and ALS: implications from novel genes and pathways. *Mol Cell Neurosci* 2013;**55**:44–9.
18. Martin LJ, Liu Z, Chen K, et al. Motor neuron degeneration in amyotrophic lateral sclerosis mutant superoxide dismutase-1 transgenic mice: mechanisms of mitochondriopathy and cell death. *J Comp Neurol* 2007;**500**(1):20–46.
19. Tadic V, Prell T, Lautenschlaeger J, Grosskreutz J. The ER mitochondria calcium cycle and ER stress response as therapeutic targets in amyotrophic lateral sclerosis. *Front Cell Neurosci* 2014;**8**:147.
20. Robberecht W, Philips T. The changing scene of amyotrophic lateral sclerosis. *Nat Rev Neurosci* 2013;**14**(4):248–64.
21. Aggarwal SP, Zinman L, Simpson E, et al. Safety and efficacy of lithium in combination with riluzole for treatment of amyotrophic lateral sclerosis: a randomised, double-blind, placebo-controlled trial. *Lancet Neurol* 2010;**9**(5):481–8.
22. Desnuelle C, Dib M, Garrel C, Favier A. A double-blind, placebo-controlled randomized clinical trial of alpha-tocopherol (vitamin E) in the treatment of amyotrophic lateral sclerosis. ALS riluzole-tocopherol Study Group. *Amyotroph Lateral Scler Other Motor Neuron Disord Off Publ World Fed Neurol Res GroupMotor Neuron Dis* 2001;**2**(1):9–18.
23. Dupuis L, Dengler R, Heneka MT, et al. A randomized, double blind, placebo-controlled trial of pioglitazone in combination with riluzole in amyotrophic lateral sclerosis. *PloS One* 2012;**7**(6):e37885.

24. Miller RG, Moore II DH, Gelinas DF, et al. Phase III randomized trial of gabapentin in patients with amyotrophic lateral sclerosis. *Neurology* 2001;**56**(7):843–8.

25. Sorenson EJ, Windbank AJ, Mandrekar JN, et al. Subcutaneous IGF-1 is not beneficial in 2-year ALS trial. *Neurology* 2008;**71**(22):1770–5.

26. A double-blind placebo-controlled clinical trial of subcutaneous recombinant human ciliary neurotrophic factor (rHCNTF) in amyotrophic lateral sclerosis. ALS CNTF Treatment Study Group. *Neurology* 1996;**46**(5):1244–9.

27. Beauverd M, Mitchell JD, Wokke JH, Borasio GD. Recombinant human insulin-like growth factor I (rhIGF-I) for the treatment of amyotrophic lateral sclerosis/motor neuron disease. *Cochrane Database Syst Rev* 2012;**11**:CD002064.

28. Cudkowicz ME, van den Berg LH, Shefner JM, et al. Dexpramipexole versus placebo for patients with amyotrophic lateral sclerosis (EMPOWER): a randomised, double-blind, phase 3 trial. *Lancet Neurol* 2013;**12**(11):1059–67.

29. Lunn JS, Hefferan MP, Marsala M, Feldman EL. Stem cells: comprehensive treatments for amyotrophic lateral sclerosis in conjunction with growth factor delivery. *Growth Factors* 2009;**27**(3):133–40.

30. Jakob H. Stem cells and embryo-derived cell lines: tools for study of gene expression. *Cell Differ* 1984;**15**(2–4):77–80.

31. Rossant J, Papaioannou VE. The relationship between embryonic, embryonal carcinoma and embryo-derived stem cells. *Cell Differ* 1984;**15**(2–4):155–61.

32. Wichterle H, Lieberam I, Porter JA, Jessell TM. Directed differentiation of embryonic stem cells into motor neurons. *Cell* 2002;**110**(3):385–97.

33. Lopez-Gonzalez R, Kunckles P, Velasco I. Transient recovery in a rat model of familial amyotrophic lateral sclerosis after transplantation of motor neurons derived from mouse embryonic stem cells. *Cell Transpl* 2009;**18**(10):1171–81.

34. Clement AM, Nguyen MD, Roberts EA, et al. Wild-type nonneuronal cells extend survival of SOD1 mutant motor neurons in ALS mice. *Science (New York, N.Y.)* 2003;**302**(5642):113–7.

35. Di Giorgio FP, Boulting GL, Bobrowicz S, Eggan KC. Human embryonic stem cell-derived motor neurons are sensitive to the toxic effect of glial cells carrying an ALS-causing mutation. *Cell Stem Cell* 2008;**3**(6):637–48.

36. Marchetto MC, Muotri AR, Mu Y, Smith AM, Cezar GG, Gage FH. Non-cell-autonomous effect of human SOD1 G37R astrocytes on motor neurons derived from human embryonic stem cells. *Cell Stem Cell* 2008;**3**(6):649–57.

37. Di Giorgio FP, Carrasco MA, Siao MC, Maniatis T, Eggan K. Non-cell autonomous effect of glia on motor neurons in an embryonic stem cell-based ALS model. *Nat Neurosci* 2007;**10**(5):608–14.

38. Nagai M, Re DB, Nagata T, et al. Astrocytes expressing ALS-linked mutated SOD1 release factors selectively toxic to motor neurons. *Nat Neurosci* 2007;**10**(5):615–22.

39. Boillee S, Yamanaka K, Lobsiger CS, et al. Onset and progression in inherited ALS determined by motor neurons and microglia. *Science (New York, N.Y.)* 2006;**312**(5778):1389–92.

40. Ilieva H, Polymenidou M, Cleveland DW. Non-cell autonomous toxicity in neurodegenerative disorders: ALS and beyond. *J Cell Biol* 2009;**187**(6):761–72.

41. Borchelt DR. Amyotrophic lateral sclerosis--are microglia killing motor neurons? *New Engl J Med* 2006;**355**(15):1611–3.

42. Evans MC, Couch Y, Sibson N, Turner MR. Inflammation and neurovascular changes in amyotrophic lateral sclerosis. *Mol Cell Neurosci* 2013;**53**:34–41.

43. Hossaini M, Cardona Cano S, van Dis V, et al. Spinal inhibitory interneuron pathology follows motor neuron degeneration independent of glial mutant superoxide dismutase 1 expression in SOD1-ALS mice. *J Neuropathol Exp Neurol* 2011;**70**(8):662–77.

44. Zagami CJ, Beart PM, Wallis N, Nagley P, O'Shea RD. Oxidative and excitotoxic insults exert differential effects on spinal motoneurons and astrocytic glutamate

transporters: implications for the role of astrogliosis in amyotrophic lateral sclerosis. *Glia* 2009;**57**(2):119–35.

45. Moloney EB, de Winter F, Verhaagen J. ALS as a distal axonopathy: molecular mechanisms affecting neuromuscular junction stability in the presymptomatic stages of the disease. *Front Neurosci* 2014;**8**:252.

46. Murdock BJ, Bender DE, Segal BM, Feldman EL. The dual roles of immunity in ALS: Injury overrides protection. *Neurobiol Dis* 2015;**77**:1–12.

47. Pansarasa O, Rossi D, Berardinelli A, Cereda C. Amyotrophic lateral sclerosis and skeletal muscle: an update. *Mol Neurobiol* 2014;**49**(2):984–90.

48. Kerr DA, Llado J, Shamblott MJ, et al. Human embryonic germ cell derivatives facilitate motor recovery of rats with diffuse motor neuron injury. *J Neurosci Off J Soc Neurosci* 2003;**23**(12):5131–40.

49. Lunn JS, Pacut C, Stern E, et al. Intraspinal transplantation of neurogenin-expressing stem cells generates spinal cord neural progenitors. *Neurobiol Dis* 2012;**46**(1):59–68.

50. Deshpande DM, Kim YS, Martinez T, et al. Recovery from paralysis in adult rats using embryonic stem cells. *Ann Neurol* 2006;**60**(1):32–44.

51. Harper JM, Krishnan C, Darman JS, et al. Axonal growth of embryonic stem cell-derived motoneurons in vitro and in motoneuron-injured adult rats. *Proc Natl Acad Sci U S A* 2004;**101**(18):7123–8.

52. Abbott A. Europe rules against stem-cell patents. *Nature* 2011;**471**(7338):280.

53. Pittenger MF, Mackay AM, Beck SC, et al. Multilineage potential of adult human mesenchymal stem cells. *Science (New York, N.Y.)* 1999;**284**(5411):143–7.

54. Uccelli A, Benvenuto F, Laroni A, Giunti D. Neuroprotective features of mesenchymal stem cells. *Best Pract Res Clin Haematol* 2011;**24**(1):59–64.

55. Uccelli A, Moretta L, Pistoia V. Mesenchymal stem cells in health and disease. *Nat Rev Immunol* 2008;**8**(9):726–36.

56. Kopen GC, Prockop DJ, Phinney DG. Marrow stromal cells migrate throughout forebrain and cerebellum, and they differentiate into astrocytes after injection into neonatal mouse brains. *Proc Natl Acad Sci U S A* 1999;**96**(19):10711–6.

57. Phinney DG, Prockop DJ. Concise review: mesenchymal stem/multipotent stromal cells: the state of transdifferentiation and modes of tissue repair--current views. *Stem Cells* 2007;**25**(11):2896–902.

58. Corti S, Locatelli F, Donadoni C, et al. Wild-type bone marrow cells ameliorate the phenotype of SOD1-G93A ALS mice and contribute to CNS, heart and skeletal muscle tissues. *Brain J Neurol* 2004;**127**(Pt 11):2518–32.

59. Ohnishi S, Ito H, Suzuki Y, et al. Intra-bone marrow-bone marrow transplantation slows disease progression and prolongs survival in G93A mutant SOD1 transgenic mice, an animal model mouse for amyotrophic lateral sclerosis. *Brain Res* 2009;**1296**:216–24.

60. Solomon JN, Lewis CA, Ajami B, Corbel SY, Rossi FM, Krieger C. Origin and distribution of bone marrow-derived cells in the central nervous system in a mouse model of amyotrophic lateral sclerosis. *Glia* 2006;**53**(7):744–53.

61. Cabanes C, Bonilla S, Tabares L, Martinez S. Neuroprotective effect of adult hematopoietic stem cells in a mouse model of motoneuron degeneration. *Neurobiol Dis* 2007;**26**(2):408–18.

62. Corti S, Nizzardo M, Nardini M, et al. Systemic transplantation of c-kit+ cells exerts a therapeutic effect in a model of amyotrophic lateral sclerosis. *Hum Mol Genet* 2010;**19**(19):3782–96.

63. Pastor D, Viso-Leon MC, Botella-Lopez A, et al. Bone marrow transplantation in hindlimb muscles of motoneuron degenerative mice reduces neuronal death and improves motor function. *Stem Cells Dev* 2013;**22**(11):1633–44.

64. Pastor D, Viso-Leon MC, Jones J, et al. Comparative effects between bone marrow and mesenchymal stem cell transplantation in GDNF expression and motor function recovery in a motorneuron degenerative mouse model. *Stem Cell Rev* 2012;**8**(2):445–58.

65. Uccelli A, Milanese M, Principato MC, et al. Intravenous mesenchymal stem cells improve survival and motor function in experimental amyotrophic lateral sclerosis. *Mol Med (Cambridge, Mass.)* 2012;**18**:794–804.
66. Boucherie C, Schafer S, Lavand'homme P, Maloteaux JM, Hermans E. Chimerization of astroglial population in the lumbar spinal cord after mesenchymal stem cell transplantation prolongs survival in a rat model of amyotrophic lateral sclerosis. *J Neurosci Res* 2009;**87**(9):2034–46.
67. Forostyak S, Jendelova P, Kapcalova M, Arboleda D, Sykova E. Mesenchymal stromal cells prolong the lifespan in a rat model of amyotrophic lateral sclerosis. *Cytotherapy* 2011;**13**(9):1036–46.
68. Kim H, Kim HY, Choi MR, et al. Dose-dependent efficacy of ALS-human mesenchymal stem cells transplantation into cisterna magna in SOD1-G93A ALS mice. *Neurosci Lett* 2010;**468**(3):190–4.
69. Vercelli A, Mereuta OM, Garbossa D, et al. Human mesenchymal stem cell transplantation extends survival, improves motor performance and decreases neuroinflammation in mouse model of amyotrophic lateral sclerosis. *Neurobiol Dis* 2008;**31**(3):395–405.
70. Zhao CP, Zhang C, Zhou SN, et al. Human mesenchymal stromal cells ameliorate the phenotype of SOD1-G93A ALS mice. *Cytotherapy* 2007;**9**(5):414–26.
71. Knippenberg S, Thau N, Dengler R, Brinker T, Petri S. Intracerebroventricular injection of encapsulated human mesenchymal cells producing glucagon-like peptide 1 prolongs survival in a mouse model of ALS. *PloS One* 2012;**7**(6):e36857.
72. Suzuki M, McHugh J, Tork C, et al. Direct muscle delivery of GDNF with human mesenchymal stem cells improves motor neuron survival and function in a rat model of familial ALS. *Mol Ther J Am Soc Gene Ther* 2008;**16**(12):2002–10.
73. Chan-Il C, Young-Don L, Heejaung K, Kim SH, Suh-Kim H, Kim SS. Neural induction with neurogenin 1 enhances the therapeutic potential of mesenchymal stem cells in an amyotrophic lateral sclerosis mouse model. *Cell Transpl* 2013;**22**(5):855–70.
74. Baek W, Kim YS, Koh SH, et al. Stem cell transplantation into the intraventricular space via an Ommaya reservoir in a patient with amyotrophic lateral sclerosis. *J Neurosurg Sci* 2012;**56**(3):261–3.
75. Karussis D, Karageorgiou C, Vaknin-Dembinsky A, et al. Safety and immunological effects of mesenchymal stem cell transplantation in patients with multiple sclerosis and amyotrophic lateral sclerosis. *Arch Neurol* 2010;**67**(10):1187–94.
76. Prabhakar S, Marwaha N, Lal V, Sharma RR, Rajan R, Khandelwal N. Autologous bone marrow-derived stem cells in amyotrophic lateral sclerosis: a pilot study. *Neurol India* 2012;**60**(5):465–9.
77. Moviglia GA, Moviglia-Brandolino MT, Varela GS, et al. Feasibility, safety, and preliminary proof of principles of autologous neural stem cell treatment combined with T-cell vaccination for ALS patients. *Cell Transpl* 2012;**21**(Suppl. 1):S57–63.
78. Mazzini L, Mareschi K, Ferrero I, et al. Stem cell treatment in amyotrophic lateral sclerosis. *J Neurol Sci* 2008;**265**(1–2):78–83.
79. Mazzini L, Mareschi K, Ferrero I, et al. Autologous mesenchymal stem cells: clinical applications in amyotrophic lateral sclerosis. *Neurol Res* 2006;**28**(5):523–6.
80. Mazzini L, Ferrero I, Luparello V, et al. Mesenchymal stem cell transplantation in amyotrophic lateral sclerosis: a Phase I clinical trial. *Exp Neurol* 2010;**223**(1):229–37.
81. Mazzini L, Mareschi K, Ferrero I, et al. Mesenchymal stromal cell transplantation in amyotrophic lateral sclerosis: a long-term safety study. *Cytotherapy* 2012;**14**(1):56–60.
82. Blanquer M, Moraleda JM, Iniesta F, et al. Neurotrophic bone marrow cellular nests prevent spinal motoneuron degeneration in amyotrophic lateral sclerosis patients: a pilot safety study. *Stem Cells* 2012;**30**(6):1277–85.
83. Blanquer M, Perez-Espejo MA, Martinez-Lage JF, Iniesta F, Martinez S, Moraleda JM. A surgical technique of spinal cord cell transplantation in amyotrophic lateral sclerosis. *J Neurosci Methods* 2010;**191**(2):255–7.

84. Deda H, Inci MC, Kurekci AE, et al. Treatment of amyotrophic lateral sclerosis patients by autologous bone marrow-derived hematopoietic stem cell transplantation: a 1-year follow-up. *Cytotherapy* 2009;**11**(1):18–25.

85. Hubel K, Engert A. Clinical applications of granulocyte colony-stimulating factor: an update and summary. *Ann Hematol* 2003;**82**(4):207–13.

86. Deng J, Zou ZM, Zhou TL, et al. Bone marrow mesenchymal stem cells can be mobilized into peripheral blood by G-CSF in vivo and integrate into traumatically injured cerebral tissue. *Neurol Sci Off J Italian Neurol Soc Italian Soc Clin Neurophysiol* 2011;**32**(4):641–51.

87. Tanaka M, Kikuchi H, Ishizu T, et al. Intrathecal upregulation of granulocyte colony stimulating factor and its neuroprotective actions on motor neurons in amyotrophic lateral sclerosis. *J Neuropathol Exp Neurol* 2006;**65**(8):816–25.

88. Zhang Y, Wang L, Fu Y, et al. Preliminary investigation of effect of granulocyte colony stimulating factor on amyotrophic lateral sclerosis. *Amyotroph Lateral Scler Off Publ World Fed Neurol Res Group Motor Neuron Dis* 2009;**10**(5-6):430–1.

89. Nefussy B, Artamonov I, Deutsch V, Naparstek E, Nagler A, Drory VE. Recombinant human granulocyte-colony stimulating factor administration for treating amyotrophic lateral sclerosis: A pilot study. *Amyotroph Lateral Scler Off Publ World Fed Neurol Res Group Motor Neuron Dis* 2010;**11**(1-2):187–93.

90. Tarella C, Rutella S, Gualandi F, et al. Consistent bone marrow-derived cell mobilization following repeated short courses of granulocyte-colony-stimulating factor in patients with amyotrophic lateral sclerosis: results from a multicenter prospective trial. *Cytotherapy* 2010;**12**(1):50–9.

91. Chio A, Mora G, La Bella V, et al. Repeated courses of granulocyte colony-stimulating factor in amyotrophic lateral sclerosis: clinical and biological results from a prospective multicenter study. *Muscle Nerve* 2011;**43**(2):189–95.

92. Cashman N, Tan LY, Krieger C, et al. Pilot study of granulocyte colony stimulating factor (G-CSF)-mobilized peripheral blood stem cells in amyotrophic lateral sclerosis (ALS). *Muscle Nerve* 2008;**37**(5):620–5.

93. Appel SH, Engelhardt JI, Henkel JS, et al. Hematopoietic stem cell transplantation in patients with sporadic amyotrophic lateral sclerosis. *Neurology* 2008;**71**(17):1326–34.

94. Janson CG, Ramesh TM, During MJ, Leone P, Heywood J. Human intrathecal transplantation of peripheral blood stem cells in amyotrophic lateral sclerosis. *J Hematother Stem Cell Res* 2001;**10**(6):913–5.

95. Martinez HR, Gonzalez-Garza MT, Moreno-Cuevas JE, Caro E, Gutierrez-Jimenez E, Segura JJ. Stem-cell transplantation into the frontal motor cortex in amyotrophic lateral sclerosis patients. *Cytotherapy* 2009;**11**(1):26–34.

96. Martinez HR, Molina-Lopez JF, Gonzalez-Garza MT, et al. Stem cell transplantation in amyotrophic lateral sclerosis patients: methodological approach, safety, and feasibility. *Cell Transpl* 2012;**21**(9):1899–907.

97. Jaing TH. Umbilical cord blood: a trustworthy source of multipotent stem cells for regenerative medicine. *Cell Transpl* 2014;**23**(4–5):493–6.

98. Flores-Guzman P, Fernandez-Sanchez V, Mayani H. Concise review: ex vivo expansion of cord blood-derived hematopoietic stem and progenitor cells: basic principles, experimental approaches, and impact in regenerative medicine. *Stem Cells Transl Med* 2013;**2**(11):830–8.

99. Yoder MC. Cord blood banking and transplantation: advances and controversies. *Curr Opin Pediatrics* 2014;**26**(2):163–8.

100. Bicknese AR, Goodwin HS, Quinn CO, Henderson VC, Chien SN, Wall DA. Human umbilical cord blood cells can be induced to express markers for neurons and glia. *Cell Transpl* 2002;**11**(3):261–4.

101. Sanchez-Ramos JR, Song S, Kamath SG, et al. Expression of neural markers in human umbilical cord blood. *Exp Neurol* 2001;**171**(1):109–15.

102. Chen R, Ende N. The potential for the use of mononuclear cells from human umbilical cord blood in the treatment of amyotrophic lateral sclerosis in SOD1 mice. *J Med* 2000;**31**(1–2):21–30.

103. Garbuzova-Davis S, Willing AE, Zigova T, et al. Intravenous administration of human umbilical cord blood cells in a mouse model of amyotrophic lateral sclerosis: distribution, migration, and differentiation. *J Hematother Stem Cell Res* 2003;**12**(3):255–70.

104. Garbuzova-Davis S, Rodrigues MC, Mirtyl S, et al. Multiple intravenous administrations of human umbilical cord blood cells benefit in a mouse model of ALS. *PloS One* 2012;**7**(2):e31254.

105. Garbuzova-Davis S, Sanberg CD, Kuzmin-Nichols N, et al. Human umbilical cord blood treatment in a mouse model of ALS: optimization of cell dose. *PloS One* 2008;**3**(6):e2494.

106. Souayah N, Coakley KM, Chen R, Ende N, McArdle JJ. Defective neuromuscular transmission in the SOD1 G93A transgenic mouse improves after administration of human umbilical cord blood cells. *Stem Cell Rev* 2012;**8**(1):224–8.

107. Bigini P, Veglianese P, Andriolo G, et al. Intracerebroventricular administration of human umbilical cord blood cells delays disease progression in two murine models of motor neuron degeneration. *Rejuvenation Res* 2011;**14**(6):623–39.

108. Habisch HJ, Janowski M, Binder D, et al. Intrathecal application of neuroectodermally converted stem cells into a mouse model of ALS: limited intraparenchymal migration and survival narrows therapeutic effects. *J Neural Transm* 2007;**114**(11):1395–406.

109. Knippenberg S, Thau N, Schwabe K, et al. Intraspinal injection of human umbilical cord blood-derived cells is neuroprotective in a transgenic mouse model of amyotrophic lateral sclerosis. *Neuro-degener Dis* 2012;**9**(3):107–20.

110. Cordes AL, Jahn K, Hass R, et al. Intramedullary spinal cord implantation of human CD34+ umbilical cord-derived cells in ALS. *Amyotroph Lateral Scler Off Publ World Fed Neurol Res Group Motor Neuron Dis* 2011;**12**(5):325–30.

111. Urban N, Guillemot F. Neurogenesis in the embryonic and adult brain: same regulators, different roles. *Front Cell Neurosci* 2014;**8**:396.

112. Liu J, Chen P, Wang Q, et al. Meta analysis of olfactory ensheathing cell transplantation promoting functional recovery of motor nerves in rats with complete spinal cord transection. *Neural Regeneration Res* 2014;**9**(20):1850–8.

113. Ramon-Cueto A, Cordero MI, Santos-Benito FF, Avila J. Functional recovery of paraplegic rats and motor axon regeneration in their spinal cords by olfactory ensheathing glia. *Neuron* 2000;**25**(2):425–35.

114. Ramon-Cueto A, Plant GW, Avila J, Bunge MB. Long-distance axonal regeneration in the transected adult rat spinal cord is promoted by olfactory ensheathing glia transplants. *J Neurosci Off J Soc Neurosci* 1998;**18**(10):3803–15.

115. Shi X, Kang Y, Hu Q, et al. A long-term observation of olfactory ensheathing cells transplantation to repair white matter and functional recovery in a focal ischemia model in rat. *Brain Res* 2010;**1317**:257–67.

116. Shyu WC, Liu DD, Lin SZ, et al. Implantation of olfactory ensheathing cells promotes neuroplasticity in murine models of stroke. *J Clin Invest* 2008;**118**(7):2482–95.

117. Teng X, Nagata I, Li HP, et al. Regeneration of nigrostriatal dopaminergic axons after transplantation of olfactory ensheathing cells and fibroblasts prevents fibrotic scar formation at the lesion site. *J Neurosci Res* 2008;**86**(14):3140–50.

118. Liu Z, Martin LJ. Pluripotent fates and tissue regenerative potential of adult olfactory bulb neural stem and progenitor cells. *J Neurotrauma* 2004;**21**(10):1479–99.

119. Martin LJ, Liu Z. Adult olfactory bulb neural precursor cell grafts provide temporary protection from motor neuron degeneration, improve motor function, and extend survival in amyotrophic lateral sclerosis mice. *J Neuropathol Exp Neurol* 2007;**66**(11):1002–18.

120. Li Y, Bao J, Khatibi NH, et al. Olfactory ensheathing cell transplantation into spinal cord prolongs the survival of mutant SOD1(G93A) ALS rats through neuroprotection and remyelination. *Anat Rec* 2011;**294**(5):847–57.

121. Morita E, Watanabe Y, Ishimoto M, et al. A novel cell transplantation protocol and its application to an ALS mouse model. *Exp Neurol* 2008;**213**(2):431–8.
122. Li Y, Chen L, Zhao Y, et al. Intracranial transplant of olfactory ensheathing cells can protect both upper and lower motor neurons in amyotrophic lateral sclerosis. *Cell Transplant* 2013(22 Suppl 1):S51–65.
123. Huang H, Chen L, Xi H, et al. Fetal olfactory ensheathing cells transplantation in amyotrophic lateral sclerosis patients: a controlled pilot study. *Clin Transplant* 2008;**22**(6):710–8.
124. Chen L, Huang H, Zhang J, et al. Short-term outcome of olfactory ensheathing cells transplantation for treatment of amyotrophic lateral sclerosis. *Zhongguo Xiu Fu Chong Jian Wai Ke Za Zhi* 2007;**21**(9):961–6.
125. Chen L, Chen D, Xi H, et al. Olfactory ensheathing cell neurorestorotherapy for amyotrophic lateral sclerosis patients: benefits from multiple transplantations. *Cell Transplant* 2012(21 Suppl 1):S65–77.
126. Piepers S, van den Berg LH. No benefits from experimental treatment with olfactory ensheathing cells in patients with ALS. *Amyotroph Lateral Scler Off Publ World Fed Neurol Res Group Motor Neuron Dis* 2010;**11**(3):328–30.
127. Gamez J, Carmona F, Raguer N, et al. Cellular transplants in amyotrophic lateral sclerosis patients: an observational study. *Cytotherapy* 2010;**12**(5):669–77.
128. Giordana MT, Grifoni S, Votta B, et al. Neuropathology of olfactory ensheathing cell transplantation into the brain of two amyotrophic lateral sclerosis (ALS) patients. *Brain Pathol (Zurich, Switzerland)* 2010;**20**(4):730–7.
129. Chew S, Khandji AG, Montes J, Mitsumoto H, Gordon PH. Olfactory ensheathing glia injections in Beijing: misleading patients with ALS. *Amyotroph Lateral Scler Off Publ World Fed Neurol Res Group Motor Neuron Dis* 2007;**8**(5):314–6.
130. Guo X, Johe K, Molnar P, Davis H, Hickman J. Characterization of a human fetal spinal cord stem cell line, NSI-566RSC, and its induction to functional motoneurons. *J Tissue Eng Regenerative Med* 2010;**4**(3):181–93.
131. Yan J, Xu L, Welsh AM, et al. Combined immunosuppressive agents or CD4 antibodies prolong survival of human neural stem cell grafts and improve disease outcomes in amyotrophic lateral sclerosis transgenic mice. *Stem Cells* 2006;**24**(8):1976–85.
132. Hefferan MP, Galik J, Kakinohana O, et al. Human neural stem cell replacement therapy for amyotrophic lateral sclerosis by spinal transplantation. *PloS One* 2012;**7**(8):e42614.
133. Xu L, Shen P, Hazel T, Johe K, Koliatsos VE. Dual transplantation of human neural stem cells into cervical and lumbar cord ameliorates motor neuron disease in SOD1 transgenic rats. *Neurosci Lett* 2011;**494**(3):222–6.
134. Xu L, Yan J, Chen D, et al. Human neural stem cell grafts ameliorate motor neuron disease in SOD-1 transgenic rats. *Transplantation* 2006;**82**(7):865–75.
135. Xu L, Ryugo DK, Pongstaporn T, Johe K, Koliatsos VE. Human neural stem cell grafts in the spinal cord of SOD1 transgenic rats: differentiation and structural integration into the segmental motor circuitry. *J Comp Neurol* 2009;**514**(4):297–309.
136. Yan J, Xu L, Welsh AM, et al. Extensive neuronal differentiation of human neural stem cell grafts in adult rat spinal cord. *PLoS Med* 2007;**4**(2):e39.
137. Boulis NM, Federici T, Glass JD, Lunn JS, Sakowski SA, Feldman EL. Translational stem cell therapy for amyotrophic lateral sclerosis. *Nat Rev Neurol* 2011;**8**(3):172–6.
138. Raore B, Federici T, Taub J, et al. Cervical multilevel intraspinal stem cell therapy: assessment of surgical risks in Gottingen minipigs. *Spine* 2011;**36**(3):E164–71.
139. Riley JP, Raore B, Taub JS, Federici T, Boulis NM. Platform and cannula design improvements for spinal cord therapeutics delivery. *Neurosurgery* 2011;**69**(2 Suppl Operative) ons147–154; discussion ons155.
140. Usvald D, Vodicka P, Hlucilova J, et al. Analysis of dosing regimen and reproducibility of intraspinal grafting of human spinal stem cells in immunosuppressed minipigs. *Cell Transplant.* 2010;**19**(9):1103–22.

141. Feldman EL, Boulis NM, Hur J, et al. Intraspinal neural stem cell transplantation in amyotrophic lateral sclerosis: phase 1 trial outcomes. *Ann Neurol* 2014;**75**(3):363–73.
142. Riley J, Glass J, Feldman EL, et al. Intraspinal stem cell transplantation in amyotrophic lateral sclerosis: a phase I trial, cervical microinjection, and final surgical safety outcomes. *Neurosurgery* 2014;**74**(1):77–87.
143. Tadesse T, Gearing M, Senitzer D, et al. Analysis of graft survival in a trial of stem cell transplant in ALS. *Ann Clin Transl Neurol* 2014;**1**(11):900–8.
144. Takahashi K, Tanabe K, Ohnuki M, et al. Induction of pluripotent stem cells from adult human fibroblasts by defined factors. *Cell* 2007;**131**(5):861–72.
145. Takahashi K, Yamanaka S. Induction of pluripotent stem cells from mouse embryonic and adult fibroblast cultures by defined factors. *Cell* 2006;**126**(4):663–76.
146. Dimos JT, Rodolfa KT, Niakan KK, et al. Induced pluripotent stem cells generated from patients with ALS can be differentiated into motor neurons. *Science (New York, N.Y.)* 2008;**321**(5893):1218–21.
147. Yao XL, Ye CH, Liu Q, et al. Motoneuron differentiation of induced pluripotent stem cells from SOD1G93A mice. *PloS One* 2013;**8**(5):e64720.
148. Burkhardt MF, Martinez FJ, Wright S, et al. A cellular model for sporadic ALS using patient-derived induced pluripotent stem cells. *Mol Cell Neurosci* 2013;**56**:355–64.
149. Egawa N, Kitaoka S, Tsukita K, et al. Drug screening for ALS using patient-specific induced pluripotent stem cells. *Sci Transl Med* 2012;**4**(145):145ra104.
150. Nicaise C, Mitrecic D, Falnikar A, Lepore AC. Transplantation of stem cell-derived astrocytes for the treatment of amyotrophic lateral sclerosis and spinal cord injury. *World J Stem Cells* 2015;**7**(2):380–98.
151. Gao J, Coggeshall RE, Tarasenko YI, Wu P. Human neural stem cell-derived cholinergic neurons innervate muscle in motoneuron deficient adult rats. *Neuroscience* 2005;**131**(2):257–62.
152. Lepore AC, O'Donnell J, Kim AS, et al. Human glial-restricted progenitor transplantation into cervical spinal cord of the SOD1 mouse model of ALS. *PloS One* 2011;**6**(10):e25968.
153. Lepore AC, Rauck B, Dejea C, et al. Focal transplantation-based astrocyte replacement is neuroprotective in a model of motor neuron disease. *Nat Neurosci* 2008;**11**(11):1294–301.
154. Rizvanov AA, Guseva DS, Salafutdinov II, et al. Genetically modified human umbilical cord blood cells expressing vascular endothelial growth factor and fibroblast growth factor 2 differentiate into glial cells after transplantation into amyotrophic lateral sclerosis transgenic mice. *Exp Biol Med* 2011;**236**(1):91–8.
155. Rizvanov AA, Kiyasov AP, Gaziziov IM, et al. Human umbilical cord blood cells transfected with VEGF and L(1)CAM do not differentiate into neurons but transform into vascular endothelial cells and secrete neuro-trophic factors to support neuro-genesis-a novel approach in stem cell therapy. *Neurochem Int* 2008;**53**(6–8):389–94.
156. Hwang DH, Lee HJ, Park IH, et al. Intrathecal transplantation of human neural stem cells overexpressing VEGF provide behavioral improvement, disease onset delay and survival extension in transgenic ALS mice. *Gene Ther* 2009;**16**(10):1234–44.
157. Park S, Kim HT, Yun S, et al. Growth factor-expressing human neural progenitor cell grafts protect motor neurons but do not ameliorate motor performance and survival in ALS mice. *Exp Mol Med* 2009;**41**(7):487–500.
158. Group A. ALSUntangled update 4: investigating the XCell-Center. *Amyotroph Lateral Scler Off Publ World Fed Neurol Res Group Motor Neuron Dis* 2010;**11**(3):337–8.
159. Group A. ALSUntangled No. 27: Precision Stem Cell. *Amyotroph Lateral Scler Frontotemp Degen* 2015:1–4.

CHAPTER

10

Gene Therapy for Spinal Muscular Atrophy

M.R. Miller, E.Y. Osman and C.L. Lorson

University of Missouri, Columbia, MO, United States

INTRODUCTION

In the past 20 years, spinal muscular atrophy (SMA) has progressed from a disease of unknown cause to a disease that is generating excitement with multiple ongoing clinical trials for disease-specific therapeutics. This rapid advancement has been beneficial not only to the SMA community, but also to researchers in other fields. SMA is sometimes considered a "model disease" as it is amenable to therapeutic interventions that are adaptable to other disease contexts. Several factors have contributed to the advances in SMA therapy development, including: (1) SMA is monogenic, caused by mutation or deletion of *SMN1*; (2) a nearly identical copy gene, *SMN2* is present in all SMA patients; and (3) SMA arises from low levels—not the complete absence—of the survival motor neuron (SMN) protein. The presence of both *SMN1* and *SMN2* provides two different targets for SMA therapy. Since SMA is caused by loss of *SMN1*, viral-based gene replacement of SMN is an attractive strategy. Alternatively, because *SMN2* is only functionally distinct from *SMN1* in its splicing patterns, antisense oligonucleotides (ASOs) or other splicing modulators designed to promote the proper splicing of *SMN2* can also result in the production of full-length SMN protein. While strategies modifying the rate of SMN2 transcription or stability have been tested, this review will focus on the development, optimization, and clinical advancement of viral and ASO-based gene therapies for SMA.

ANIMAL MODELS

Due to the complex interplay between gene therapy vectors, tissues, and the immune system, developing strategies for gene therapy must be performed on an organismal level. While less complex models such as fruit flies (*Drosophila melanogaster*), nematodes (*Caenorhabditis elegans*), and zebrafish (*Danio rerio*) have been helpful in understanding the molecular roles of SMN, most therapeutic development has been performed in transgenic mouse models. These animal models have been essential as a means of confirming the SMA disease mechanism as well as testing potential therapeutics. *SMN2* is unique to humans;[1] thus to truly model the disease, transgenic animal models must be developed in the lab. When animal experiments first began in SMA, the debate on whether *SMN1* was the SMA causative gene had been mostly settled. However, after several iterations of genetic manipulation, researchers developed models that displayed characteristic SMA features and further solidified the link between *SMN1/2* and SMA. The first mouse SMN (*mSmn*) knockout demonstrated that complete loss of *mSmn* is embryonic lethal, highlighting the critical importance of SMN in fulfilling basic cellular needs.[2] This experiment failed to create a model of SMA because knockout mice lacked the low levels of SMN protein present in SMA patients that support the viability of the embryo. Integration of the full *SMN2* gene into an embryonic lethal *mSmn* knockout model demonstrated that animals with one to two copies of *SMN2* fell into two groups: most of the SMA animals died within 6 hours of birth while 25% lived up to 6 days.[3] This mouse model represents a very severe form of SMA, and only a limited number of interventions can be carried out on such a short-lived animal. In contrast, eight integrated copies of SMN2 yielded healthy pups that lived in excess of 18 months.[3,4] Another research group designed a mouse model by integrating a larger segment of chromosome 5q13 including SMN2 as well as parts of its neighboring regions.[5] This resulted in mice with variable disease severity: mice termed "type 1" died by day 10, "type 2" died within 2–4 weeks, and "type 3" were long-lived and suffered only from a blunt tail.[5]

To summarize the previous experiments, loss of mSmn is embryonic lethal, the addition of two copies of SMN2 rescued embryonic lethality but the mice die early in life, and eight copies yielded complete rescue. Thus, it was hypothesized that more copies of SMN2 would result in progressively milder symptoms, possibly providing a spectrum of SMA mouse models available to study. Osborne et al. produced an SMN2 allelic series of transgenic mice, replacing the mSmn locus with either 0, 1, 2, 3, 4, 5, 6, or 8 copies of SMN2.[6] However, this allelic series did not demonstrate a strong linear effect between SMN2 and gross disease phenotype. The line receiving two copies of SMN2 was embryonic lethal, whereas addition of three or more copies produced mice that were very healthy with only a moderate behavioral and physiological semblance to SMA even though SMN was restored only to a modest 30–40% of wild-type protein levels.[6] These SMN2-based models elucidated several concepts: (1) enhancing SMN2 may yield profound results in vivo; (2) increasing SMN levels to ~30–40% of normal may be sufficient for phenotypic improvements; and (3) titrating levels of SMN may be difficult using transgenic means.

A number of additional mouse models have been developed for SMA. The most widely utilized mouse is known as the SMNΔ7 model. This model was created to settle the debate as to whether the truncated SMNΔ7 protein was simply functionally insufficient, or whether it carried a dominant negative effect, exacerbating SMA symptoms. Toward this end, researchers added many copies of the SMNΔ7 cDNA into a severe $mSMN^{-/-}$, $SMN2^{+/+}$ model and found that SMNΔ7 addition extended the model's lifespan from ~7 days to ~14 days. This lifespan provides adequate time for testing therapeutics before and after symptom onset. Around 8 days of age, the mice experience a sharp decline in muscle strength followed by a very predictable and reproducible death curve, which has been ideal for detecting therapeutic effects. The SMNΔ7 model has been instrumental in advancing the ASO and AAV-based therapies currently in clinical trials. While many other severe, intermediate, and mild SMA models have been developed, they have mostly been used in very early preclinical testing and will not be discussed at length here.

STRATEGIES OF GENE THERAPY IN SPINAL MUSCULAR ATROPHY

The genetic situation of SMA allows for two possible routes of therapy. Early SMA research recognized that *SMN2* is a natural therapeutic target. Thus, researchers have aimed at increasing *SMN2* expression, stability, and splicing inclusion of exon 7. Many therapeutic strategies, including bifunctional RNAs, HDAC inhibitors, small molecules, repurposed drugs, aminoglycosides, and ASOs, have been used toward this end with varying degrees of risk and efficacy.

The second route for therapy is based on delivering a replacement for the patient's missing or nonfunctional *SMN1* gene. This is accomplished using a viral vector to deliver the gene of interest so that it can be continually expressed. Safe and effective delivery of viral-based therapeutics can be challenging however, this strategy has quickly developed from a mouse-based preclinical investigation to a clinical trial. Further, delivery of SMN as well as other genes of interest has allowed researchers to probe the pathology and physiology of SMA.

Antisense Oligonucleotides

The genetic context of SMA provides unique opportunities for therapeutic development, many of which are focused on the presence of *SMN2*. This gene is an appealing target because it is present in all patients, it is often found in multiple copies, and it is capable of encoding a wild-type SMN protein. This gene fails to rescue the loss of *SMN1* because it has the tendency to exclude exon 7 during mRNA splicing; consequently reducing the amount of full-length SMN it produces.[7] Shifting the splicing preferences of the *SMN2* pre-mRNA such that it includes exon 7 could, in theory, provide sufficient SMN to replace the function of patients' missing *SMN1* genes. This strategy is highly attractive because modulating the native *SMN2* pre-mRNA to produce full-length SMN protein makes it less likely that SMN will accumulate to superphysiologic levels. There is no known toxicity related to overexpression of SMN,[8] but considering the irreversible nature of gene therapies, the utilization of a native SMN transcript provides an additional layer of spatiotemporal regulation that may increase safety. One such therapeutic approach utilizes ASO sequences that target gene-specific splicing modulators.

ASOs are short nucleic acid sequences able to bind a target sequence with a high degree of specificity. When designing specific ASOs for SMA therapeutics, several criteria are considered, including their ability to increase exon retention with high efficiency and specificity, low toxicity, long-lasting resistance to cellular degradation, and high permeability to target cells like motor neurons within the central nervous system (CNS). In SMA, a variety of sequences have been targeted by various chemistries as illustrated in Fig 10.1.

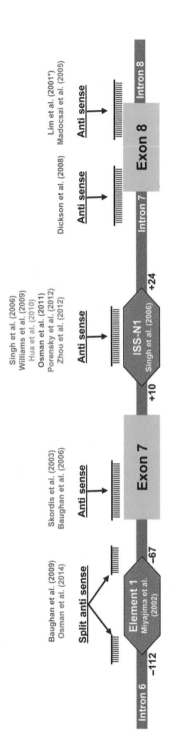

FIGURE 10.1 Schematic of antisense oligonucleotides with various chemical modifications utilized in spinal muscular atrophy (SMA) research. The intron and exon regions (blue and green, respectively) within the SMN2 exon 7 region are indicated, as well as potent repressors of SMN2 exon 7 splicing, Element 1 and ISS-N1 (red). Relative positions of previously published antisense oligonucleotides that increase SMN2 exon 7 splicing are shown above each region.

Alternative splicing plays an important role in gene expression and abnormal splicing has been recognized as a cause of an increasing number of diseases. In SMA, nearly 90% of SMN2-derived transcripts encode a truncated protein that lacks the final coding exon 7.[7,9] Pre-mRNA splicing of SMN exon 7 is regulated by a variety of factors that bind to pre-mRNA sequences called exonic or intronic splicing enhancers (ESEs) and silencers (ESSs).[10] The role of these enhancer and silencer motifs is to promote or repress the splice-site decision. It has been previously shown that there are a multitude of alternative splice signals within and flanking SMN exon 7.

Several ASOs have been designed utilizing different chemistry backbones[11] and studied in cell and animal models. Two research groups used 2'-O-methyl (2'-OMe) and 2'-O-methoxyethyl (2'-MOE) ASOs targeting the putative ESE abolished by the C to T transition, combined with a tailed splice-factor recruiting ASO.[12,13] Researchers have also attempted to redirect positive splicing factors to the vicinity of exon 7 using an RNA molecule that can bind to the SMN2 pre-mRNA and recruit splicing factors.[13] These sequences are known as bifunctional RNAs due to the presence of these two functional domains: an RNA sequence that is complementary to the target RNA and inhibits its function (e.g., SMN exon 7, intron 6); and an RNA segment that serves as a sequence-specific binding platform for cellular splicing factors. In SMA, bifunctional RNAs have been designed to inhibit intronic repressors through the antisense sequence and to recruit SR proteins promoting specific ESEs.

Two in vivo studies have examined the efficacy of bifunctional RNAs.[14,15] In both studies, the bifunctional RNA was able to increase the production of SMN in the CNS. The treated animals exhibited improved motor function, and lifespan was increased by 2–7 days. Nevertheless, due to the unstable chemistry of native RNA, bifunctional RNAs will need to be delivered continuously. Gene therapy vectors expressing bifunctional RNAs driven by a ubiquitous promoter have the potential to overcome this limitation by providing a continuous source of therapeutic RNA.[16]

The mode of action with morpholino-modified oligonucleotides resembles that of ASOs. Morpholino oligonucleotides are synthetic molecules similar in structure to natural nucleic acids, made by substituting morpholine and phosphorodiamidate for ribose and phosphate in the backbone of the molecule. Due to these modifications, morpholinos exhibit greater stability in vivo and do not activate toll-like receptors. Like RNA-based ASOs, morpholinos have been used to block sites on RNA to obstruct cellular processes by binding to its selected target site and hindering access of RNA processing machinery. By blocking sites involved in splicing pre-mRNA, morpholinos can be used to modify normal splicing events. Morpholino-based ASOs targeting different intronic splicing repressors have been shown to be effective at modulating SMN2 splicing in vivo.[17–19]

Some strategies have utilized ASOs to target the intron 7/exon 8 junction with the goal of reducing recognition of the exon 8 3'-splice-site in favor of the 3'-splice-site of exon 7, increasing exon 7 inclusion.[20,21] Additionally, employing a bifunctional ASO that recruits negative splice effectors (hnRNP A1/A2) can further facilitate splice suppression.[22,23] ASO-mediated therapy can also be enhanced by incorporating the ASO into a small nuclear ribonucleic protein-like (snRNP-like) complex. The U7 snRNP is normally involved in the maturation of replication-dependent histone mRNAs, using its guide RNA to target the mRNA for cleavage. This RNA has two regions of interest, one that mediates binding with RNA regulatory proteins called Smith Core proteins (Sm)/Lsm and one that complements the histone pre-mRNA (guide sequence). Altering the Sm/Lsm sequence of the U7 snRNP RNA creates a particle that can bind to the mRNA and competitively inhibit wild-type U7 snRNP activity. When combined with an alternative guide sequence, the new snRNP can inhibit the binding of other RNA binding proteins to a target RNA.[24] For instance, a U7 snRNA containing a sequence complementary to the 3'-splice-site of SMN exon 8 can prevent splicing factors from accessing this portion of the SMN pre-mRNA, increasing the rate of exon 7 inclusion in the mature mRNA.[25]

Though each of these strategies demonstrated the ability to increase exon 7 inclusion in the mature SMN mRNA, vector-based delivery of splicing modulating RNAs has not advanced into animal models. However, treatments employing ASOs have applications, not only for SMA therapy, but also for a number of diseases where altering splicing may be therapeutic, such as Duchenne muscular dystrophy, Huntington's disease, and myotonic dystrophy.

Utilizing Viruses for Spinal Muscular Atrophy Research

Viral vectors have been utilized for a variety of purposes in SMA research. For SMA therapeutics, viral delivery has provided a means to achieve stable expression of otherwise transient RNA- or protein-based therapies. Researchers have also used viruses to deliver SMN or other genes of interest in order to investigate biological properties of SMA, providing an inexpensive alternative to creating transgenic animals. This approach can provide insight into the spatiotemporal requirements of SMN and may inform SMA treatment development in the future.

Creating Transgenics

Viral-based knockdown of SMN has been able to compensate where transgenic models are unavailable. For instance, preclinical testing and assessment of biomarkers for SMA would benefit from an *SMN2*-based large-animal model. Pigs are an attractive candidate for this purpose due to their physiological similarities to humans,[26] and initial studies have shown the feasibility of an SMN2-based swine model.[26,27] As an alternative approach, self-complementary AAV9 (scAAV9) has been used to deliver shRNAs to knock down the expression of the endogenous SMN.[28] SMN expression in pigs was reduced to a clinically relevant level throughout the CNS, and these animals exhibited SMA-like symptoms such as hindlimb weakness and atrophy. Animals became nonambulant within weeks following delivery of the shRNA vector. This model was subsequently used to examine distribution and efficacy of a scAAV9-SMN vector.[28]

Determining the Spatial Requirements for SMN

The mechanism through which the ubiquitously expressed SMN protein can cause a seemingly motor-neuron-specific disease has been a pressing question within the SMA community. Many have wondered whether this disease is actually specific to motor neurons or whether inadequate SMN levels adversely affect other types of neurons or peripheral tissues. Interestingly, SMN depletion in motor neurons does not cause an overt SMA phenotype,[29] and expression exclusively in motor neurons is not sufficient to rescue the disease in mice.[30–34] Other transgenic models demonstrated that pan-neuronal restoration of SMN is much more effective than muscle-specific restoration.[35] These findings led to two questions: which other tissue types are affected in SMA and which cell types contribute to the degradation of motor neurons? Viral gene delivery has allowed researchers to probe disease progression in well-established transgenic models. Recently SMN was delivered specifically to astrocytes in the SMNΔ7 and SMN$^{2B/-}$ models[36] using a scAAV vector delivering SMN under the control of a GFAP promoter. Specificity was ensured by including binding sites for neuronally expressed miRNAs in the 3′-UTR.[36] Astrocyte-specific expression of SMN in the SMNΔ7 model led to a two-fold improvement in lifespan. The milder SMN$^{2B/-}$ animals all survived past 80 days while their untreated counterparts lived for ~35 days.[36] This is one example of how researchers have been able to utilize gene therapy technology to investigate SMA pathogenesis.

Delivering Neuroprotectants

Viral delivery has been used to test therapeutic candidates that address the symptoms of SMA rather than the molecular cause of the disease. Two candidates for this approach have come from the field of amyotrophic lateral sclerosis.[37,38] Because neurotrophic factors protect motor neurons under stressful conditions,[39] researchers investigated whether these proteins could alter the disease trajectory of SMA. Supplementing SMA mice with the neuroprotectant cardiotrophin 1 ameliorated the phenotypes associated with SMA, though overall extension of life span was modest.[40] Another group utilized insulin-like growth factor-1 to treat a mild SMA mouse model. While this therapy yielded a reduction in motor neuron death, it did not produce a phenotypic improvement.[41]

Replacing SMN1 in Spinal Muscular Atrophy

As SMA is monogenic, vector-mediated gene replacement has been an intriguing area of research for over a decade. The initial proof-of-concept experiments in SMA involved the development of an adenovirus vector that expressed the complete SMN cDNA. In cell-based models, expression of *SMN1* restored SMN protein levels, increased the levels of SMN-associated factors such as Gemin proteins, and dramatically increased the nuclear foci referred to as SMN-positive "gems."[42] However, this vector was not examined in an animal model of disease to test for efficacy.[42]

The first in vivo experiments followed shortly thereafter using a lentiviral platform.[43] Utilizing the natural tropism of the rabies G (Rab-G) protein for neuronal populations, the investigators incorporated a ubiquitously expressed SMN transgene into a Rab-G pseudotyped equine infectious anemia virus. In SMA patient fibroblasts, the vector dramatically elevated levels of SMN and SMN-positive gems. For delivery of the vector to the mouse nervous system, intramuscular (i.m.) injections were performed into the hindlimb gastrocnemius, diaphragm, intercostal, facial, and tongue muscles. The goal of this route was to transduce motor neurons innervating these muscles through retrograde transport, the process responsible for shuttling cellular cargoes from the nerve terminal to the soma.[44] Vector-derived SMN expression was present in spinal cord motor neurons following i.m. injections, and total motor neuron numbers were increased in lenti-SMN-treated animals. These improvements were met with only a mild effect on the gross phenotype. Lenti-SMN extended survival of severe SMA mice from ~13 days to ~18 days while a control virus expressing LacZ extended survival to ~15 days.[43] Despite a fairly small sample size ($n=5$), each of these improvements was statistically significant.[43] Though the life extension was modest, this work provided a powerful proof-of-concept demonstrating that SMA was a candidate for gene therapy. This allowed the field to optimize vector type, delivery, and timing for SMA therapeutics.

Adeno-associated viral vectors have yielded substantial improvements in disease progression in animal models of SMA. When delivered directly into the CNS via a combination of intracerebroventricular and lumbar intrathecal injections, ssAAV8 and scAAV8 vectors expressing SMN were able to significantly improve the phenotype of a severe model of SMA. ssAAV8 increased median survival to 50 days of age, while scAAV8 increased survival to 150 days and prevented motor neuron loss until close to endpoint.[45]

Intravenous (i.v.) delivery is an attractive injection method due to the minimal risks associated with the procedure as well as its potential to reach the entire body. However, the vasculature of the brain is not permeable to most macromolecules, viruses, and bacteria, preventing them from reaching the brain from the bloodstream. This blood–brain barrier (BBB), therefore, presents a significant obstacle for most i.v. administered gene therapy vectors. One AAV serotype, scAAV9, has been shown to efficiently cross the BBB in multiple species, including mice, adult cats, and nonhuman primates (NHPs), allowing for CNS transduction following intravenous injection (Fig. 10.2).[28,46–50] Due to its ability to transit the BBB,

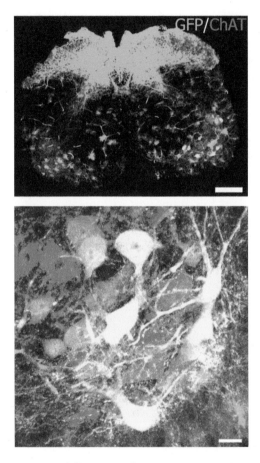

FIGURE 10.2   Intravenous delivery transduces motor neurons in a murine spinal cord. Intravenous injections into postnatal day 1 mice with a self-complementary AAV9 vector containing a chicken β-actin promoter driving GFP expression results in robust gene expression in the spinal cord. Animals received 3×10^{11} vector genomes and were sacrificed 2 weeks postinjection. Spinal cords were sectioned on a vibratome into 40-mm-thick sections and stained for GFP (green) and the motor neuron marker choline acetyltransferase (ChAT). Yellow indicates colocalization. The high-magnification image is a compressed z-stack of a lumbar motor neuron pool. Whole section scale bar=100μm. MN cluster scale bar=20μm. *Source: Courtesy of Kevin Foust, PhD, Nationwide Children's Hospital, Columbus, OH.*

many SMA preclinical studies have employed i.v. AAV9.[47,51,52] In 2010, researchers intravenously delivered scAAV9 expressing SMN under the control of a ubiquitous (chicken β-actin or CB) promoter.[47] This therapy, scAAV9.CB.SMN, extended the life of SMNΔ7 mice from ~14 days to an unprecedented range of 90–250+ days.[47] Soon after, another group published similar results with a codon-optimized *SMN*.[50,51] Improvements in lifespan were associated with increased motor neuron survival and the preservation of innervated neuromuscular junctions (Fig. 10.3).

Although SMA is considered a motor neuron disease, many research groups have demonstrated defects in other organ systems, including hyperglycemia and glucose resistance, heart defects, and defects in muscle development.[53–55] Consistent with these observations, some preclinical studies observed an increased benefit when therapeutics were delivered to the periphery as well as to the CNS.[19,56] Thus, the robust efficacy observed with i.v. injection of scAAV9.CB.SMN may be due, in part, to correction of these non-CNS abnormalities. Due to the fact that these peripheral defects may be overshadowed by problems in the CNS, clinicians involved in clinical trials for SMA will need to be cognizant that these defects may become uncovered and represent a second hurdle that needs to be addressed.

Defining the Therapeutic Window of Opportunity

Transgenic models and viral expression of SMN are both valuable tools for probing temporal requirements of this protein in vivo. While AAV acts relatively quickly, in certain circumstances one must consider the delay between AAV injection and transgene expression. Once injected, viral

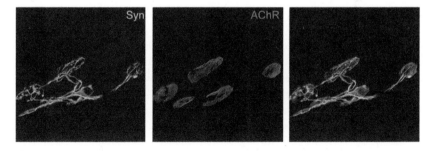

FIGURE 10.3   Mice treated with scAAV.CB.SMN have restored neuromuscular junctions. SMNΔ7 mice (mSMN$^{-/-}$; SMN2$^{+/+}$; Δ7$^{+/+}$) received 10^{11} vector genomes into the central nervous system (CNS) through intracerebroventricular injection on postnatal day 7. Animals were euthanized and preserved in PFA through cardiac infusion on postnatal day 12. Longissimus capitis muscles were isolated and stained; presynapse was stained using antineurofilament and antisynaptophysin (green) and postsynapse was stained with alpha bungarotoxin (red). Confocal images were taken at 40× magnification in 1-µm intervals and displayed in compressed z-stack form.

entry, gene transcription, and translation must occur before transgenes can begin to impact the course of disease. For AAV, there is a modest delay from the time of delivery to transgene expression for the two types of AAV: ~2 days for scAAV and ~9 days for ssAAV. In the context of human therapies, this delay is modest. However, within the context of the murine models of SMA, this delay was initially envisioned to be a significant hurdle since many of the severe SMA models live less than 14 days.[57]

The question of when to deliver therapeutics has been an important concern in the SMA field. It has become apparent that a relatively narrow window is available for therapeutic intervention in SMA models. Several studies utilizing various therapeutic modalities support the idea that 'earlier is better'.[17,32,46,47,50,58,59] Using a single i.v. delivery, scAAV9-SMN at postnatal day 2 (PND2) resulted in a near-complete rescue of the SMA mice. Delaying this injection until PND5 produced a dramatically reduced benefit, resulting in an average life span of only ~15 days. Delivery at PND10 failed to alter the disease course.[47] Using i.c.v. delivery of scAAV9-SMN, similar results were shown in which cohorts were treated at different time points ranging from PND2 to PND8. While in each cohort disease severity was significantly reduced, earlier treatment corresponded with a greater rescue of the phenotype.[58] Data from large-animal experiments support these results. In the shRNA-based pig model, neonatal swine were initially injected with the scAAV9-shRNA vector to induce disease and then subsequently treated with the scAAV9-SMN vector at either PND6 or when they became symptomatic (PND33-36). In nearly all phenotypic parameters examined, presymptomatic treatment resulted in a profound rescue of the SMA-like phenotype, whereas symptomatic treatment had a greatly reduced benefit.[28]

For gene therapy, biology may also provide hurdles separate from disease progression. Some evidence suggests that targeting motor neurons with AAV9 is particularly time-sensitive. In mice, neonatal treatment with intravenous AAV9 yields high motor neuron transduction, while transduction shifts toward astrocyte populations when delivered in older animals.[48] High levels of astrocyte transduction have been found in NHPs as well.[60]

CLINICAL TRIALS

Ionis-SMN$_{Rx}$

Ionis Pharmaceuticals, USA, and Biogen Idec have developed a proprietary molecule, Ionis-Smn$_{Rx}$. As of July 2014, this drug is in phase III clinical trials. Ionis-Smn$_{Rx}$ is a 2'-O-methoxyethyl-modified ASO targeting the ISS-N1 repressor site located in intron 7. Systemic administration of the drug in animals increased SMN2 exon 7 inclusion in a dose- and time-dependent manner in liver, kidney, and skeletal muscle, but not CNS tissues.[61] Administration of the ASO into the lateral ventricles of the brain resulted in a dose-dependent increase in exon 7 containing transcripts in motor neurons and other cells in the CNS.[62] The ASO was well tolerated at doses that promoted almost complete exon 7 inclusion.[62]

SMA type II and III patients enrolled in the Phase II SMN_{Rx} dose-escalation study were injected intrathecally every 6 months with 3, 6, 9, or 12 mg of the drug.[63] Similar trials in infants with SMA type I evaluated doses of 6 and 12 mg. Results in the SMA type I trial indicated an increased time of incident-free longevity.[63] Children with SMA type II and type III also saw benefits. Whereas the natural history for patients with Type II and III would predict a slow decline in strength tests over time, the patients enrolled in this trial have been gaining strength as measured by the Hammersmith Functional Motor Scale-Expanded, the 6-minute walk test, and the upper-limb module test for those patients who are nonambulatory.[64] Given the promising results of the Phase II trial, patients proceeded into an open label extension study in which all patients received a 12-mg dose (Clinicaltrials.gov ID NCT02052791, NCT01780246).[64] Phase III trials of Ionis-Smn$_{Rx}$ for both SMA type I and SMA type II/III will be randomized, double-blind, and sham procedure controlled.[65]

AveXis-ChariSMA

The development of scAAV9.CB.SMN, called ChariSMA, is being pioneered by AveXis Inc., the Sophia's Cure Foundation, and investigators at Nationwide Children's Hospital, Ohio under the direction of Dr. Jerry R. Mendell. AveXis Inc. has also developed a European subsidiary in anticipation of European trials.[66]

ChariSMA trials have already begun in the United States. A Phase I/II dose-escalation trial delivered scAAV9.CB.SMN through i.v. injection to Type I patients at 6–9 months of age (Clinicaltrials.gov ID: NCT02122952). Certain criteria preclude patients from participating, including the need for ventilator support and the presence of high levels of neutralizing antibodies against AAV9. In 2014, the trial initiated with the treatment of a 6 month old SMA patient.[67] This was the first patient of the low-dose cohort receiving 6.7×10^{13} vg/kg (vector genomes per kilogram body weight), followed by an intermediate-dose cohort (2.0×10^{14} vg/kg) and a high-dose cohort (3.3×10^{14} vg/kg). After 6 months the group announced that the low-dose cohort had completed treatment and had no adverse effects.[67]

While the i.v.-based trial is ongoing, this group is preparing to test an alternate approach, delivering scAAV9-CB-SMN to the cerebrospinal fluid (CSF). This type of delivery was preclinically validated using i.c.v. injection in an SMA mouse model. At the highest dose evaluated, median lifespan of the mouse was increased from 18 days to 282 days.

Although substantial, this improvement in life span was probably not quite as robust as that observed in i.v. scAAV9-CB-SMN-treated animals. In 1-year-old cynomolgus macaques, CSF delivery was accomplished via intrathecal injections in the lumbar spinal cord. To improve transduction in the cervical cord, animals were placed in the Trendelenburg position, placing the animals so that their bodies were tilted head-down, increasing CSF flow toward the brain. This delivery method successfully transduced 55% of cervical motor neurons and 80% of lumbar motor neurons.[68]

CLOSING REMARKS

While advances from a patient perspective can never come soon enough, the SMA field has been extremely fortunate to advance to a stage where multiple exciting therapeutics are being examined. Without the support of the SMA community at large and the commitment from grass roots foundations, much of this work would not be possible. Additionally, while the genetics of SMA are complex, the fact that the disease is monogenic and low levels of functional SMN are present creates an ideal environment for vector-mediated gene replacement. Clinical trials are not merely confirmatory studies of the preclinical work. Rather, the ongoing trials are pioneering first-in-human experiments. A certain level of trepidation is required. However, the basic and translational research underpinning the current gene therapy trials in SMA provides a solid foundation of hope, not just for SMA patients, but for patients suffering from many other rare diseases that could benefit from a highly targeted vector-mediated gene replacement strategy.

Acknowledgments

We would like to thank members of the Lorson lab for assisting in the review of the manuscript and Dr. Kevin Foust for the unpublished immunofluorescence images. This work was supported by an NIH Training Grant (M.R.M.) T32 GM0008396, and a research grant from the Muscular Dystrophy Association (C.L.L.).

References

1. Rochette CF, Gilbert N, Simard LR. SMN gene duplication and the emergence of the SMN2 gene occurred in distinct hominids: SMN2 is unique to Homo sapiens. *Hum Genet* 2001;**108**(3):255–66.
2. Schrank B, Gotz R, Gunnersen JM, et al. Inactivation of the survival motor neuron gene, a candidate gene for human spinal muscular atrophy, leads to massive cell death in early mouse embryos. *Proc Natl Acad Sci U S A* 1997;**94**(18):9920–5.
3. Monani UR, Sendtner M, Coovert DD, et al. The human centromeric survival motor neuron gene (SMN2) rescues embryonic lethality in Smn(−/−) mice and results in a mouse with spinal muscular atrophy. *Hum Mol Genet* 2000;**9**(3):333–9.

4. Monani UR, Coovert DD, Burghes AH. Animal models of spinal muscular atrophy. *Hum Mol Genet* 2000;**9**(16):2451–7.

5. Hsieh-Li HM, Chang JG, Jong YJ, et al. A mouse model for spinal muscular atrophy. *Nat Genet* 2000;**24**(1):66–70.

6. Osborne M, Gomez D, Feng Z, et al. Characterization of behavioral and neuromuscular junction phenotypes in a novel allelic series of SMA mouse models. *Hum Mol Genet* 2012;**21**(20):4431–47.

7. Lorson CL, Hahnen E, Androphy EJ, Wirth B. A single nucleotide in the SMN gene regulates splicing and is responsible for spinal muscular atrophy. *Proc Natl Acad Sci USA* 1999;**96**(11):6307–11.

8. Goulet BB, McFall ER, Wong CM, Kothary R, Parks RJ. Supraphysiological expression of survival motor neuron protein from an adenovirus vector does not adversely affect cell function. *Biochem Cell Biol* 2013;**91**(4):252–64.

9. Monani UR, Lorson CL, Parsons DW, et al. A single nucleotide difference that alters splicing patterns distinguishes the SMA gene SMN1 from the copy gene SMN2. *Hum Mol Genet* 1999;**8**(7):1177–83.

10. Bebee TW, Gladman JT, Chandler DS. Splicing regulation of the survival motor neuron genes and implications for treatment of spinal muscular atrophy. *Front Biosci (Landmark Ed)* 2010;**15**:1191–204.

11. Porensky PN, Burghes AH. Antisense oligonucleotides for the treatment of spinal muscular atrophy. *Hum Gene Ther* 2013;**24**(5):489–98.

12. Cartegni L, Krainer AR. Disruption of an SF2/ASF-dependent exonic splicing enhancer in SMN2 causes spinal muscular atrophy in the absence of SMN1. *Nat Genet* 2002;**30**(4):377–84.

13. Skordis LA, Dunckley MG, Yue B, Eperon IC, Muntoni F. Bifunctional antisense oligonucleotides provide a trans-acting splicing enhancer that stimulates SMN2 gene expression in patient fibroblasts. *Proc Natl Acad Sci USA* 2003;**100**(7):4114–9.

14. Baughan TD, Dickson A, Osman EY, Lorson CL. Delivery of bifunctional RNAs that target an intronic repressor and increase SMN levels in an animal model of spinal muscular atrophy. *Hum Mol Genet* 2009;**18**(9):1600–11.

15. Osman EY, Yen PF, Lorson CL. Bifunctional RNAs targeting the intronic splicing silencer N1 increase SMN levels and reduce disease severity in an animal model of spinal muscular atrophy. *Mol Ther J Am Soc Gene Ther* 2012;**20**(1):119–26.

16. Baughan T, Shababi M, Coady TH, Dickson AM, Tullis GE, Lorson CL. Stimulating full-length SMN2 expression by delivering bifunctional RNAs via a viral vector. *Mol Ther J Am Soc Gene Ther* 2006;**14**(1):54–62.

17. Porensky PN, Mitrpant C, McGovern VL, et al. A single administration of morpholino antisense oligomer rescues spinal muscular atrophy in mouse. *Hum Mol Genet* 2012;**21**(7):1625–38.

18. Andreassi C, Jarecki J, Zhou J, et al. Aclarubicin treatment restores SMN levels to cells derived from type I spinal muscular atrophy patients. *Hum Mol Genet* 2001;**10**(24):2841–9.

19. Osman EY, Miller MR, Robbins KL, et al. Morpholino antisense oligonucleotides targeting intronic repressor Element1 improve phenotype in SMA mouse models. *Hum Mol Genet* 2014;**23**(18):4832–45.

20. Lim SR, Hertel KJ. Modulation of survival motor neuron pre-mRNA splicing by inhibition of alternative 3′ splice site pairing. *J Biol Chem* 2001;**276**(48):45476–83.

21. Madocsai C, Lim SR, Geib T, Lam BJ, Hertel KJ. Correction of SMN2 Pre-mRNA splicing by antisense U7 small nuclear RNAs. *Mol Ther J Am Soc Gene Ther* 2005;**12**(6):1013–22.

22. Gendron D, Carriero S, Garneau D, et al. Modulation of 5′ splice site selection using tailed oligonucleotides carrying splicing signals. *BMC Biotechnol* 2006;**6**:5.

23. Dickson A, Osman E, Lorson CL. A negatively acting bifunctional RNA increases survival motor neuron both in vitro and in vivo. *Hum Gene Ther* 2008;**19**(11):1307–15.

24. Schumperli D, Pillai RS. The special Sm core structure of the U7 snRNP: far-reaching significance of a small nuclear ribonucleoprotein. *Cell Mol Life Sci* 2004;**61**(19–20):2560–70.
25. Geib T, Hertel KJ. Restoration of full-length SMN promoted by adenoviral vectors expressing RNA antisense oligonucleotides embedded in U7 snRNAs. *PloS One* 2009;**4**(12):e8204.
26. Lorson MA, Spate LD, Prather RS, Lorson CL. Identification and characterization of the porcine (Sus scrofa) survival motor neuron (SMN1) gene: an animal model for therapeutic studies. *Dev Dyn Off Publ Am Assoc Anatomists* 2008;**237**(8):2268–78.
27. Lorson MA, Spate LD, Samuel MS, et al. Disruption of the survival motor neuron (SMN) gene in pigs using ssDNA. *Transgenic Res* 2011;**20**(6):1293–304.
28. Duque SI, Arnold WD, Odermatt P, et al. A large animal model of spinal muscular atrophy and correction of phenotype. *Ann Neurol* 2015;**77**(3):399–414.
29. Park G-H, Maeno-Hikichi Y, Awano T, Landmesser LT, Monani UR. Reduced survival of motor neuron (SMN) protein in motor neuronal progenitors functions cell autonomously to cause spinal muscular atrophy in model mice expressing the human centromeric (SMN2) gene. *J Neurosci* 2010;**30**(36):12005–19.
30. Gogliotti RG, Quinlan KA, Barlow CB, Heier CR, Heckman CJ, DiDonato CJ. Motor Neuron rescue in spinal muscular atrophy mice demonstrates that sensory-motor defects are a consequence, not a cause, of motor neuron dysfunction. *J Neurosci* 2012;**32**(11):3818–29. 11/18/received 01/18/revised 01/25/accepted.
31. Paez-Colasante X, Seaberg B, Martinez TL, Kong L, Sumner CJ, Rimer M. Improvement of Neuromuscular synaptic phenotypes without enhanced survival and motor function in severe spinal muscular atrophy mice selectively rescued in motor neurons. *PloS One* 2013;**8**(9):e75866.
32. Lutz CM, Kariya S, Patruni S, et al. Postsymptomatic restoration of SMN rescues the disease phenotype in a mouse model of severe spinal muscular atrophy. *J Clin Invest* 2011;**121**(8):3029–41.
33. Martinez TL, Kong L, Wang X, et al. Survival motor neuron protein in motor neurons determines synaptic integrity in spinal muscular atrophy. *J Neurosci* 2012;**32**(25):8703–15.
34. Lee AJH, Awano T, Park G-H, Monani UR. Limited phenotypic effects of selectively augmenting the SMN protein in the neurons of a mouse model of severe spinal muscular atrophy. *PloS One* 2012;**7**(9):e46353.
35. Gavrilina TO, McGovern VL, Workman E, et al. Neuronal SMN expression corrects spinal muscular atrophy in severe SMA mice while muscle-specific SMN expression has no phenotypic effect. *Hum Mol Genet* 2008;**17**(8):1063–75.
36. Rindt H, Feng Z, Mazzasette C, et al. Astrocytes influence the severity of spinal muscular atrophy. *Hum Mol Genet* 2015;**24**(14):4094–102.
37. Bordet T, Schmalbruch H, Pettmann B, et al. Adenoviral cardiotrophin-1 gene transfer protects pmn mice from progressive motor neuronopathy. *J Clin Invest* 1999;**104**(8):1077–85. 01/11/received 09/10/accepted.
38. Kaspar BK, Llado J, Sherkat N, Rothstein JD, Gage FH. Retrograde viral delivery of IGF-1 prolongs survival in a mouse ALS model. *Science* 2003;**301**(5634):839–42.
39. Mattson MP. Apoptosis in neurodegenerative disorders. *Nat Rev Mol Cell Biol* 2000;**1**(2):120–30.
40. Lesbordes JC, Cifuentes-Diaz C, Miroglio A, et al. Therapeutic benefits of cardiotrophin-1 gene transfer in a mouse model of spinal muscular atrophy. *Hum Mol Genet* 2003;**12**(11):1233–9.
41. Tsai LK, Chen YC, Cheng WC, et al. IGF-1 delivery to CNS attenuates motor neuron cell death but does not improve motor function in type III SMA mice. *Neurobiol Dis* 2012;**45**(1):272–9.
42. DiDonato CJ, Parks RJ, Kothary R. Development of a gene therapy strategy for the restoration of survival motor neuron protein expression: implications for spinal muscular atrophy therapy. *Hum Gene Ther* 2003;**14**(2):179–88.

43. Azzouz M, Le T, Ralph GS, et al. Lentivector-mediated SMN replacement in a mouse model of spinal muscular atrophy. *J Clin Invest* 2004;**114**(12):1726–31. 08/04/received 10/18/accepted.
44. Griffin JW, Watson DF. Axonal transport in neurological disease. *Ann Neurol* 1988;**23**(1):3–13.
45. Passini MA, Bu J, Roskelley EM, et al. CNS-targeted gene therapy improves survival and motor function in a mouse model of spinal muscular atrophy. *J Clin Invest* 2010;**120**(4):1253–64.
46. Duque S, Joussemet B, Riviere C, et al. Intravenous administration of self-complementary AAV9 enables transgene delivery to adult motor neurons. *Mol Ther* 2009;**17**(7):1187–96.
47. Foust KD, Wang X, McGovern VL, et al. Rescue of the spinal muscular atrophy phenotype in a mouse model by early postnatal delivery of SMN. *Nat Biotechnol* 2010;**28**(3):271–4.
48. Foust KD, Nurre E, Montgomery CL, Hernandez A, Chan CM, Kaspar BK. Intravascular AAV9 preferentially targets neonatal neurons and adult astrocytes. *Nat Biotechnol* 2009;**27**(1):59–65.
49. Gray SJ, Matagne V, Bachaboina L, Yadav S, Ojeda SR, Samulski RJ. Preclinical differences of intravascular AAV9 delivery to neurons and glia: a comparative study of adult mice and nonhuman primates. *Mol Ther J Am Soc Gene Ther* 2011;**19**(6):1058–69.
50. Valori CF, Ning K, Wyles M, et al. Systemic delivery of scAAV9 expressing SMN prolongs survival in a model of spinal muscular atrophy. *Sci Transl Med* 2010;**2**(35) 35ra42.
51. Dominguez E, Marais T, Chatauret N, et al. Intravenous scAAV9 delivery of a codon-optimized SMN1 sequence rescues SMA mice. *Hum Mol Genet* 2011;**20**(4):681–93.
52. Glascock JJ, Osman EY, Coady TH, Rose FF, Shababi M, Lorson CL. Delivery of therapeutic agents through intracerebroventricular (ICV) and intravenous (IV) injection in mice. *J Vis Exp* 2011;**56**:2968.
53. Bowerman M, Swoboda KJ, Michalski JP, et al. Glucose metabolism and pancreatic defects in spinal muscular atrophy. *Ann Neurol* 2012;**72**(2):256–68.
54. Shababi M, Habibi J, Yang HT, Vale SM, Sewell WA, Lorson CL. Cardiac defects contribute to the pathology of spinal muscular atrophy models. *Hum Mol Genet* 2010;**19**(20):4059–71.
55. Hamilton G, Gillingwater TH. Spinal muscular atrophy: going beyond the motor neuron. *Trends Mol Med* 2013;**19**(1):40–50.
56. Hua Y, Liu YH, Sahashi K, Rigo F, Bennett CF, Krainer AR. Motor neuron cell-nonautonomous rescue of spinal muscular atrophy phenotypes in mild and severe transgenic mouse models. *Genes Dev* 2015;**29**(3):288–97.
57. Bebee TW, Dominguez CE, Chandler DS. Mouse models of SMA: tools for disease characterization and therapeutic development. *Hum Genet* 2012;**131**(8):1277–93.
58. Robbins KL, Glascock JJ, Osman EY, Miller MR, Lorson CL. Defining the therapeutic window in a severe animal model of spinal muscular atrophy. *Hum Mol Genet* 2014;**23**(17):4559–68.
59. Le TT, McGovern VL, Alwine IE, et al. Temporal requirement for high SMN expression in SMA mice. *Hum Mol Genet* 2011;**20**(18):3578–91.
60. Passini MA, Bu J, Richards AM, et al. Translational fidelity of intrathecal delivery of self-complementary AAV9-survival motor neuron 1 for spinal muscular atrophy. *Hum Gene Ther* 2014;**25**(7):619–30.
61. Hua Y, Vickers TA, Okunola HL, Bennett CF, Krainer AR. Antisense masking of an hnRNP A1/A2 intronic splicing silencer corrects SMN2 splicing in transgenic mice. *Am J Hum Genet* 2008;**82**(4):834–48.
62. Hua Y, Sahashi K, Hung G, et al. Antisense correction of SMN2 splicing in the CNS rescues necrosis in a type III SMA mouse model. *Genes Dev* 2010;**24**(15):1634–44.
63. Pharmaceuticals I. Isis pharmaceuticals reports data from ISIS-SMN Rx Phase 2 Study in infants with spinal muscular atrophy: PR Newswire 2015.

64. Pharmaceuticals I. Isis pharmaceuticals reports data from ISIS-SMN Rx in children with spinal muscular atrophy. PR Newswire 2015.
65. Pharmaceuticals I. Isis pharmaceuticals Earns $2.15M for advancing ISIS-SMN Rx in children with spinal muscular atrophy: PR Newswire 2015.
66. AveXis I, Bowman J. *AveXis announces the formation of its European subsidiary.* Dallas, Texas: AveXis EU, Ltd; 2015. <http://avexisinc.com/category/press-releases/>.
67. AveXis I, Bowman J. *AveXis zannounces the completion of dosing of the low dose cohort in US clinical trial for spinal muscular atrophy.* Dallas, Texas: AveXis EU, Ltd; 2014. <http://avexisinc.com/category/press-releases/>.
68. Meyer K, Ferraiuolo L, Schmelzer L, et al. Improving single injection CSF delivery of AAV9-mediated gene therapy for SMA: a dose-response study in mice and nonhuman primates. *Mol Ther J Am Soc Gene Ther* 2014;**23**(3):477–87.

11

Cellular Therapy for Spinal Muscular Atrophy: Pearls and Pitfalls

I. Faravelli[1,2] *and S. Corti*[1,2]

[1]University of Milan, Milan, Italy [2]IRCCS Foundation Ca' Granda
Ospedale Maggiore Policlinico, Milan, Italy

INTRODUCTION

Spinal muscular atrophies (SMAs) include a series of neuromuscular genetic diseases in which spinal motor neurons degenerate leading to progressive paralysis with proximal muscular atrophy.[1] The most frequent condition is caused by mutations of the survival motor neuron 1 (*SMN1*) gene on chromosome 5q13[2]. The human genome also contains a highly homologous gene on the same chromosome, called *SMN2*. It differs from *SMN1* by the presence of a C→T nucleotidic replacement in a critical position on exon 7.[3] This results in an altered splicing of the transcript leading to the production of a very small amount of SMN full-length functional protein.[4]

The 5q form of SMA represents the most common genetic cause of death during childhood affecting about 1 in 10,000 newborns.[5] SMA clinical manifestations include muscular weakness and atrophy especially affecting lower limbs and proximal segments.[6] Muscular wasting leads to severe difficulties in breathing and swallowing; death usually occurs due to cardiorespiratory failure.[6] The number of copies of the paralogous gene, *SMN2*, which can contribute only a small amount of SMN full-length protein, appears to determine the severity of the clinical picture.[7] The SMA type 1 form represents the most frequent and severe phenotype; affected children manifest the disorder within 5–6 months after birth, and never reach motor-developmental milestones.[1] They cannot sit or walk autonomously, and require early ventilator support. SMA type 2 displays a milder clinical phenotype; patients become symptomatic between 6 and 18 months from birth and are able to sit independently. SMAs 3 and 4 become overt later (the latter during adulthood), and while patients display symptoms related to muscular weakness and altered motor performance, their life expectancy is comparable to the general population.[8] There is no effective treatment clinically available for SMA patients, who require a 24-h assistance with specific sanitary devices.[9] A multidisciplinary environment in which clinicians, physiotherapists, psychologists, and speech therapists work collectively currently represents the most proper supportive approach.[9] Due to its high frequency, monogenic etiology, and severe clinical and social burden, SMA represents a valuable target for experimental therapeutic approaches.

POTENTIAL THERAPEUTIC EFFECTS OF STEM CELLS ON SPINAL MUSCULAR ATROPHY DISEASE MECHANISMS

Although the genetic origin of SMA has been revealed during the last decade, many questions still remain about downstream *SMN1* mutation molecular pathogenic mechanisms. The precise physiological role of SMN protein is the focus of several ongoing researches. SMN appears to complex with Gemin proteins forming a structure that cooperates into the biogenesis of uridine-rich small nuclear ribonuclear proteins. They are a crucial contributor to spliceosome formation and splicing process.[10] Other functions have been attributed to SMN including a role in axon maturation and myelination.[11,12] The rather selective vulnerability of motor neurons to SMN reduction remains a major issue to be addressed.

Several research groups have exploited the murine model to investigate pathological mechanisms involved in SMA development. One model has been obtained by introducing the human *SMN2* transgene on a Smn-null background.[13] The so-called SMNΔ7 mice mimic SMA1 human phenotype in terms of neuromuscular waste, with a rather selective involvement of spinal motor neurons.[13] However, SMNΔ7 animals display a systemic multiorgan involvement with distal tissue necrosis, which is more severe than it appears to be in humans, and it significantly affects mouse life span.[14] This and other differences are rather species-specific and need to be taken into account during the analyses of the experimental data. The discovery of human pluripotent stem cells and their use in disease-modeling-studies has led to the precious possibility of investigating pathogenic mechanisms peculiar to human pathology.[15]

Several studies on murine models and human pluripotent stem cells have highlighted widespread axonal and synaptic alterations in SMA motor neurons, including scarcity of dendrites and spines, impairment of calcium metabolism and dysfunctional remodeling ability.[16–18] SMA human pluripotent stem cells are able to differentiate into motor neurons but derived cells are characterized by a reduced life span; a crucial role of apoptosis is noted in the disease onset, with increased caspase-8 and-3 activation.[19]

Moreover, other cell types beside motor neurons have been recently implied in SMA onset. Interneurons and sensory neurons have been shown to be affected in the disease course thus suggesting the SMN loss could lead to a widespread perturbation of the neural network.[20]

Non-cell-autonomous mechanisms due to a toxic activation of glia cells are emerging as a central contributor in motor neuron death. Astrogliosis has been known to play a pathogenetic role in amyotrophic lateral sclerosis (ALS), and recent discoveries have pointed out that SMA astrocytes could trigger motor neuron degeneration, both by losing their trophic function and acquiring inflammatory features.[21,22] Overall, SMA pathogenesis appears to involve complex molecular mechanisms in multiple cell types highly interactive in creating a pathological microenvironment.

In this context, stem-cell-based therapeutic approaches could counteract several disease mechanisms at the same time. Pluripotent stem cells are able to give rise to mature motor neurons reacting to exogenous signals and thus replacing lost cells after engraftment, similar to what happens during physiological neurogenesis after injury.[23-26] Stem cells are also able to differentiate in glia lineage and substitute toxic astrocytes with healthy ones, which are able to provide neuroprotection to endogenous motor neurons.[27] Indeed, transplanting neural stem cells (NSCs) could be more effective than transplanting more mature cell subtypes. NSCs are more robust and intrinsically plastic; they can give rise to different subpopulations of healthy cells to replace and sustain endogenous ones.[28] They could contribute to alleviate the neural network impairment by building alternative circuitry and stimulating the formation of new synapses. Furthermore, the role of oligodendrocytes and myelination dysfunction in SMA pathogenesis as potential therapeutic targets has to be further investigated. As a consequence, stem cell-based therapeutic approaches could be clinically effective through a multifactorial action consisting of modulation of glial impairment, trophic sustainment of the endogenous cells, and enrichment of the unhealthy microenvironment.

Indeed, several methods exploit stem cells in order to give rise to de novo production and exogenous import of neurotrophic molecules to the diseased spinal cord. Human NSCs are able to express several trophic factors and can be manipulated to produce specific substances. Insulin-like growth factor-I (IGF-I), brain-derived neurotrophic factor (BDNF), glial-derived neurotrophic factor (GDNF), and vascular endothelial growth factor (VEGF) have been demonstrated to alleviate neurodegeneration and provide environmental support in a series of models of neurodegenerative diseases.[29] Moreover, the SMA nervous system showed an impairment in IGF1 signaling, which could represent a valid target for this approach in stem cell-mediated therapy.[30,31]

Overall, these considerations support the hypothesis that cell-based therapies may function in supporting SMA motor neurons, especially by providing a trophic and protective environment within the spinal cord, and thus counteracting the multifactorial pathogenetic mechanisms of neuronal death. Stem cell transplants could represent a complementary approach to SMA molecular and gene therapies. The latter are able to precisely target the loss of SMN restoring its function, but they could have a minor impact on later-diagnosed patients, while cell therapy could be useful in managing symptomatic phases of the disease. However, it is still worth considering that mechanisms downstream of stem cell activity are poorly understood, as are the molecular pathways underlying neural repair. Elucidating these aspects will provide invaluable insight into SMA pathogenesis and possibilities for therapy, which could be suitable also for other motor neuron diseases.

THE SELECTION OF CELL TYPES TO BE TRANSPLANTED

During the last decades, a few relevant preclinical studies have been conducted on the use of different cell subtypes for SMA therapy.[17,32–35] Cells have been screened for their ability to target the injured area, survive, and correctly engraft exerting a therapeutic action.

Many of these studies employed NSCs or derived motor neurons. A series of relevant preclinical studies have been reported (see Table 11.1). Hopefully, these results could open the way to clinical translation, even though more experiments will be necessary to accurately assess safety and effectiveness of stem cell based approaches.

TABLE 11.1 Preclinical Studies on Cell Transplantation for Spinal Muscular Atrophy Therapy

Study	Cell source	Animal model	Conclusions
Wyatt et al. 2011 [35]	Human embryonic stem cell-derived motor neuron progenitors	SMNΔ7 mice	Transplanted cells were able to survive and differentiate secreting active growth factors. It resulted in an increase of the number of rescued endogenous cells
Corti et al.[17]	Human iPSCs derived motor neurons	SMNΔ7 mice	Treated SMA mice presented an increased survival and motor performance
Corti et al. 2008 [32]	Murine ALDH(hi) SSC(lo) NSCs	SMNΔ7 mice	SMA mice showed ameliorated functional (neuromuscular function and life span) and histological features
Corti et al. 2010 [33]	Murine embryonic stem cell-derived NSCs	SMNΔ7 mice	Cell transplants could be combined with pharmacological therapy to select suitable cells enhancing the beneficial effects

SMA, spinal muscular atrophy; iPSCs, induced pluripotent stem cells; NSCs, neural stem cells.

EMBRYONIC STEM CELL DERIVED MOTOR NEURON PRECURSORS FOR SPINAL MUSCULAR ATROPHY

A novel technique to obtain highly enriched human motor neuron precursor (hMNP) cultures was established by California Stem Cell Inc.[35,36] Hmnps were derived from human embryonic stem cell (ESC) lines that were expanded on Matrigel for the first 3 weeks in a media supplemented with basic fibroblast growth factor (bFGF)[36]. Cells were then transferred to ultralow binding dishes and suspended in MN differentiation media, including Glutamax, B27, insulin, sodium selenite, transferrin, MgSO4, and bFGF.[36] Cells were exposed to the differentiation media for 5 days, supplemented with retinoic acid (RA), and then fully characterized. The transplantation of derived cells in the spinal cord of SMA models resulted in a proper engraftment and survival.[37] Moreover, transplanted hMNP were able to produce trophic factors in situ (i.e., neurotrophin-3 (NT-3) and nerve growth factor). The number of endogenous motor neurons was significantly increased.[37] Given these results, the research group designed a clinical trial, but until now, the FDA has not approved the study.

INDUCED PLURIPOTENT STEM CELL DERIVED MOTOR NEURONS AS A CELL SOURCE FOR TRANSPLANTATION

Spinal motor neurons appear to degenerate rather selectively during the course of SMA and this background led to the experimental trial to transplant healthy motor neurons in SMA murine models.

Human motor neurons can be obtained by differentiating human embryonic or induced pluripotent stem cells (iPSCs). ESCs can be isolated from the blastocyst and are characterized by the ability to give rise to all the three germ layers. iPSCs derive from patients' somatic cells (Fig. 11.1), which have been reprogrammed to an embryonic stage.[38] Both ESCs and iPSCs can be differentiated toward a motor neuron lineage with established protocols including the use of RA and Sonic hedgehog, which play a fundamental role in physiological motor neuron development.[39] Derived motor neurons express specific markers (i.e., choline acetyltransferase (ChAT) and HB9) and spread axons to neuromuscular junctions after in vivo transplantation.[40]

We firstly transplanted a purified motor neuron population differentiated from LeX+ stem cells derived from embryonic spinal cords into an SMA respiratory distress type 1 (SMARD1) transgenic model.[41] LeX is a developmentally regulated tetrasaccharide carbohydrate, which is expressed in embryonic and adult NSCs. It can be exploited to select primitive stem cell fractions.[42] A pharmacological treatment aiming to promote axonal growth directed at the neuromuscular junction was administered in combination. More specifically, isolated motoneurons were treated with a solution containing dbcAMP to enhance cell survival and axonal growth. Animals were also systemically injected with rolipram, a pharmacological compound modulating phosphodiesterase type 4 activity with the aim of counteracting the inhibitory effects of myelin proteins on axonal outgrowth. Moreover, GDNF was administered to the animals with intramuscular injections in order to direct donor axons toward their targets. Analyses at the end stage of the disease revealed that transplanted cells were integrated into the host anterior horn and their axons were directed toward the muscular target. The disease phenotype of SMARD1 mice was improved with better motor performance and increased survival. Moreover, neuroinflammation within the host spinal cord was significantly reduced after treatment.[41]

Our research group exploited iPSC-derived motor neurons as a tool for study SMA pathology and as an experimental therapeutic approach after genetic manipulation[17] (Fig. 11.1). We derived human iPSCs from SMA patients' fibroblasts, reprogramming them with the use of nonintegrating and nonviral episomal vectors. This reprogramming strategy is slightly less efficient than the viral reprogramming method but could ensure a more rapid translation to clinical trials. Generated cells were treated with single-stranded DNA oligonucleotides designed to genetically edit *SMN2* inducing a substitution of a single nucleotide within exon 7, thus ensuring the inclusion of the exon within the transcript and as a consequence the production of a *SMN1*-like functional protein.

Engineered iPSCs were then differentiated toward motor neurons employing a multistage differentiation protocol already validated for human ESCs.[17] After 1 month, cells in culture expressed motor neuron markers (i.e., Hb9, ISLET1), pan-neuronal markers (TuJ1, Neurofilament, and MAP2), or spinal cord progenitor markers (OLIG2). Hb9- and ISLET1-positive cells could also be stained for ChAT and SMI32, thus showing a motor neuronal lineage.[17] Uncorrected and modified SMA iPSC-derived motor neurons displayed significant differences in terms of morphology and survival. SMA cell phenotype was characterized by reduced axonal length and neuromuscular junction formation with a decreased survival in culture.[17] These features were substantially rescued in motor neurons derived from corrected SMA iPSCs. An in-depth analysis of the transcriptome of both cell populations (i.e., treated and untreated ones with oligonucleotides) revealed a difference in a specific group of genes responsible for RNA metabolism, motor neuronal and axonal development.[17] SMA iPSC-derived motor neurons were then transplanted into the spinal cord of SMA transgenic model and they were able to survive and

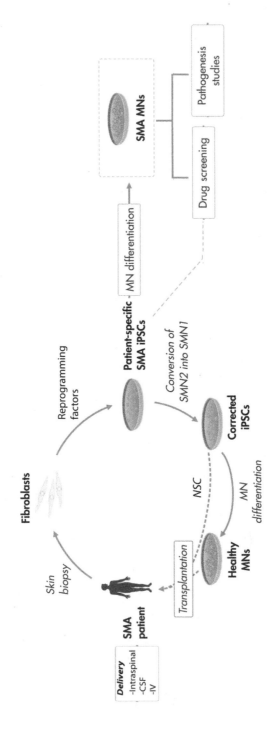

FIGURE 11.1 Overview on the potential use of human pluripotent stem cells for spinal muscular atrophy (SMA) modeling and therapy. SMA patients' fibroblasts can be reprogrammed into induced pluripotent stem cells and genetically corrected to be used for autologous cell therapy. Induced pluripotent stem cells (iPSC)-derived motor neurons (MNs) represent an invaluable source for disease-modeling-studies. *CSF*, cerebrospinal fluid; *IV*, intravenous; *NSC*, neural stem cells; *SMN*, survival motor neuron.

properly engraft into the host microenvironment without modifications of the motor neuronal phenotype.[17] An amelioration of SMA mice phenotype could be observed after transplant; treated SMA mice presented an increased survival (about 50%) in comparison to vehicle-treated mice.[17] The beneficial effect was more evident with the transplant of corrected iPSC-derived motor neurons. Moreover, transplanted SMA iPSC motor neurons showed a decreased life span with a reduced ability to spread their axons when compared to motor neurons generated from oligonucleotide-corrected stem cells.[17]

Overall, these data showed a way to exploit transplanted SMA-iPSC motor neurons to create an in vivo model of the disease. Moreover, the possibility of effectively correcting the SMA motor neuron phenotype by using molecular strategies could pave the way for experimental therapeutic approaches with autologous cells. It is worth considering that the proper engraftment and integration within the host microenvironment of motor neurons differentiated from oligonucleotide-treated SMA iPSCs appeared to be modulated by the time of correction. A very early treatment with oligonucleotides was more effective in ensuring beneficial effects, thus reinforcing the hypothesis of a "window of opportunity" for molecular therapeutic approaches.[43–46]

In our study, transplanted human iPSC-derived motor neurons were able to provide a protective environment for endogenous cells by secreting neurotrophins and growth factors.[17] This beneficial action appeared to be related to a successful engraftment in a proportional way. Indeed, motor neurons obtained from genetically manipulated SMA iPSCs displayed a major therapeutic effect on endogenous motor neurons, which presented an increase in cell size with enhanced axonal length and a prolonged life span.[17]

Taken together, the results from our study provided data on the possibility of deriving SMA patients' iPSCs, which could be genetically manipulated and used to derive precious insight into SMA pathogenesis, and experimental stem cell based therapeutic approaches.

NEURAL STEM CELLS

NSCs represent the progenitors of all the three neuroectodermal progenies within the central nervous system (CNS). NSCs can be directly derived from embryonic or mature neural tissue, or can be obtained through differentiation of ESCs or iPSCs using well-established protocols. When transplanted, NSCs are able to improve the phenotype in different transgenic models of motor neuron disease.[32,33,37,42,47–49] Our research group investigated, for the first time, the possibility of transplanting NSCs to ameliorate the disease phenotype in SMA and SMARD1 animal models.

TRANSPLANTATION OF SPECIFIC
NEURAL STEM CELL SUBPOPULATIONS

NSCs represent a group of heterogeneous cells, which can be selected for their biological potential to survive within the host environment after transplant, migrate to the injury site and properly engraft, and finally differentiate into the relevant phenotypes. Our research group has shown that choosing specific subpopulations of NSCs could be advantageous for therapeutic purposes and for clinical application in motor neuron disorders such as SMA. In our first work, we studied the therapeutic action of a primary NSC population selected for the levels of aldehyde dehydrogenase activity and low side scatter (ALDHhiSSClo cells), and their ability to modulate the disease course of the *nmd* mouse, a SMARD1 transgenic model. ALDH activity has been used as a parameter to isolate hematopoietic stem cells, and multiple research studies have described that high ALDH activity is related to the stem cell state.[50,51]

Intrathecal injection of derived stem cells in *nmd* mice resulted in delayed disease progression, protection of motor neurons, and increased survival.[47] Transplanted motor neuron precursors correctly migrated to the ventral horns of the host spinal cord. In mutated and treated *nmd* spinal cord, we showed with microarray and RT-PCR analyses a reduction of genes involved in glutamate damage and oxidative toxicity, and an overexpression of genes linked to the chromatin structure. These genes might be involved in SMARD1 development. Spinal cord of *nmd*-treated animals expressed high transcript levels of genes involved in neurogenesis (i.e., LIS1, doublecortin (DCX) and drebrin).[47,52] The detection of DCX-positive cells within adult *nmd* spinal cord led to the speculation that both exogenous and endogenous neurogenesis could contribute to the *nmd* mice phenotypic improvement. We then tested the same selected subpopulation derived from spinal cord murine NSCs on SMA mice.[32] ALDHhiSSClo cells were derived from fetal murine spinal cord neurospheres and cultured with specific priming and motor neuron differentiation protocols.[42,47] GFP+ ALDHhiSSClo-derived cells exhibited morphological and histochemical characteristics of fully differentiated motor neuron precursors in vitro. After ALDHhiSSClo cell intrathecal injection, 1-day-old SMA mice showed ameliorated functional (neuromuscular function and life span) and histological features (motor neuron cell number and dimension and neuromuscular junction size) compared to untreated SMA mice.[32]. SMA mice treated with primed ALDHhiSSClo NSCs survived significantly longer also compared to SMA mice injected with undifferentiated ALDHhiSSClo cells, ALDHhiSSClo-derived astrocytes, and murine primary fibroblasts. Transplanted cells (HB9-GFP+) resembled true motor neurons morphologically, migrated, and engrafted along the recipient spinal cords. Full gene expression analyses of laser-capture-microdissected motor neurons derived from transplanted animals through microarray analysis showed an amelioration of SMA-diseased features in the direction of the wild-type pattern (for primary data refer to the GEO website: http://www.ncbi. nlm.nih.gov/geo/; GEO accession number: GSE10224). Moreover, with Luminex multianalyte profiling technology, we detected different factors secreted by NSCs including VEGF, granulocyte colony-stimulating factor (G-CSF), and several neurotrophins (BDNF, GDNF, NT-3, and transforming growth factor alpha (TGF-α)), which may exert neuroprotective effects on motor neurons after ALDHhiSSClo cell transplantation.[32] These data demonstrated the feasibility and therapeutic potential of NSC transplantation in the SMA mouse model.

In a second study, we investigated the hypothesis that NSCs obtained from murine pluripotent stem cells could improve the SMA mouse model phenotype.[33] We used NSCs derived from wild-type murine ESCs and drug-selectable (with ganciclovir and G418) ESC lines, to promote neuronal differentiation and increase cell safety.[33,53] Pluripotent-derived NSCs migrated into the spinal cord of SMA mice after intrathecal transplantation and improved phenotype and survival of SMA mice through replacement of motor neuron cells and conferring neuroprotection.

To differentiate drug-selectable mouse ESCs, the differentiation medium was supplemented with G418 and ganciclovir to select for neuroepithelial cells and against undifferentiated ESCs.[33,53] We obtained motor neurons with classical phenotype and expressing correct neuroectodermal factors and cholinergic molecules using various signaling molecules in culture. The drug-selected NSCs promoted longer survival than the wild-type ESCs and the primary NSCs of our previous report.[32] Moreover, we demonstrated that the differentiation toward motor neurons could be promoted once NSC phenotype had been established and reinforced.[33] Combining cell transplants with pharmacological or gene therapy could enhance the therapeutic effectiveness of this strategy to a clinically relevant threshold.

In a more recent study, we transplanted murine LewisX+; CXCR4+ NSCs directed toward a motor neuron fate into the spinal cord of SMARD1 transgenic mice.[42] Treated animals presented a prolonged survival with increased motor neuron numbers within the spinal cord. About 18% of transplanted cells expressed motor neuronal markers at the end stage of the disease and some of them have spread their axons toward the anterior roots.

GLIAL CELLS

Astrocytes play a key role in maintaining a healthy and trophic environment for motor neurons.[54,55] SMA astrocytes seem to lose their protective action and undergo phenotypical and functional modifications.[22]

McGivern's group showed that SMA astrocytes displayed morphological changes suggestive of overactivation even before motor neuron degeneration. Activated cells can secrete inflammatory molecules provoking neuronal death through the apoptotic signaling.[22] Moreover, astrocytes derived from SMA patients' iPSCs produced less GDNF (a key growth factor) in culture.[22] Impaired astrocyte function can trigger motor neuron degeneration and, complementary, transplantation of Glial-Restricted Progenitors has been demonstrated to give rise to enriched healthy astrocytes within an ALS animal model (SOD1G93A rats).[56] This resulted in delay of the disease progression.[56] The beneficial effect was due in part to the restoration of astrocyte GLT1, which is essential for glutamate balance in the extracellular fluid.[56,57] The site of transplantation was chosen around cervical respiratory motor neurons, which are responsible for respiratory failure in patients affected by motor neuron diseases.[58,59] These data could open the path to the employment of astrocytes for SMA therapy, addressing neuroinflammation within the spinal cord, which has not been performed up to now. Complementarily, the role of oligodendrocytes in contributing for SMA pathogenesis has been considered; alterations in myelination have been reported in SMA murine models.[12] More studies will be needed to clearly define the role of these modifications and the potential therapeutic effects of stem cell derived oligodendrocyte transplant.

MUSCLE CELLS

Skeletal muscle is affected in SMA both as an event secondary to the denervation as well as an independent pathogenic phenomenon.[14,60] In particular, myogenesis and terminal differentiation problems of skeletal muscle myofibers have been described in animal models and SMA patients.[14,60] In this respect, the transplantation of skeletal muscle stem/precursor cells, the so-called, satellite cells, could provide therapeutic benefit by promoting enhanced muscle regeneration. In fact, engrafted satellite cells can contribute to the muscle regeneration by fusing with endogenous fibers.[61] However, given that local transplantation of muscle tissue is inefficient, it may not lead to any functional improvement. Further data in SMA mice are required to quantitatively confirm the effectiveness of this approach. As an alternative to local injection, systemic delivery of satellite cells or stem cell-derived skeletal myocytes or mesangioblasts, a particular subset of stem cells, can be hypothesized to allow a more efficient and widespread distribution to the muscles even if the clinical feasibility and efficacy of this approach have to be optimized.[14] Overall SMA is a disease of both nerve and muscle tissues; therefore, the attempt to target both tissues could present a therapeutic efficacy.[14,60]

MINIMALLY INVASIVE STRATEGIES OF ADMINISTRATION TO EASE THE CLINICAL TRANSLATION: THE CEREBROSPINAL FLUID AND SYSTEMIC ROUTES

The strategy of cell administration is a crucial point for clinical translation of a stem cell approach in SMA. The ideal delivery route should achieve the greatest therapeutic efficacy with minimal invasiveness using a protocol appropriate for human patients, and especially children who represent a large proportion of SMA cases. For the treatment to be effective, introduced cells must be dispersed along the CNS and be capable of migrating in degenerating areas of the brainstem and spinal cord. Noninvasive methods, such as intrathecal and systemic intravenous injections, could allow repeated cell administrations and represent, therefore, effective approaches to SMA treatment. A practicable administration protocol requires the selection of a proper cell source able to migrate from CSF and/or blood to CNS and the identification of the most appropriate stem cell subpopulation in terms of therapeutic efficacy. Recent studies have demonstrated that minimally invasive transplantation of NSCs (via intravenous or intrathecal injections) positively impact the course of neurological disease models, including multiple sclerosis and motor neuron disorders.[27,62–65] These methods are capable of allowing the extensive distribution of cells in the CNS, which represents an indicator of potential therapeutic efficacy. Damaged zones within the CNS are targeted by NSCs, which tend to migrate into the diseased areas.[65] The inflammatory molecules resulting from neuronal death in SMA disease can exert a chemoattractant role on NSCs.[65] Murine and human NSCs express a series of adhesion surface molecules, chemokine, and other immune receptors on the membrane surface.[66–70] This feature makes NSCs similar in some way to leukocytes and permits them to interact with endothelial and blood–brain barrier cells targeting the zone of active degeneration.[65,71,72] These molecular interactions are prerequisites for both systemic and CSF delivery.[29,65] Preclinical studies in SMA mice confirmed the migration of NSC in the right areas of the diseased spinal cord following intrathecal injection, which could sustain the practicability of CSF administration and, likely, of intravenous protocols,[27,72] thus combining therapeutic effects with minimal invasiveness.[32]

There is growing evidence that SMA is the motor neuron manifestation of a broader, multiorgan systemic disease. Alterations in skeletal muscle, skin, and several metabolic pathways have been described in SMA patients.[1] These findings support the utility of a systemic approach, when an appropriate stem cell population able to target these events is available. For instance, transplanted NSCs can exert an immune-regulatory action. From the blood, they can migrate to lymph nodes and spleen and assemble near blood vessel borders where they interact with lymphocytes and antigen-presenting cells, influencing their activity.[65] A peripheral therapeutic effect is obtained, also, through the secretion of neuroprotective molecules in skeletal muscle tissue that can contribute to preserving neuromuscular junctions. Moreover, systemic injections can be repeated, thus obtaining an extensive distribution of sufficient cells in the CNS to deliver a therapeutic effect, which is likely related to the number of stem cells reaching the CNS. Although several aspects need further investigation prior to clinical application, a noninvasive approach, either intrathecal or systemic, could offer great promise of bringing cell-mediated strategies to the bedside. We focused our research on these aspects in one of our recent studies, in which we selected a specific NSC population from human iPSCs based on high aldehyde dehydrogenase activity, low side scatter, and integrin VLA4 positivity (ALDHhiSSCloVLa4+ NSCs). VLA4 is a membrane protein that modulates transendothelial migration and is present on T lymphocytes, where it regulates cell migration through the blood–brain barrier.[73] As shown in multiple sclerosis animal studies, NSCs positive for VLA4 can bypass the blood–brain barrier, particularly when inflammatory processes or a damaged barrier are present.[65] We evaluated the therapeutic efficacy of these selected NSCs on ALS phenotype in mice following intrathecal and intravenous injections.[27] The NSCs were obtained by plating iPSCs on neuronal medium supplemented with minimum essential medium nonessential amino acid solution, N2, and heparin to promote neuroepithelial commitment. Cells were then selected for their ALDHhiSSClo properties using FACS[32,47] and for their VLA4 positivity. To test their capacity to acquire a motor neuronal phenotype, cell differentiation was induced by plating them on polylysine/laminin-coated dishes and adding at different time points a combination of growth factors. When transplanted into CSF or in the tail vein of ALS rodent models, these cells migrate into the CNS where they demonstrated robust survival and did not trigger side effects. Further examination revealed that they localized throughout the entire spinal cord, specifically in the regions near degenerating endogenous motor neurons. Donor ALDHhiSSCloVLA4+ cells preserved their stem cell/precursor phenotypes or differentiated into neuronal and glial cells, enriching the local cell populations within the host tissue microenvironment. Remarkably, the ALDHhiSSCloVLA4+ NSC-treated SOD1 mice showed an improvement in the motor neuron phenotype in their neuromuscular function tests and prolonged life span. This effect was more evident with the systemic administration of NSCs. These positive events were associated with the neurotrophic factor production by NSCs, the reduction of micro- and macro-gliosis, as well as by other still unknown actions. Other studies of noninvasive administration either intrathecal or intravenous, of ALDHhiSSCloVLA4+ NSCs in SMA rodent models are planned in our lab. Taken together, these results reinforce the hypothesis that the selection and isolation of the best-performing stem cell subpopulation could be essential for clinical application, potentially being the most advantageous approach to cell-mediated therapy, including the chance of minimally invasive administration.

FROM BENCHTOP TO CLINICAL TRANSLATION: ISSUES TO OVERCOME

Some of the specific characteristics of SMA render this disease a particularly appropriate target for stem cell based therapeutic strategies.[74,75] While spinal cord injury, multiple sclerosis, and other neurodegenerative disorders represent multilayered diseases in which several cell subtypes are evenly affected, in the context of SMA, motor neurons appear rather selectively vulnerable to the disease process, thus representing a valuable therapeutic target.[76,77] Moreover, the milieu of the spinal cord in SMA patients could be more receptive to the transplanted cells.[76,77] In fact, SMA-affected children could present a more plastic CNS compared with the aging tissues of neurodegenerative patients. Because the spinal cord is still in a "growing phase," it possesses a proportionally proper amount of blood flow and nourishment and high levels of growth factors, which could support grafted cells.[76,77] In addition, transplanted cells once differentiated into motor neurons could elongate their axons to the muscular target for relatively smaller distances within SMA children's bodies, thus forming early healthy synapses at the neuromuscular junctions. On the other hand, considering that SMA involves mainly infants and children, there are also many practical and ethical issues to handle. Cell transplantation trials involve surgical operation and its related risks.[75] Cell transplantation methods for SMA patients require protocols optimized for the efficient and minimally invasive delivery of stem cells into children's spinal cords. Therefore, appropriate cell delivery to the CNS should first be defined, including standard operating procedures that specify the proper "amount" of cells and the scheduling and number of injections.[75,78] In addition, details about the permissive milieu or the growth substances injected have to be determined. Cells need to be accurately studied in terms of proper migration and engraftment into the host tissues: proof of donor cell survival is essential to ensure the therapeutic efficacy. The demonstration of the presence of human cells in in vivo models has been achieved by immunohistochemical analyses detecting human-specific antigens.[79] More advanced methods, however, will be needed to determine the status of grafted cells in the clinical setting.

Immunorejection of donor grafted cells represents an issue to overcome before clinical translation. In this perspective, the transplanted stem cells could be autologous patient-derived iPSC-NSCs and thus not targeted by the immune system. However, this assumption has not been fully investigated and some experiments suggest that reprogrammed cells may induce an immune reaction, even causing immune rejection, in animal models.[80] There are different reasons for iPSC-derived cell immunogenicity and some may be linked to the techniques of cell preparation: stem cells generated using retroviral vectors seem to elicit the strongest host immune response.[81] Moreover, random viral integration can negatively influence endogenous gene expression. Nonviral methods such as plasmids, RNAs, or protein reprogramming could represent alternative strategies to overcome these problems.[81] iPSC-derived cells offer hope for the treatment of some genetic diseases, like SMA; however, this strategy is feasible only if the causative mutation is corrected in the patient's autologous cells *ex vivo*. It has to be considered that the presence of proteins unknown to the patient's immune system may still provoke an immune reaction. It is crucial that the reprogramming of the cells is complete and that they eliminate any epigenetic memory of the original cell phenotype (i.e., fibroblasts). Chromatin status could inhibit the developmental potential and, possibly, hamper the ability to derive completely differentiated and functional neurons. An epigenetic memory could also cause an aberrant cell-surfacing antigen expression when iPSCs are differentiated into other cell types. Furthermore, these cell-surface abnormalities could induce an immunoresponse.[82] These questions have to be experimentally addressed increasing our knowledge in the reprogramming process.

Another imperative question is the oncogenic risk of transplanted cells.[83–85] The employment of differentiated cells, such as neural-committed progenitors or mature neurons, could alleviate this problem as more committed cells would hold a limited potential of further/aberrant proliferation. An alternative option includes the insertion of apoptotic genes into stem cells, which prompt cell death under certain conditions. Thus, grafted cells could be selectively targeted, if needed, by the administration of a specific drug.[86,87]

Furthermore, even though protocols for the directed differentiation of stem cells into specific neural types have been described, and their morphologies and functions characterized, further research is needed to define these events and their temporal evolution. Indeed, several technical differences are present between various reprogramming and differentiation methods that have to be further evaluated and optimized.[29,88] Other significant issues to be addressed remain the ideal therapeutic window for cell transplantation, the optimum range between maturation and maintenance of stem cell identity within the host environment, and the beneficial effect to be fostered in a balance between direct cell replacement and/or trophic support. It has to be determined whether a therapeutic window for cell therapy exists in SMA, as well as for therapies aiming for SMN restoration, and this timing needs to be defined. This is fundamental for planning future human clinical trials. A crucial milestone in the successful clinical application of molecular and cellular approaches will also be the careful design of human clinical trials, with a proper selection of patients in which a specific therapeutic approach could be beneficial.

Furthermore, while the results of early NSC administration in rodent SMA seem positive, the next step regards confirming these evidences in larger animal models to define the feasibility and safety of this therapeutic approach. In conjunction with the safety studies, it has to be assessed whether donor motor neurons can connect with their targets over extended distances. Regarding these aspects, the pig (Sus scrofa) *SMN* gene has been discovered[89] and swine models have been recently developed and tested for gene therapy development.[90]

Overall, the upcoming therapeutic uses of patient-derived stem cells should be encouraged while considering the multitude of questions presented above. The application of this approach must also be inspired by ethical considerations and appropriate selection of patients to enroll in clinical trials.

STEM CELLS FOR DRUG DISCOVERY

In addition to the fascinating promises of therapeutic application, human stem pluripotent cells hold a great potential in the field of disease modeling. Standardized methods for human stem cell differentiation have made it possible to generate a consistent number of human motor neurons as a precious source for disease modeling and drug screening.[29,91] Ebert and Svendsen gave the first demonstration of the possibility of reprogramming SMA fibroblasts.[15] Employing a lentiviral transfection protocol, the authors obtained SMA and wild-type iPSCs from fibroblasts from a 3-year-old patient diagnosed with SMA type 1 and his healthy mother. They showed that SMA iPSCs had markedly decreased expression of *SMN* full-length transcripts with respect to healthy subjects. In the early steps of motor neuron differentiation, no relevant disparity in the number and dimensions between the affected and unaffected iPSC-derived motor neurons was reported, thus pointing out that no defect in motor neuron early development underlies SMA pathogenesis. However, after differentiation, SMA motor neurons displayed a selective reduction in number and size with respect to wild-type cells. Based on these data, it was hypothesized that the lack of SMN1 gene permits a normal development of SMA motor neurons, but causes a secondary selective degeneration, suggesting that the pathogenesis of

SMA resembles a neurodegenerative process. These experiments represented the first demonstration that SMA pathology specifically affects human cell survival. It also revealed the crucial role that iPSCs could play as reliable disease models recapitulating at least some features of the human disorder.[15] In a second study, Sareen and Svendsen's group investigated the activation of the apoptotic cascade as a possible mechanism underlying SMA motor neuron death.[19] They demonstrated that during motor neuron differentiation, SMA iPSCs present an upregulation in Fas ligand-mediated apoptosis together with an overactivation of caspase-8 and-3. Remarkably, this mechanism could be modulated by exposure with either a Fas-blocking antibody or a caspase-3 inhibitor. Our research group has demonstrated the utility of SMA iPSC lines and their differentiated motor neurons as a precious tool for disease modeling. In particular, we described the possibility of genetically engineering iPSCs from SMA patients to counteract the disease phenotype in differentiated motor neurons.[17] We presented the results obtained with SMA iPSCs and motor neurons in vitro and in vivo above, in this chapter. Our study confirmed that molecular therapy with oligonucleotides successfully increased SMN and rescued the cellular damage and neuropathological features of SMA.[17]

SMA iPSCs could be also exploited to screen potential therapeutic compounds in the field of small-molecule drug research. SMN is a ubiquitously expressed protein, which assembles within cells into dot-like nuclear structures called "gems," for "Gemini of the coiled bodies."[92] The number of identified gems correlates with the disease severity in fibroblasts. Thus, gem expression can also be a possible marker for the efficacy of therapeutic drugs. Small molecules can modify the expression of gems in patients' iPSCs and their derived motor neurons, thus representing a useful tool in the development of potential therapies for SMA.[93] For instance, Tobramycin and Valproic acid have been demonstrated to significantly upregulate the number of gems in SMA iPSCs with respect to control cells.[15] Novel small molecules have been also recently tested in SMA iPSCs, which are able to shift the balance of SMN2 splicing toward the production of full-length SMN2 messenger RNA with high selectivity.[94] Moreover, iPSCs could contribute to the spectrum of diagnostic tools for SMA. The quantification of gems in SMA iPSC-derived motor neurons may represent a powerful screening test with a high specificity, also having a role in predicting SMA disease progression. This can be useful also for the stratification of patients in clinical trials as well as for the evaluation of the outcome measure of SMA candidate drugs. In this perspective, the evaluation of SMN2 transcripts by quantitative RT-PCR in SMA motor neurons could be another biomarker that may aid in selecting those patients who will benefit most from treatment.[95,96]

CONCLUSIONS

No effective treatment exists for SMA. Currently, the main therapeutic strategies consist of symptomatic relief and palliative treatments. Overall, results from different preclinical studies provide evidence that stem cells may represent a potential approach for the treatment of SMA. Therapeutic effects derived from stem cell transplantation are probably due to multiple positive events on different pathogenic mechanisms of the disease; this approach is essential for complex neurodegenerative diseases such as SMA. The possibility to exploit patients' specific cells, such as iPSCs after *ex vivo* gene correction, and to administer them using minimally invasive methods will encourage investigations into further clinical applications. By focusing investigations on selecting stem cell populations that demonstrate the best biological attributes in terms of migration to the CNS and ability to engraft, minimally invasive administration protocols will emerge that are highly suitable for clinical applications. Cell therapy could be exploited in combination with complementary therapeutic strategies (i.e., drugs, molecular targeted approaches, gene therapy) that counteract different pathogenic mechanisms of the disease. It could be speculated that with this approach the possibility of delivering clinically meaningful therapy for SMA will increase. In terms of improvement to phenotype and survival, it is evident that the results achieved with stem cell based therapies are less impactful than those observed using gene and/or molecular strategies. Indeed, gene therapy and molecular approaches have been demonstrated to prolong life span with a statistically significant percentage, which reaches the 1500%; this result could vary on the basis of the viral vector and antisense oligonucleotides exploited, but it appears always clinically meaningful.[75,97–101] Conversely, our group found a life span increase of ~40% in animals treated with cellular therapy. Consequently, the most beneficial effect of stem cells, like other non-SMN-targeted therapeutic approaches, might be their application in affected children who present with clinical symptoms. The purpose in this case would be to avoid further motor neuron degeneration and support endogenous motor neurons through the release of neuroprotective and trophic molecules by donor stem cells. Stem cell based molecular and gene therapies could be applied in combination to achieve the most beneficial effect for patients. Indeed, in the case of SMA, the best-performing therapeutic strategy could be represented by specifically targeting the genetic alteration, exploiting gene therapy or antisense oligonucleotides, and complementarily ameliorating SMA phenotype using other approaches, such as stem cell based therapies. There are potential additive effects between the molecular and cellular experiments; however, this possibility has not yet been attempted in animal models at the preclinical level. Understanding the molecular mechanisms of stem cell action could allow simulation of the same effects with small molecules without the need for complex transplantation procedures. Together, parallel advances in our knowledge of motor neuron disorder specific pathways and stem cell biology will expand the reliable experimental data and, simultaneously, contribute to developing suitable cell sources. This research will open the way for the development of novel effective therapies for SMA.

References

1. D'Amico A, Mercuri E, Tiziano FD, Bertini E. Spinal muscular atrophy. *Orphanet J Rare Dis* 2011;**6**:71. http://dx.doi.org/10.1186/1750-1172-6-71.
2. Brzustowicz LM, Lehner T, Castilla LH, et al. Genetic mapping of chronic childhood-onset spinal muscular atrophy to chromosome 5q11.2-13.3. *Nature* 1990;**344**(6266): 540–1. http://dx.doi.org/10.1038/344540a0.
3. Monani UR, Lorson CL, Parsons DW, et al. A single nucleotide difference that alters splicing patterns distinguishes the SMA gene SMN1 from the copy gene SMN2. *Hum Mol Genet* 1999;**8**(7):1177–83.
4. Lorson CL, Hahnen E, Androphy EJ, Wirth B. A single nucleotide in the SMN gene regulates splicing and is responsible for spinal muscular atrophy. *Proc Natl Acad Sci U S A* 1999;**96**(11):6307–11.
5. Prior TW, Snyder PJ, Rink BD, et al. Newborn and carrier screening for spinal muscular atrophy. *Am J Med Genet A* 2010;**152A**(7):1608–16. http://dx.doi.org/10.1002/ajmg.a.33474.
6. Munsat TL, Davies KE. International SMA consortium meeting. (26-28 June 1992, Bonn, Germany). *Neuromuscul Disord NMD* 1992;**2**(5–6):423–8.
7. Prior TW, Swoboda KJ, Scott HD, Hejmanowski AQ. Homozygous SMN1 deletions in unaffected family members and modification of the phenotype by SMN2. *Am J Med Genet A* 2004;**130A**(3):307–10. http://dx.doi.org/10.1002/ajmg.a.30251.
8. Piepers S, van den Berg LH, Brugman F, et al. A natural history study of late onset spinal muscular atrophy types 3b and 4. *J Neurol* 2008;**255**(9):1400–4. http://dx.doi.org/10.1007/s00415-008-0929-0.
9. Wang CH, Finkel RS, Bertini ES, et al. Consensus statement for standard of care in spinal muscular atrophy. *J Child Neurol* 2007;**22**(8):1027–49. http://dx.doi.org/10.1177/0883073807305788.
10. Meister G, Hannus S, Plöttner O, et al. SMNrp is an essential pre-mRNA splicing factor required for the formation of the mature spliceosome. *EMBO J* 2001;**20**(9):2304–14. http://dx.doi.org/10.1093/emboj/20.9.2304.
11. Burghes AHM, Beattie CE. Spinal muscular atrophy: why do low levels of smn make motor neurons sick? *Nat Rev Neurosci* 2009;**10**(8):597–609. http://dx.doi.org/10.1038/nrn2670.
12. Hunter G, Aghamaleky Sarvestany A, Roche SL, Symes RC, Gillingwater TH. SMN-dependent intrinsic defects in Schwann cells in mouse models of spinal muscular atrophy. *Hum Mol Genet.* 2014;**23**(9):2235–50. http://dx.doi.org/10.1093/hmg/ddt612.
13. Monani UR, Sendtner M, Coovert DD, et al. The human centromeric survival motor neuron gene (SMN2) rescues embryonic lethality in Smn(-/-) mice and results in a mouse with spinal muscular atrophy. *Hum Mol Genet* 2000;**9**(3):333–9.
14. Iascone DM, Henderson CE, Lee JC. Spinal muscular atrophy: from tissue specificity to therapeutic strategies. *F1000Prime Rep* 2015;**7**:04. http://dx.doi.org/10.12703/P7-04.
15. Ebert AD, Yu J, Rose FF, et al. Induced pluripotent stem cells from a spinal muscular atrophy patient. *Nature* 2009;**457**(7227):277–80. http://dx.doi.org/10.1038/nature07677.

16. Ruiz R, Casañas JJ, Torres-Benito L, Cano R, Tabares L. Altered intracellular Ca2+ homeostasis in nerve terminals of severe spinal muscular atrophy mice. *J Neurosci Off J Soc Neurosci* 2010;**30**(3):849–57. http://dx.doi.org/10.1523/JNEUROSCI.4496-09.2010.

17. Corti S, Nizzardo M, Simone C, et al. Genetic correction of human induced pluripotent stem cells from patients with spinal muscular atrophy. *Sci Transl Med* 2012;**4**(165) 165ra162. http://dx.doi.org/10.1126/scitranslmed.3004108.

18. Cherry JJ, Osman EY, Evans MC, et al. Enhancement of SMN protein levels in a mouse model of spinal muscular atrophy using novel drug-like compounds. *EMBO Mol Med* 2013;**5**(7):1035–50. http://dx.doi.org/10.1002/emmm.201202305.

19. Sareen D, Ebert AD, Heins BM, McGivern JV, Ornelas L, Svendsen CN. Inhibition of apoptosis blocks human motor neuron cell death in a stem cell model of spinal muscular atrophy. *PloS One* 2012;**7**(6):e39113. http://dx.doi.org/10.1371/journal.pone.0039113.

20. Imlach WL, Beck ES, Choi BJ, Lotti F, Pellizzoni L, McCabe BD. SMN is required for sensory-motor circuit function in Drosophila. *Cell* 2012;**151**(2):427–39. http://dx.doi.org/10.1016/j.cell.2012.09.011.

21. Papadimitriou D, Le Verche V, Jacquier A, Ikiz B, Przedborski S, Re DB. Inflammation in ALS and SMA: sorting out the good from the evil. *Neurobiol Dis* 2010;**37**(3):493–502. http://dx.doi.org/10.1016/j.nbd.2009.10.005.

22. McGivern JV, Patitucci TN, Nord JA, Barabas M-EA, Stucky CL, Ebert AD. Spinal muscular atrophy astrocytes exhibit abnormal calcium regulation and reduced growth factor production. *Glia* 2013;**61**(9):1418–28. http://dx.doi.org/10.1002/glia.22522.

23. Quintard H, Heurteaux C, Ichai C. Adult neurogenesis and brain remodelling after brain injury: from bench to bedside? *Anaesth Crit Care Pain Med* 2015 http://dx.doi.org/10.1016/j.accpm.2015.02.008.

24. Yang H, Lu P, McKay HM, et al. Endogenous neurogenesis replaces oligodendrocytes and astrocytes after primate spinal cord injury. *J Neurosci Off J Soc Neurosci* 2006;**26**(8):2157–66. http://dx.doi.org/10.1523/JNEUROSCI.4070-05.2005.

25. Gulino R, Parenti R, Gulisano M. Novel mechanisms of spinal cord plasticity in a mouse model of motoneuron disease. *Bio Med Res Int* 2015;**2015**:654637. http://dx.doi.org/10.1155/2015/654637.

26. Rolfe A, Sun D. Stem cell therapy in brain trauma: implications for repair and regeneration of injured brain in experimental TBI models. In: Kobeissy FH, editor. *Brain neurotrauma: molecular, neuropsychological, and rehabilitation aspects. Vol frontiers in neuroengineering*. Boca Raton (FL): CRC Press; 2015. http://www.ncbi.nlm.nih.gov/books/NBK299210/. [accessed 19.08.15].

27. Nizzardo M, Simone C, Rizzo F, et al. Minimally invasive transplantation of iPSC-derived ALDHhiSSCloVLA4+ neural stem cells effectively improves the phenotype of an amyotrophic lateral sclerosis model. *Hum Mol Genet* 2014;**23**(2):342–54. http://dx.doi.org/10.1093/hmg/ddt425.

28. Conti L, Cattaneo E, Papadimou E. Novel neural stem cell systems. *Expert Opin Biol Ther* 2008;**8**(2):153–60. http://dx.doi.org/10.1517/14712598.8.2.153.

29. Faravelli I, Bucchia M, Rinchetti P, et al. Motor neuron derivation from human embryonic and induced pluripotent stem cells: experimental approaches and clinical perspectives. *Stem Cell Res Ther* 2014;**5**(4):87. http://dx.doi.org/10.1186/scrt476.

30. Locatelli D, Terao M, Fratelli M, et al. Human axonal survival of motor neuron (a-SMN) protein stimulates axon growth, cell motility, C-C motif ligand 2 (CCL2), and insulin-like growth factor-1 (IGF1) production. *J Biol Chem* 2012;**287**(31):25782–94. http://dx.doi.org/10.1074/jbc.M112.362830.

31. Krieger F, Metzger F, Jablonka S. Differentiation defects in primary motoneurons from a SMARD1 mouse model that are insensitive to treatment with low dose PEGylated IGF1. *Rare Dis Austin Tex* 2014;**2**:e29415. http://dx.doi.org/10.4161/rdis.29415.

32. Corti S, Nizzardo M, Nardini M, et al. Neural stem cell transplantation can ameliorate the phenotype of a mouse model of spinal muscular atrophy. *J Clin Invest* 2008;**118**(10):3316–30. http://dx.doi.org/10.1172/JCI35432.
33. Corti S, Nizzardo M, Nardini M, et al. Embryonic stem cell-derived neural stem cells improve spinal muscular atrophy phenotype in mice. *Brain J Neurol* 2010;**133** (Pt 2):465–81. http://dx.doi.org/10.1093/brain/awp318.
34. Wyatt TJ, Keirstead HS. Stem cell-derived neurotrophic support for the neuromuscular junction in spinal muscular atrophy. *Expert Opin Biol Ther* 2010;**10**(11):1587–94. http:// dx.doi.org/10.1517/14712598.2010.529895.
35. Wyatt TJ, Rossi SL, Siegenthaler MM, et al. Human motor neuron progenitor transplantation leads to endogenous neuronal sparing in 3 models of motor neuron loss. *Stem Cells Int* 2011;**2011**:207230. http://dx.doi.org/10.4061/2011/207230.
36. Nistor G, Siegenthaler MM, Poirier SN, et al. Derivation of high purity neuronal progenitors from human embryonic stem cells. *PloS One* 2011;**6**(6):e20692. http://dx.doi. org/10.1371/journal.pone.0020692.
37. Rossi SL, Nistor G, Wyatt T, et al. Histological and functional benefit following transplantation of motor neuron progenitors to the injured rat spinal cord. *PloS One* 2010;**5**(7):e11852. http://dx.doi.org/10.1371/journal.pone.0011852.
38. Takahashi K, Yamanaka S. Induction of pluripotent stem cells from mouse embryonic and adult fibroblast cultures by defined factors. *Cell* 2006;**126**(4):663–76. http://dx.doi. org/10.1016/j.cell.2006.07.024.
39. Wichterle H, Lieberam I, Porter JA, Jessell TM. Directed differentiation of embryonic stem cells into motor neurons. *Cell* 2002;**110**(3):385–97.
40. Lee H, Shamy GA, Elkabetz Y, et al. Directed differentiation and transplantation of human embryonic stem cell-derived motoneurons. *Stem Cells Dayt Ohio* 2007;**25**(8): 1931–9. http://dx.doi.org/10.1634/stemcells.2007-0097.
41. Corti S, Nizzardo M, Nardini M, et al. Motoneuron transplantation rescues the phenotype of SMARD1 (spinal muscular atrophy with respiratory distress type 1). *J Neurosci Off J Soc Neurosci* 2009;**29**(38):11761–71. http://dx.doi.org/10.1523/ JNEUROSCI.2734-09.2009.
42. Corti S, Locatelli F, Papadimitriou D, et al. Neural stem cells LewisX+ CXCR4+ modify disease progression in an amyotrophic lateral sclerosis model. *Brain J Neurol* 2007;**130**(Pt 5):1289–305. http://dx.doi.org/10.1093/brain/awm043.
43. Valori CF, Ning K, Wyles M, et al. Systemic delivery of scAAV9 expressing SMN prolongs survival in a model of spinal muscular atrophy. *Sci Transl Med* 2010;**2**(35) 35ra42. http://dx.doi.org/10.1126/scitranslmed.3000830.
44. Passini MA, Bu J, Richards AM, et al. Antisense oligonucleotides delivered to the mouse CNS ameliorate symptoms of severe spinal muscular atrophy. *Sci Transl Med* 2011;**3**(72) 72ra18. http://dx.doi.org/10.1126/scitranslmed.3001777.
45. Foust KD, Wang X, McGovern VL, et al. Rescue of the spinal muscular atrophy phenotype in a mouse model by early postnatal delivery of SMN. *Nat Biotechnol* 2010;**28**(3):271–4. http://dx.doi.org/10.1038/nbt.1610.
46. Dominguez E, Marais T, Chatauret N, et al. Intravenous scAAV9 delivery of a codon-optimized SMN1 sequence rescues SMA mice. *Hum Mol Genet* 2011;**20**(4):681–93. http://dx.doi.org/10.1093/hmg/ddq514.
47. Corti S, Locatelli F, Papadimitriou D, et al. Transplanted ALDHhiSSClo neural stem cells generate motor neurons and delay disease progression of nmd mice, an animal model of SMARD1. *Hum Mol Genet* 2006;**15**(2):167–87. http://dx.doi.org/10.1093/ hmg/ddi446.
48. Teng YD, Benn SC, Kalkanis SN, et al. Multimodal actions of neural stem cells in a mouse model of ALS: a meta-analysis. *Sci Transl Med* 2012;**4**(165) 165ra164. http:// dx.doi.org/10.1126/scitranslmed.3001777.

49. Boulis NM, Federici T, Glass JD, Lunn JS, Sakowski SA, Feldman EL. Translational stem cell therapy for amyotrophic lateral sclerosis. *Nat Rev Neurol* 2011;**8**(3):172–6. http://dx.doi.org/10.1038/nrneurol.2011.191.

50. Hess DA, Meyerrose TE, Wirthlin L, et al. Functional characterization of highly purified human hematopoietic repopulating cells isolated according to aldehyde dehydrogenase activity. *Blood* 2004;**104**(6):1648–55. http://dx.doi.org/10.1182/blood-2004-02-0448.

51. Cai J, Cheng A, Luo Y, et al. Membrane properties of rat embryonic multipotent neural stem cells. *J Neurochem* 2004;**88**(1):212–26.

52. Francis F, Koulakoff A, Boucher D, et al. Doublecortin is a developmentally regulated, microtubule-associated protein expressed in migrating and differentiating neurons. *Neuron* 1999;**23**(2):247–56. http://dx.doi.org/10.1016/S0896-6273(00)80777-1.

53. Billon N, Jolicoeur C, Ying QL, Smith A, Raff M. Normal timing of oligodendrocyte development from genetically engineered, lineage-selectable mouse ES cells. *J Cell Sci* 2002;**115**(18):3657–65. http://dx.doi.org/10.1242/jcs.00049.

54. Khakh BS, Sofroniew MV. Diversity of astrocyte functions and phenotypes in neural circuits. *Nat Neurosci* 2015;**18**(7):942–52. http://dx.doi.org/10.1038/nn.4043.

55. Ben Haim L, Carrillo-de Sauvage M-A, Ceyzériat K, Escartin C. Elusive roles for reactive astrocytes in neurodegenerative diseases. *Front Cell Neurosci* 2015;**9**:278. http://dx.doi.org/10.3389/fncel.2015.00278.

56. Lepore AC, O'Donnell J, Kim AS, et al. Human glial-restricted progenitor transplantation into cervical spinal cord of the SOD1 mouse model of ALS. *PloS One* 2011;**6**(10):e25968. http://dx.doi.org/10.1371/journal.pone.0025968.

57. Li K, Javed E, Scura D, et al. Human iPS cell-derived astrocyte transplants preserve respiratory function after spinal cord injury. *Exp Neurol* 2015;**271**:479–92. http://dx.doi.org/10.1016/j.expneurol.2015.07.020.

58. Nichols NL, Van Dyke J, Nashold L, Satriotomo I, Suzuki M, Mitchell GS. Ventilatory control in ALS. *Respir Physiol Neurobiol* 2013;**189**(2):429–37. http://dx.doi.org/10.1016/j.resp.2013.05.016.

59. Shimizu T, Komori T, Kugio Y, Fujimaki Y, Oyanagi K, Hayashi H. Electrophysiological assessment of corticorespiratory pathway function in amyotrophic lateral sclerosis. *Amyotroph Lateral Scler Off Publ World Fed Neurol Res Group Mot Neuron Dis* 2010;**11**(1–2):57–62. http://dx.doi.org/10.1080/17482960903207385.

60. Monani UR. Spinal muscular atrophy: a deficiency in a ubiquitous protein; a motor neuron-specific disease. *Neuron* 2005;**48**(6):885–96. http://dx.doi.org/10.1016/j.neuron.2005.12.001.

61. Nicole S, Desforges B, Millet G, et al. Intact satellite cells lead to remarkable protection against Smn gene defect in differentiated skeletal muscle. *J Cell Biol* 2003;**161**(3):571–82. http://dx.doi.org/10.1083/jcb.200210117.

62. Chu K, Kim M, Jeong S-W, Kim SU, Yoon B-W. Human neural stem cells can migrate, differentiate, and integrate after intravenous transplantation in adult rats with transient forebrain ischemia. *Neurosci Lett* 2003;**343**(2):129–33.

63. Kim SU, de Vellis J. Stem cell-based cell therapy in neurological diseases: a review. *J Neurosci Res* 2009;**87**(10):2183–200. http://dx.doi.org/10.1002/jnr.22054.

64. Jeong S-W, Chu K, Jung K-H, Kim SU, Kim M, Roh J-K. Human neural stem cell transplantation promotes functional recovery in rats with experimental intracerebral hemorrhage. *Stroke J Cereb Circ* 2003;**34**(9):2258–63. http://dx.doi.org/10.1161/01.STR.0000083698.20199.1F.

65. Pluchino S, Cossetti C. How stem cells speak with host immune cells in inflammatory brain diseases. *Glia* 2013;**61**(9):1379–401. http://dx.doi.org/10.1002/glia.22500.

66. Martino G, Pluchino S. The therapeutic potential of neural stem cells. *Nat Rev Neurosci* 2006;**7**(5):395–406. http://dx.doi.org/10.1038/nrn1908.

67. Martino G, Franklin RJM, Baron Van Evercooren A, Kerr DA. Stem cells in multiple sclerosis (STEMS) consensus group. Stem cell transplantation in multiple sclerosis:

current status and future prospects. *Nat Rev Neurol* 2010;**6**(5):247–55. http://dx.doi.org/10.1038/nrneurol.2010.35.

68. De Feo D, Merlini A, Laterza C, Martino G. Neural stem cell transplantation in central nervous system disorders: from cell replacement to neuroprotection. *Curr Opin Neurol* 2012;**25**(3):322–33. http://dx.doi.org/10.1097/WCO.0b013e328352ec45.

69. Kokaia Z, Martino G, Schwartz M, Lindvall O. Cross-talk between neural stem cells and immune cells: the key to better brain repair? *Nat Neurosci* 2012;**15**(8):1078–87. http://dx.doi.org/10.1038/nn.3163.

70. Ottoboni L, De Feo D, Merlini A, Martino G. Commonalities in immune modulation between mesenchymal stem cells (MSCs) and neural stem/precursor cells (NPCs). *Immunol Lett* 2015 http://dx.doi.org/10.1016/j.imlet.2015.05.005.

71. Hermann DM, Peruzzotti-Jametti L, Schlechter J, Bernstock JD, Doeppner TR, Pluchino S. Neural precursor cells in the ischemic brain - integration, cellular crosstalk, and consequences for stroke recovery. *Front Cell Neurosci* 2014;**8**:291. http://dx.doi.org/10.3389/fncel.2014.00291.

72. Donegà M, Giusto E, Cossetti C, Schaeffer J, Pluchino S. Systemic injection of neural stem/progenitor cells in mice with chronic EAE. *J Vis Exp JoVE* 2014(86) http://dx.doi.org/10.3791/51154.

73. Steinman L. A molecular trio in relapse and remission in multiple sclerosis. *Nat Rev Immunol* 2009;**9**(6):440–7. http://dx.doi.org/10.1038/nri2548.

74. O'Hare E, Young PJ. Childhood spinal muscular atrophy and stem cell research: is cellular replacement therapy the answer? (Review). *Mol Med Rep* 2009;**2**(1):3–5. http://dx.doi.org/10.3892/mmr_00000052.

75. Donnelly EM, Boulis NM. Update on gene and stem cell therapy approaches for spinal muscular atrophy. *Expert Opin Biol Ther* 2012;**12**(11):1463–71. http://dx.doi.org/10.1517/14712598.2012.711306.

76. Kostova FV, Williams VC, Heemskerk J, et al. Spinal muscular atrophy: classification, diagnosis, management, pathogenesis, and future research directions. *J Child Neurol* 2007;**22**(8):926–45. http://dx.doi.org/10.1177/0883073807305662.

77. Swoboda KJ, Kissel JT, Crawford TO, et al. Perspectives on clinical trials in spinal muscular atrophy. *J Child Neurol* 2007;**22**(8):957–66. http://dx.doi.org/10.1177/0883073807305665.

78. O'Connor DM, Boulis NM. Cellular and molecular approaches to motor neuron therapy in amyotrophic lateral sclerosis and spinal muscular atrophy. *Neurosci Lett* 2012;**527**(2):78–84. http://dx.doi.org/10.1016/j.neulet.2012.04.079.

79. Gordon P, Corcia P, Meininger V. New therapy options for amyotrophic lateral sclerosis. *Expert Opin Pharmacother* 2013;**14**(14):1907–17. http://dx.doi.org/10.1517/14656566.2013.819344.

80. Cao J, Li X, Lu X, Zhang C, Yu H, Zhao T. Cells derived from iPSC can be immunogenic - yes or no? *Protein Cell* 2014;**5**(1):1–3. http://dx.doi.org/10.1007/s13238-013-0003-2.

81. Miyazaki S, Yamamoto H, Miyoshi N, et al. Emerging methods for preparing iPS cells. *Jpn J Clin Oncol* 2012;**42**(9):773–9. http://dx.doi.org/10.1093/jjco/hys108.

82. Lister R, Pelizzola M, Kida YS, et al. Hotspots of aberrant epigenomic reprogramming in human induced pluripotent stem cells. *Nature* 2011;**471**(7336):68–73. http://dx.doi.org/10.1038/nature09798.

83. Nori S, Okada Y, Nishimura S, et al. Long-term safety issues of iPSC-based cell therapy in a spinal cord injury model: oncogenic transformation with epithelial-mesenchymal transition. *Stem Cell Rep* 2015;**4**(3):360–73. http://dx.doi.org/10.1016/j.stemcr.2015.01.006.

84. Kobayashi Y, Okada Y, Itakura G, et al. Pre-evaluated safe human iPSC-derived neural stem cells promote functional recovery after spinal cord injury in common marmoset without tumorigenicity. *PloS One* 2012;**7**(12):e52787. http://dx.doi.org/10.1371/journal.pone.0052787.

85. Bayart E, Cohen-Haguenauer O. Technological overview of iPS induction from human adult somatic cells. *Curr Gene Ther* 2013;**13**(2):73–92.
86. Naujok O, Kaldrack J, Taivankhuu T, Jörns A, Lenzen S. Selective removal of undifferentiated embryonic stem cells from differentiation cultures through HSV1 thymidine kinase and ganciclovir treatment. *Stem Cell Rev* 2010;**6**(3):450–61. http://dx.doi.org/10.1007/s12015-010-9148-z.
87. Lim T-T, Geisen C, Hesse M, Fleischmann BK, Zimmermann K, Pfeifer A. Lentiviral vector mediated thymidine kinase expression in pluripotent stem cells enables removal of tumorigenic cells. *PloS One* 2013;**8**(7):e70543. http://dx.doi.org/10.1371/journal.pone.0070543.
88. Nizzardo M, Simone C, Falcone M, et al. Human motor neuron generation from embryonic stem cells and induced pluripotent stem cells. *Cell Mol Life Sci CMLS* 2010;**67**(22):3837–47. http://dx.doi.org/10.1007/s00018-010-0463-y.
89. Lorson MA, Spate LD, Prather RS, Lorson CL. Identification and characterization of the porcine (Sus scrofa) survival motor neuron (SMN1) gene: an animal model for therapeutic studies. *Dev Dyn Off Publ Am Assoc Anat* 2008;**237**(8):2268–78. http://dx.doi.org/10.1002/dvdy.21642.
90. Duque SI, Arnold WD, Odermatt P, et al. A large animal model of spinal muscular atrophy and correction of phenotype. *Ann Neurol* 2014 http://dx.doi.org/10.1002/ana.24332.
91. Frattini E, Ruggieri M, Salani S, et al. Pluripotent stem cell-based models of spinal muscular atrophy. *Mol Cell Neurosci* 2015;**64**:44–50. http://dx.doi.org/10.1016/j.mcn.2014.12.005.
92. Liu Q, Dreyfuss G. A novel nuclear structure containing the survival of motor neurons protein. *EMBO J* 1996;**15**(14):3555–65.
93. Makhortova NR, Hayhurst M, Cerqueira A, et al. A screen for regulators of survival of motor neuron protein levels. *Nat Chem Biol* 2011;**7**(8):544–52. http://dx.doi.org/10.1038/nchembio.595.
94. Naryshkin NA, Weetall M, Dakka A, et al. Motor neuron disease. SMN2 splicing modifiers improve motor function and longevity in mice with spinal muscular atrophy. *Science* 2014;**345**(6197):688–93. http://dx.doi.org/10.1126/science.1250127.
95. Garbes L, Heesen L, Hölker I, et al. VPA response in SMA is suppressed by the fatty acid translocase CD36. *Hum Mol Genet* 2013;**22**(2):398–407. http://dx.doi.org/10.1093/hmg/dds437.
96. Wirth B, Garbes L, Riessland M. How genetic modifiers influence the phenotype of spinal muscular atrophy and suggest future therapeutic approaches. *Curr Opin Genet Dev* 2013;**23**(3):330–8. http://dx.doi.org/10.1016/j.gde.2013.03.003.
97. Nizzardo M, Simone C, Salani S, et al. Effect of combined systemic and local morpholino treatment on the spinal muscular atrophy Δ7 mouse model phenotype. *Clin Ther* 2014;**36**(3):340–56. e5. http://dx.doi.org/10.1016/j.clinthera.2014.02.004.
98. Osman EY, Miller MR, Robbins KL, et al. Morpholino antisense oligonucleotides targeting intronic repressor Element1 improve phenotype in SMA mouse models. *Hum Mol Genet* 2014;**23**(18):4832–45. http://dx.doi.org/10.1093/hmg/ddu198.
99. Zhou H, Janghra N, Mitrpant C, et al. A novel morpholino oligomer targeting ISS-N1 improves rescue of severe spinal muscular atrophy transgenic mice. *Hum Gene Ther* 2013;**24**(3):331–42. http://dx.doi.org/10.1089/hum.2012.211.
100. Benchaouir R, Robin V, Goyenvalle A. Gene and splicing therapies for neuromuscular diseases. *Front Biosci Landmark Ed* 2015;**20**:1190–233.
101. Faravelli I, Nizzardo M, Comi GP, Corti S. Spinal muscular atrophy-recent therapeutic advances for an old challenge. *Nat Rev Neurol* 2015;**11**(6):351–9. http://dx.doi.org/10.1038/nrneurol.2015.77.

12

Clinical Trials to Date

B.J. Mader and N. Boulis

Emory University, Atlanta, GA, United States

Molecular and Cellular Therapies for Motor Neuron Diseases.
DOI: http://dx.doi.org/10.1016/B978-0-12-802257-3.00012-2

277

SITUATION AND INTRODUCTION

The characteristic symptoms of neurological disorders are caused by the dysfunction or death of cells required for normal function of the central and/or peripheral nervous system. Neural pathways, originating from the highest brain levels, send out complex signals through a series of neuronal synaptic connections arranged through the spinal cord and on through peripheral nerves, and must be maintained in order to preserve normal neurophysiology. For example, amyotrophic lateral sclerosis (ALS) is a devastating degenerative neuromuscular disease believed to occur as a result of cellular dysfunction at the spinal cord level and/or the neuromuscular junction, a specific location where the electrochemically propagated nerve impulse regulates muscle contraction and tone.[1] There is evidence that the motor neurons start to malfunction at a time that precedes detection of motor neuron death.[2] Unfortunately, the sobering reality is that in many neurological diseases there remains a deep chasm that is stalling the development of effective therapies due to a general lack of understanding of the underlying pathological mechanisms driving the disease processes.

The vast majority of preclinical and clinical studies are focused on those disease processes affecting the largest patient populations. The two most funded research areas for neurological disorders are in Alzheimer's and Parkinson's, diseases for which there are over 3000 clinical trials listed on ClincalTrials.gov. Alzheimer's and Parkinson's diseases are the two most common neurodegenerative diseases and were among the first to be considered for clinical trials involving gene- or cell-based therapies. The advanced development of gene based therapeutics in these disorders compared to others is due to a slightly better understanding of the disease process. For example, researchers have determined that Parkinson's disease symptoms occur due to dopamine deficiency caused by neuronal cell death within the substantia nigra pars compacta, an area of the brain that regulates body posture and movement by signaling other neurons through synaptic release of dopamine.[3] Therefore, clinical trial strategies have been designed and implemented to elevate or replace dopamine levels to improve functional movement. Some of these strategies have improved the standard of care for many Parkinson's disease patients, but still no cure has been found. For many of these age-related neurodegenerative diseases, the key to development of effective therapeutics remains in achieving a stronger understanding of the disease process.

The National Institutes of Health defines motor neuron disease (MND) as a group of neurological disorders where the critical cells that control muscle activity are destroyed. These essential cells are the motor neurons that control breathing, swallowing, speaking, and walking. Upper motor neurons are found within the brain and lower motor neurons are found within the brainstem and spinal cord. Currently there exists no cure for MND. While treatment options are available for MND they only provide the patient temporary symptomatic relief or maintain quality of life. The most prevalent MND is ALS. There have been no therapies to date shown to provide long-term benefit for those suffering from this devastating disease. Currently the best option for ALS patients is a drug called riluzole, and it has been shown to increase life expectancy by 2–3 months, albeit on a ventilator. Neurodegenerative diseases such as MND greatly reduce a patient's quality of life and can be life threatening. Patients who have exhausted all other treatment options are prime candidates for gene- or cell-based therapeutics.[4,5]

Currently there are nearly 1000 clinical trials that have been launched worldwide to study MND clinical pathologies, or to validate the safety and efficacy of promising MND therapeutics. Even with this high level of scientific investigation and clinical interest there are no cures or therapies proven to provide long-term benefits.[6] While the majority of ALS diagnoses are sporadic, there are many patients who have a genetic link associated with the familial form of ALS.[7] Research into familial ALS has uncovered over 150 mutations in the gene encoding superoxide dismutase 1 (SOD1) that are linked to the disease.[8] These mutations are associated with either a loss of the SOD1 enzymatic activity or a toxic gain of function.[9,10] These potential causes of ALS lead to increased oxidative stress, protein misfolding and aggregation, mitochondrial dysfunction, and excitotoxicity. Identification of these mutations can be used to study and validate therapeutic targets to treat those patients diagnosed with a specific mutation. While these are exciting possibilities, the fact remains that the majority of ALS patients are diagnosed without a known genetic component.

While the vast majority of MND is diagnosed in patients over the age of 40, an MND called spinal muscular atrophy (SMA) occurs in patients during the early stages of life. Ninety-eight percent of patients with SMA have an alteration of the gene SMN1 through deletion, rearrangement, or mutation.[11] The discovery of alterations within this gene in 1995 spawned many SMA studies focused on the structure of the SMN1 gene. It is believed that the SMN protein protects and sustains the motor neuron population through the maintenance of axonal transport and neuromuscular junctions. Patients presenting with SMA-like symptoms are tested for homozygous deletion for the SMN1 gene, which if confirmed, is 100% specific for developing SMA. Therefore strategies aiming to replace the deleted gene within these motor neurons are believed to very likely provide therapeutic relief. Preclinical studies using SMA models have shown that viral-vector-mediated expression of SMN1 in skeletal muscle and in the brain increases motor function and motor neuron survival.[12–15] Following up on the preclinical success of this gene therapy strategy, a clinical trial has been launched to treat SMA using modified viral vector expressing the native SMN1 protein (NCT02122952). This viral-mediated gene therapy is intended to halt or delay the disease progression, thus offering a better quality of life for patients suffering from this terrible disease. Diseases with a strong genetic link, such as SMA, are prime candidates for the implementation of gene- or cell-based therapies, which are designed to replenish or correct aberrant gene expression.

Therapeutic Technology

For decades, researchers and clinicians have advanced gene- and cell-based therapies to the cusp of fundamentally altering the way we treat human disease. The ability to transplant multipotent or pluripotent cells or to constitutively express a therapeutic gene in the target diseased tissue provides an attractive alternative to continuous pharmacological intervention. Cellular therapies have been approved by the U.S. Food and Drug Administration (FDA) to treat cosmetic flaws (Laviv®, Fibrocell Technologies Inc.), blood disorders (Ducord, Duke University), cartilage deficiencies (Carticel, Genzyme Biosurgery), dental pathologies (Gintuit, Organogenesis Incorporated), and cancer (Provenge®, Dendreon Corporation). There is widespread research in numerous diseases that utilize viral-vector-mediated therapies. A great many preclinical trials have shown efficacy with viral-mediated gene delivery in animal models of disease. Gene-based therapeutics have been shown to be generally safe for human application, however there were serious complications on the path to acceptability. Specifically, a patient died from a therapeutically induced catastrophic immune response in an early clinical trial. The death of this patient caused a significant delay in the development of gene-based technology as the safety of viral vectors was called into question.[16] Some cellular therapeutic products are already approved for public use but gene therapy products are not yet approved for market release. A particular clinical trial (NCT00643890) using viral-mediated gene therapy in Parkinson's disease did show efficacy above the placebo range but the study was terminated because it did not elevate the current standard of care.[17] Modern generations of viral vectors have been demonstrated to be safe in multiple trials but their efficacy in clinical trials rarely surpass either the placebo-effect range or the current standard of care.

Viral-Mediated Gene Therapy

Gene therapy uses viral or synthetic vectors to deliver exogenous genetic material to living cells and tissues to elicit therapeutic benefit. Preclinical studies commonly use modified viral vectors derived from naturally occurring viruses that infect human cells. Modified viruses include retroviruses, such as lenti and HIV; Herpes simplex virus 1 presents as cold sores; adenovirus presents as respiratory sickness; or adeno-associated viruses, which have no known pathogenicity in humans. Researchers use the naturally evolved mechanisms of infectious viruses to transduce diseased cells and express a potentially therapeutic gene. For example, the adeno-associated virus serotype 9 (AAV9) is particularly useful for peripheral administration (i.e., intravenous) because it can cross the blood–brain barrier to express exogenous gene products within the central nervous system (CNS).[18–20] While some particular genetic characteristics of these viral vectors make them attractive for therapeutic engineering, other pathogenic viral genes must be removed to make them safe, and therefore useful for therapeutic development.

Stem Cell Therapy

Stem cells are extremely promising therapeutics that have the potential to support or replace cells vulnerable to dysfunction and/or death in a wide range of human diseases. The term "stem cell" was first used by German scientist Ernst Haeckel to describe the fertilized egg that will give rise to a complicated organism.[21] Stem cells demonstrate these three general characteristics: (1) they can divide and multiply for prolonged periods; (2) they are unspecialized cells; and (3) they can differentiate into specific cell types depending on stimuli (http://stemcells.nih.gov/info/basics/pages/basics2.aspx). There are different types of stem cells depending on how they were isolated and from what tissue they were taken. Stem cells after transplantation have been shown to align with and become tissue-specific cells given specific physiological or experimental stimuli. Other stem cell strategies involve transplanting cells that will express growth factors or neurotrophic factors to support viability or function in vulnerable cell populations.[22] Although to date there have been many stem cell-based clinical trials launched to treat neurological diseases, the field remains in the early stages of development and application. Many preclinical studies have shown great efficacy in a large number of wide-ranging disease models, but successful implementation into human therapies have been incremental.

The advancement of stem cell therapy has been relatively slow because early stem cell lines were embryo-derived, a stem cell source that has remained a national ethical dilemma from its inception. Many nations have government-imposed regulations on the development and use of stem cells in biomedical research. The controversial issues surrounding stem cell therapeutics span ethical, moral, and political debates.[23]

The two main types of stem cells are human embryonic derived stem cells (hESCs) and somatic (adult) stem cells (ASCs) that are found in and isolated from specific body tissues. A particular advantage of hESCs is their ability to differentiate to become any cell found within the human body, which is termed "pluripotency."[24] Pluripotent cells are amenable to a broad range of therapeutic applications given their ability to divide for long periods of time in addition to their vast differentiation potential. Conversely, isolated ASCs do not readily produce large quantities of stem cells and may only be able to differentiate into the limited cell types found in the tissue of origin.[25] Within the bone marrow, a common tissue where stem cells are isolated from, there are two types of stem cells: hematopoietic (HSCs; give rise to all blood cells) and mesenchymal stem cells (MSCs; give rise to bone, cartilage, and fat cells).[26] In the 1990s it was discovered that neural stem cells (NSCs) inhabited the adult brain and could give rise to the three major brain cell types: astrocytes, oligodendrocytes, and neurons.[27,28] This discovery was of particular interest to those researching neurological disease in that it could allow for isolation of neural precursors to develop novel strategies for advanced treatments.

The deleterious effect of the host's immune system on stem cell survival after transplantation is of particular concern. Patients that receive allogeneic stem cells (usually hESCs) are likely to need continuous administration of immunosuppressive drugs, which can leave the patient vulnerable to infection. Stem cell transplantations using "autologous" stem cells are believed to reduce the risk of host immune-response mediated rejection as the patient is receiving their own cells and not cells from a foreign donor. While a patient is less likely to have a robust immune response after autologous cell transplantation, when compared to donor cell transplantation, the risk remains that newly transplanted autologous cells within diseased tissue will not function as they did within the tissue of origin. There is also potential that the transplanted cells isolated from a diseased patient could be in a diseased state as well. Since many millions of cells are needed for cell-mediated therapy, ESCs are more amenable to growing large numbers of therapeutic cells for the treatment of disease. Therefore, more research is warranted on the specific conditions that are required to achieve successful transplantation and subsequent engraftment.

Another type of stem cell, termed the induced pluripotent stem cell (iPSC) has recently been elevated as the new improved stem cell therapy that has characteristics of both hESCs and MSCs. iPSCs are pluripotent because they have been reprogrammed to replicate gene expression observed in ESCs. Human iPSCs express embryonic stem cell markers and are capable of generating cells characteristic of all three germ layers.[29] This technology enables the transplantation of reprogrammed autologous cells that maintain somatic genetic homogeneity with the patient (as with autologous MSCs) but with the potential to differentiate into any cell found in the body (as with hESCs). While these cells have been met with great excitement in the therapeutic arena, a concerning issue with advancing the clinical use of iPSCs is their propensity to form tumors.[30] As with any new medical technology, a patient's safety must be paramount.

In recent years the FDA has approved many trials that aim to use stem cell therapy to treat many devastating diseases. These trials include evaluation of stem cell therapies in diseases such as neurodegenerative disorders, cardiovascular disorders, genetic disorders, cancer, musculoskeletal disorders, and degenerative eye conditions. Moreover, HSCs have been included in blood transfusions for almost half a century, and transplantations of isolated and expanded HSCs have been conducted (http://stemcells.nih.gov/info/scireport/pages/chapter5.aspx).

Review of Clinical Trial Process

Clinical trials are used to determine whether or not novel therapies that exhibit therapeutic benefit in preclinical studies can be "translated" into effective treatments for human disease. In the United States, the FDA has regulated the clinical trial process since the 1970s. Prior to the FDA approving the commencement of human clinical trials, they first require that these clinical trials must be designed to comply with established good clinical practices (GCPs) to ensure the safety of human trial subjects. There are now international guidelines established in concert with the FDA to standardize the GCPs used to conduct clinical trials throughout the world. Therapeutics should only advance to early-stage clinical trials if there is sufficient preclinical data supporting the probable safe usage in human subjects. Therapies validated by the clinical trial process can be chemical, biological, thermal electrical, or behavioral. Depending on the nature of the therapy to be tested, there is a department within the FDA responsible for approval and oversight of the trial. Studies in gene and cellular therapies are specifically regulated by the FDA's Center for Biologics Evaluation and Research (CBER). In addition to establishing safe effective parameters for the development of therapeutic biologics, the CBER also educates the public on the safe and appropriate usage of biological products.

The clinical trial process (http://www.fda.gov/ForPatients/Approvals/Drugs/ucm405622.htm) utilizes a series of trials to scientifically establish parameters for use of an investigational therapeutic pertaining to safety, dosage, efficacy, and side effects. In phase I trials, a team of investigators will test a new therapeutic in a relatively limited group of people to establish safety data. In addition to safety, the dosing range and potential efficacy may be monitored prior to launching the phase II trial. Phase I trials are usually conducted in healthy subjects to ensure that the therapeutic itself causes no adverse effects in the absence of a disease state. However, in cell- and gene-based therapeutics, phase I trials usually evaluate safety as a primary measure, and efficacy as a secondary measure in diseased patients. This diversion from enrolling healthy participants in phase I clinical trials is due to the potential for long-term or permanent adverse effects associated with gene- or cell-based therapies when compared to transient small-molecule drug therapies. Phase I trials are usually open-label, which means that the patient is aware of the potential therapeutic outcome. This means that there is a likely placebo effect that must be taken into consideration when looking to move therapeutics into the next trial phase.

The phase II trial remains committed to monitoring safety of treatment in a larger cohort while determining whether the therapeutic represents an effective treatment for the relevant disease. If possible, these larger cohorts will be conducted under "closed-label" and "blinded" conditions, meaning that that trial subjects are not told what therapeutic they are given or even whether or not they have been given the therapeutic or the inert placebo. This gives the trial more validity if there is evidence of efficacy in the treated group versus the control because neither group knows what they had received. Phase I and II clinical trials involving terminally ill patients or complex invasive procedures may not include a true "control or placebo" group. While the ideal trial involves blinding the researchers and the patients to the treatments it is unlikely that there will be true control groups in late-stage neurodegenerative disease studies, unless they advance to phase III, due to the invasive nature of intraparenchymal or intrathecal CNS delivery. In cases such as these, efficacy is usually established by a comparison with established disease progression rates.

If the therapeutic shows efficacy in a phase II trial then it will be moved into the phase III trial to be administered in multiple-center trials to confirm efficacy, generate more safety and dosage data, and compare it with currently available treatments. Phase III trials have large control and therapeutic cohorts (usually minimum of 100 participants) where randomized patients and investigational clinicians are blinded to who gets what. Finally, if the novel therapeutic drug or biologic advances successfully through the first three clinical trial phases, and it exceeds the current standard of care, it will be brought to market for distribution and clinical use. During this open-market period, ongoing phase IV studies continue to evaluate the therapeutic efficacy in diverse populations and determine long-term effects of use. These studies are of particular use to determine future applications for new therapeutics.

STEM CELL CLINICAL TRIALS IN AMYOTROPHIC LATERAL SCLEROSIS

There are close to 5000 clinical trials related to stem cell therapies listed on ClincalTrials.gov, a US government website that lists proposed, ongoing, or completed clinical trials. Numerous clinical trials using stem cell therapy to treat ALS have been launched worldwide. These ALS clinical trials have shown the use of stem cell therapies to be safe for delivery to affected muscle regions, cerebrospinal fluid (CSF),[31] and direct CNS parenchyma.[32,33] As many of the following studies have yet to release their trial data, much of the information provided is taken from ClincalTrials. gov. However, the FDA cautions that while Clinicaltrials.gov makes available information provided by the trial sponsors, this information is neither reviewed nor confirmed by the FDA. These stem cell trials are organized by geographical location, and trials that are conducted by the same group of investigators are listed with one another (Table 12.1).

North America

Pennsylvania

One of the first documented uses of stem cell therapy to treat ALS was described in 2001 by researchers at Thomas Jefferson University's CNS Gene Therapy Center in a three-patient case study that described an "off-label" usage of HSCs isolated from peripheral blood.[44] The stem cell isolation protocol was FDA-approved for use in bone marrow transplantations. In a 51-year-old male patient (Patient 1) CD34+ stem cells were intrathecally infused (1×10^8 cells) into the lumbar region over a 2-day period. A 2-hour transient loss of sensation in the lower limbs was reported during the procedure, but subsided quickly. Patient 1 had improvements to speech but demonstrated no other neurological advantages. Patient 2 was a male who received cells transplanted into the cisterna magna and at the level L3–L4 with a fluoroscopically guided needle. A total of 20 million CD34+ PBS cells were infused with 1/3 injected in the cisterna. In this patient, no inflammatory markers were detected and no adverse clinical or serological effects were seen after the procedure. Patient 3 was a female who received 100 million cells with 40% injected into the intrathecal space at C1/C2 and the remaining 60% into the lumbar region. Patient 3 had reported gains in leg and neck strength through 6 months of follow-up. This preliminary case study showed no side effects at 6- and 12-month periods during the follow-up period. However, it was stated that none of the patients that underwent the procedure elected to repeat the therapy.[44]

United States (Multicenter): Neuralstem

The biotech company Neuralstem launched an open-label phase I clinical trial (NCT01348451) to establish the feasibility and safety for treating ALS patients with human spinal cord-derived neural stem cells (HSSCs). The HSSCs are derived from 8-week fetal tissue obtained by an elective

TABLE 12.1 Amyotrophic Lateral Sclerosis Stem Cell Trials Listed on Clinicaltrials.gov

Sponsor/trial location	Nomenclature (estimated start)	Trial phase: outcome measures	Stem cell type	Delivery technique	Dosages tested	Published results (PMID)
NORTH AMERICA						
Neuralstem/ Emory University (Atlanta, GA)	NCT01348451 (2009)	Phase I	Human spinal cord-derived (fetal)	Direct injections into the cervical and/or lumbar region of spinal cord	500,000 to 1.5×10^6 cells total	Riley et al.[34] (22565043) Glass et al.[49] (22415942) Feldman et al.[35] (24510776) Boulis et al.[22] (22158518) Tadesse et al.[36] (25540804)
Neuralstem/Emory University (Atlanta, GA)	NCT01730716 (2013)	Phase II			500,000 to 1.5×10^6 cells total	N/A
Mass general (Boston, MA)						
U. of Michigan (Ann Arbor, MI)						
Mayo clinic/ (Rochester, MN)	NCT01142856 (2010)	Phase I	Autologous adipose- derived mesenchymal (adult)	Intrathecal infusion	0.6×10^6 cells/kg	http://www. cellr4.org/ article/530
Mayo clinic/ (Rochester, MN)	NCT01609283 (2012)	Phase I			1×10^7 to 2×10^8 cells total	N/A

Institution/Location	NCT number (Year)	Phase	Cell type	Delivery method	Cell dose	Reference
TCA cellular therapy/Covington, LA	NCT01082653 (2010)	Phase I	Autologous bone-marrow-derived mesenchymal (adult)	Intrathecal infusion	N/A	N/A
Q Therapeutics, Inc./(Salt Lake City, UT)	NCT02478450 (2015)	Phase I/II	Glial restricted progenitor cells	Direct injections into the cervical or lumbar region of spinal cord	N/A	N/A
Brainstorm-cell therapeutics/(Mass General, UMass, Mayo Clinic)	NCT02017912 (2013)	Phase II	Autologous bone marrow–derived mesenchymal that express neurotrophic factors	Intrathecal infusion and intramuscular injection	N/A	N/A
Hospital Universitario/(Monterrey, Mexico)	NCT01933321 (2012)	Phase I	Autologous hematopoietic	Intrathecal infusion	N/A	N/A
EUROPE						
Azienda Ospedaliera/(Santa Maria, Italy)	NCT01254539 (2001)	Phase I/II	Autologous bone-marrow-derived mesenchymal	Intrathecal infusion	8×10^7 to 1.5×10^8 cells total	Mazzini et al.[37] (19682989) Mazzini et al.[38] (21954839)
Foundation for health research and training/(Murcia, Spain)	NCT01640067 (2011)	Phase I	Human neural (fetal)	Unilateral or bilateral direct spinal cord injection	2.25×10^6 to 5.5×10^6 cells total	Mazzini et al.[39] (25889343)
Foundation for health research and training/(Murcia, Spain)	NCT00855400 (2009)	Phase I/II	Autologous bone-marrow-derived mesenchymal (described as mononuclear)	Direct injections into the thoracic region of spinal cord	1.38×10^8 to 6.03×10^8 cells total	Blanquer et al.[40] (22415951)

(*Continued*)

TABLE 12.1 Amyotrophic Lateral Sclerosis Stem Cell Trials Listed on Clinicaltrials.gov (Continued)

Sponsor/trial location	Nomenclature (estimated start)	Trial phase: outcome measures	Stem cell type	Delivery technique	Dosages tested	Published results (PMID)
Andalusian initiative for advanced therapies/ (Multi-Locations, Spain)	NCT02290886 (2014)	Phase I/II	Adipose-derived mesenchymal	Intravenous injection	1×10^6 to 4×10^6 cells total	N/A
Red de Terapia Celular/(Murcia, Spain)	NCT02286011 (2014)	Phase I	Autologous bone-marrow-derived mesenchymal (described as mononuclear)	Intramuscular infusion	1×10^8 to 1.2×10^9 cells total	N/A
Pomeranian medical uni./(Szczecin, Poland)	NCT02193893	Phase I	Autologous bone-marrow-derived mesenchymal	Intrathecal infusion	N/A	N/A
ASIA						
Haddassah medical Org./(Jerusalem, Israel)	NCT01051882 (2010)	Phase I/II	Autologous bone-marrow-derived mesenchymal that express neurotrophic factors	Intrathecal infusion and intramuscular injection	2.4×10^7 to 6×10^7 cells total	Karussis et al.[41] (20937945)
	NCT01777646 (2012)	Phase II			9.4×10^7 to 1.4×10^8 cells total	N/A
Neurogen Brain and Spine Institute/ (Mumbai, India)	NCT02242071 (2008)	Phase I	Autologous bone-marrow-derived mesenchymal	Intrathecal infusion and intramuscular injection	$<8 \times 10^7$ cells total	Sharma et al.[42] (25973331)
	NCT01984814 (2008)	Phase II			N/A	N/A

Institution/(Location)	NCT (Year)	Phase	Cell type	Route	Dose	Reference
Royan institute/ (Tehran, Iran)	NCT01759784 (2012)	Phase I	Autologous bone-marrow–derived mesenchymal	Intraventricular injection	N/A	N/A
	NCT01759797 (2012)	Phase I		Intravenous injection	N/A	N/A
	NCT01771640 (2013)	Phase I		Intrathecal infusion	N/A	N/A
	NCT02492516 (2014)	Phase I	Autologous adipose–derived mesenchymal	Intravenous injection	N/A	N/A
Alzahra hospital/ (Isfahan, Iran)	NCT02116634 (2014)	Phase I/II	Autologous bone-marrow–derived mesenchymal	Intrathecal infusion	1×10^8 cells total	N/A
Corestem Hanyang University/ (Seoul, Republic of Korea)	NCT01363401 (2011)	Phase I/II	Autologous bone-marrow-derived mesenchymal	Intrathecal injection	1×10^6 cells/kg	Oh et al.[43] (25934946)
	NCT01758510 (2012)	Phase I	Allogenic bone-derived mesenchymal	Intrathecal infusion	2.5×10^5 to 1×10^6 cells/kg	N/A

abortion. Their strategy was based on post mortem analysis of ALS patients demonstrating an abnormal amount of glutamate accumulation potentially leading to motor neuron toxicity within the brain and spinal cord.[45,46] HSSCs are known to produce amino acid transporters and could potentially reduce toxic levels of glutamate within the ALS CNS. Additionally, they hypothesized that the expression of neurotrophic factors from these cells could provide increased motor neuron support to delay or halt ALS disease progression.

The surgical procedure involved the transplantation of the HSSCs into the ALS patient spinal cord parenchyma after the protective lamina bone was removed through a procedure called a laminectomy. Following the laminectomy and exposure of the spinal cord, a specialized platform was used to minimize potential risk of spinal cord damage.[47,48] Injections were made in 5–10 locations along the cord. The study design included 18 patients presenting with ALS divided into 5 different groups. The groups were laid out in a risk-escalation model where safety in earlier groups was validated before moving on to successive riskier groups.

Group "A" included six patients with advanced-stage ALS who are unable to walk regardless of breathing status. This group was divided into two subsets. The first subset of three patients received injections unilaterally into the lumber cord. The second subset of three patients received bilateral injections into the lumbar cord.

Group "B" included three ambulatory ALS patients who received injections unilaterally in the lumbar cord, whereas group "C" also included three ambulatory patients who received bilateral injections in the lumbar cord. Group "D" included ambulatory patients experiencing upper-limb dysfunction, and five received unilateral cervical injections. Finally, group "E" included ambulatory patients from group "C" who received unilateral cervical injections, in addition to their bilateral lower spine injections.

In total, the trial enrolled 15 patients (13 male, 2 female) who received either HSSC lumbar injections, cervical injections, or both.[33,35] The age range of the trial participants was 35–66 years. This study followed a "risk escalation" paradigm in order to minimize the potential risk for harming the patient. All injections were administered through a specialized device and each injection consisted of a 10-μL volume containing 100,000 cells. Over the course of the study, 18 surgeries were performed to transplant between 500,000 and 1.5 million cells (Group "E").

The result of this phase I safety trial confirmed that the procedure was well-tolerated and exhibited minimal perioperative or postoperative adverse events.[33,34,49] These results successfully established the safety of the procedure and the usage of HSSCs in single or dual intraparenchymal injections in the cervical and lumbar cord regions. At the time of this publication, 7 of the 15 participants had died. Six of those patients died due to respiratory complications associated with ALS progression, and one died of congenital heart defect.[33] Following autopsy, investigators found no evidence of hemorrhage, cyst formation, or inflammatory response. A nest of transplanted cells were believed to be found within a subject (#14) that underwent cervical injections suggesting engraftment within the host tissue.[36] None of the first 12 subjects were reported to have an acceleration of ALS disease progression. Alternatively, one subject in group C (#11) showed modest gains in neurological testing (ALS functional rating score (ALSFRS-R), hand-held dynamometry (HHD), and electrical impedance myography measurements).

Of the group D patients (#13–15) receiving cervical injections, one (#13) developed cervical kyphosis and died 20 months later, and the other (#14) died 200 days later. Patient #15 was reported to demonstrate a delayed progression of the disease following transplantation. Clinical progression was also monitored for these patients but at the time of this publication has yet to be reported.

Following the success of Neuralstem's phase I trial, a phase II study was launched in September of 2013 and surgical procedures to deliver the stem cells to the cord were completed in the July of 2014. This open-label phase II Neuralstem clinical trial (NCT01730716) was initiated to determine the safety and tolerability of human spinal-derived stem-cell transplantation in a dose-escalation trial. This trial will determine the maximum tolerated dose of HSSCs that can be used to treat ALS. Patients will be followed through 24 months postsurgery. Patients enrolled will be ambulatory but suffer from extremity weakness and/or spasticity caused by ALS progression. This trial's endpoints are expected to be completed by late 2016.

Minnesota: Mayo Clinic

In the summer of 2010, a trial was launched by the Mayo Clinic (NCT01142856) to test the safety of the cell-harvesting and stem cell delivery protocol. For this study a 62-year-old male ALS patient underwent a bone marrow aspiration for the expansion of autologous MSCs in a good manufacturing practices (GMP) facility. These cells were expanded *ex vivo* to increase the population of MSCs. Following successful expansion, these cells were to be delivered via lumbar puncture and intrathecal infusion into the CSF. However, the cell aspirate expansion produced a large portion of mature osteoclasts or their progenitors while containing very few MSCs. Time-extended cultures demonstrated that the MSCs were nondividing. The treatment was unable to go forward after a second attempt at MSC expansion failed, and by protocol guidelines the patient was withdrawn from the study. The patient was instead treated with allogeneic MSCs from a donor (http://www.cellr4.org/article/530). The patient will be followed up with and observed for 2 years postprocedure to confirm safety and identify changes to neurological disability score.

As the primary outcome of the study was to establish safety for the "cell product infusion procedure," donor MSCs did not prohibit evaluation of the procedure. After confirming the viability of the cell product and that it contained no known endotoxin, mycoplasma, or bacterial contamination, the cells were injected into the lumbar region intrathecal space. A total infusion volume of 4 mL containing 0.6×10^6 MSCs/kg of the patient's body weight was administered. The patient was observed for vital signs and the injection site for 48 h after the procedure, and the patient was followed for 12 months to determine neurological alterations. No adverse events or side effects relating to the cell treatment were observed up to 12 months of clinical observation.

Investigators reported evidence of mild efficacy, although this positive outcome must be taken in the context of a one-patient study. Forced vital capacity (FVC), a measure of respiratory function, remained mostly unchanged. ALS-FRS score exhibited an early improvement that was mostly sustained over the 12-month follow-up. Bilateral improvement was observed in the upper limb grip strength HHD. However, only the right grip strength improvement was sustained while the left grip strength eventually dropped below baseline after 12 months. By telephone assessment, the patient was believed to have achieved a "remarkable repossession of quality of life" (http://www.cellr4.org/article/530). This study claims to be the first to use a noninvasive procedure to infuse adult stem cells in contrast to studies that have infused laboratory-differentiated stem cells for ALS therapeutic intervention.

The Mayo Clinic launched a follow-on phase I clinical trial (NCT01609283) in May of 2012. The purpose of this clinical trial in the treatment of ALS was to determine the safety of intrathecal delivery of MSCs in a dose-escalation study. Twenty-five adult ALS patients who were not yet using breathing assistance were to be enrolled. Cells were isolated from the patient's adipose tissue and expanded *ex vivo* for 8 weeks prior to their therapeutic administration.

There were five treatment groups, five patients/group. In this dose-escalation study, subsequent groups were not recruited until at least 60% of the previous treatment group had undergone the therapeutic intervention and experienced no toxic or adverse events 1 month postoperation. Patients were observed through 2 years postsurgery to determine safety endpoints. Group "1" was to receive 1×10^7 cells; Group "2," 5×10^7 cells; Group "3," 5×10^7 cells per dose, 1 month between doses; Group "4," 1×10^8 cells in a single dose; Group "5," two doses of 1×10^8 cells, 1 month between doses. Patients participated in the initial clinical follow-up for 1 month by providing weekly sampling of blood and CSF in addition to magnetic resonance imaging (MRI) testing. Following the 1-month follow-up period, patients returned for clinical evaluations every 3 months.

Louisiana: TCA Cellular

An open-label phase I clinical trial (NCT01082653) was launched by TCA Cellular in March of 2010 for the treatment of ALS. This trial's primary outcome was focused on establishing safety and efficacy for the intrathecal infusion of autologous bone marrow derived stem cells (MSCs). The TCA Cellular trial was conducted at a single center in Covington, LA, and aimed to enroll a total of six patients, either male or female. Investigators wanted to ensure that no complications were found at the site of infusion and that no new neurological deficits unattributed to ALS were observed. In addition to safety implications, the study was secondarily interested in establishing the efficacy of the treatment. Investigators were interested in determining whether there are any preliminary therapeutic benefits observed. These benefits were assessed by slowing the decline of FVC, ALS-FRS, maximum voluntary isometric contraction-arm (MVIC-arm), and MVIC-grip scores. The trial was suspended in May of 2014 due to noncompliance with approved protocols FDA.

Mexico: Servicio Hematologia Hospital Universitario

In Monterrey, Mexico at the Servicio Hematologia Hospital Universitario, a phase II/III trial (NCT01933321) was launched to evaluate whether intrathecal infusion of autologous HSCs could slow or stop clinical progression of ALS. This study aimed to enroll 14 patients beginning in 2013 and was scheduled to complete in 2014. This study's primary and secondary outcome measures were to establish the procedural safety of intrathecal infusion of autologous HSCs, both during the procedure and at 12 months postoperatively.

There are no results published for this study. Of particular note, the description in ClincalTrials.gov does not list dosage or number of cells numbers to be infused intrathecally. Also a statement found in no other study description section indicates that HSCs would be administered intravenously as well as intrathecally. Finally, the classification of this study as a phase II/III is confusing because there are no "efficacy" outcome measures stated in the study description, which are always included as outcome measures for phase III clinical trials.

EUROPE

Italy: Azienda

The first instance of stem cell transplantation into the spinal cord was performed in Italy as two prospective, open-phase I safety studies (Italian

Registration #12947-29.3 and #16454-pre21-823). This group conducted the first studies to transplant autologous MSCs, as well as was the first to conduct a surgical trial in ALS.[37,38] Launched in 2001, the study enrolled 19 patients (8 female, 11 male) who were diagnosed with ALS in two separate phase I trials. The first group consisted of patients who experienced severe lower-limb dysfunction to mitigate potential risk associated with the therapy. This group had injections at the T7–T9 cord levels.[38] The second group was comprised of patients demonstrating higher levels of autonomy and function, and they were injected at the T4–T6 cord levels.[37] Prior to the surgical transplantation of MSCs, patients were assessed every 3 months for 9 months to determine the progression rate of the disease. The patients were also monitored by a clinical psychologist to ensure that their mental health was accurately assessed. Electromyography was used to assess the abductor digiti minimi muscle function. Sensory function was monitored before and after surgery through detection of somatosensory-evoked potentials that were stimulated at the tibial nerve. Brain and spinal cord MRIs were performed 15 days prior to surgery and intermittently for 2 years postsurgery.[37,38]

Patients had about 10 mL of bone marrow aspirate removed from the posterior iliac crest. Cells were then expanded for four passages. For a patient's continued inclusion in the study, their stem-cell expansion must yield minimum of 110 million cells. Cells were expanded in a GMP facility. The day of the procedure, the cells were prepared and re-suspended in 1 mL of the patient's own CSF. Cells were infused at a constant flow into the patient's CSF, and then injected directly into the spinal cord. Cells were injected in rows 3 mm apart. All procedures were conducted by the same neurosurgeon.[37,38]

This study reported no clinical, laboratory, or radiological evidence that this therapy caused any serious treatment-related events. These phase I safety trials were deemed successful as this study showed that neither the surgical procedure performed on ALS patients nor the stem cell therapy cause additional lesions at the cord. Common adverse symptoms were postoperative pain that lasted for 10 days in the longest-affected patient. Patients also reported sensory anomalies that lasted from 3 to 180 days postprocedure, although none were classified as severe. Patients did not receive the same number of cells transplanted and also had different numbers of injections. The optimal number of cells to transplant was not established and was stated as a goal for future studies. Because of these questions and inconsistencies, efficacy was not determined.

This study reported no definitive clinical efficacy, but showed long-term safety for transplantation of stem cells into the spinal cord parenchyma. The authors of this study also extend their support for future investigation into the potentially therapeutic use of ASCs in neurodegenerative disorders.[38]

The same Italian researchers launched a phase I study (NCT01640067) sponsored by Azienda Ospedaliera in 2011 to establish the safety and tolerability of expanded human fetal-derived NSCs administered through a "microsurgery" for intra-spinal cord delivery. In this study six nonambulatory patients were enrolled and treated.[39] The fetal neural cells used were obtained from fetuses that had undergone natural *in utero* death.[39] The use of neural tissue from natural miscarriages alleviated the ethical concerns commonly associated with the usage of fetal tissue. Two groups of three received either unilateral or bilateral lumbar cord injections (three injections per side) using a mild immune suppression regimen. Each injection totaled 15 microliters containing 750,000 NSCs. Following surgery patients were evaluated monthly for the first year and then every 3 months for the 2nd and 3rd years.

No patients demonstrated increased progression of ALS as a result of the treatment at 18 months postsurgery.[39] Two patients were reported to improve their scores on the ALS-FRS-R, whereas a third patient showed functional improvement that lasted for 7 months at the tibialis anterior. The other three patients refused invasive ventilation and all died at 8 months postprocedure due to respiratory failure. While this study did not show efficacy in halting ALS progression it did show that a much larger number of stem cells can be injected into the spinal cord without causing detectible toxicity.[39]

Spain: Hospital Universitario Virgen de la Arrixaca

A phase I/II clinical trial was conducted at the Hospital Universitario Virgen de la Arrixaca in Spain from 2007 to 2010. This open-label safety study was approved by the Spanish Agency of Medicines and enrolled 11 patients (5 male, 6 female). The phase I arm of the trial aimed to establish the safety of a laminectomy procedure (removal of lamina bone from vertebra) and spinal injection for autologous MSCs. The phase II arm of the trial was for the collection of the autologous bone marrow stem cells. Primary outcome measures were to determine changes to FVC every 3 months. The secondary outcome measures were changes to ALS-FRS, MRC, and Norris scale scores.

The procedure began with the sedation of the patient and then harvesting of 60 mL of bone marrow through multiple aspirations at the iliac crest. Ficoll density gradient was used to isolate the MSCs. Flow cytometry was used to quantify populations of different cell types to approximate a ratio of MSCs injected into the spine. After MSCs were expanded, transplantation first required a laminectomy to expose the dura followed by intraspinal implantation between the T3 and T4 vertebrae. One milliliter of cell suspension was injected over a 3-minute span with a 22-gauge lumbar puncture needle into the most avascular pial surface on the posterior spinal funiculus, ranging from 1 to 2.5mm from midline. Some received ipsilateral or contralateral injections of the remaining 1 mL of cell suspension. This depended on the pial vasculature structure. Eleven patients were included in the trial. Median age was 46 years with a range from 32 to 61 years. Patients received infusion on an average of 21 months postdiagnosis (range 11–40 months). Median scores at the time of entry into the study FVC 105% (79–121); ALS-FRS 30 (24–38); Norris 74 (54–95); and MRC 46 (35–54).[40]

The quantity of cells infused ranged from 1.38×10^8 to 6.03×10^8, with viability reported to be at approximately 80%. The solitary adverse event relating to the procedure lasted for a 10-minute period and was exhibited by a patient suffering a 50% reduction in spinal somatosensory evoked potentials (SSEPs), a measure of nerve conductance at the spinal cord. While six severe adverse events were reported in this study, they were not believed to occur due to the procedure. Fifty-one percent of the adverse events occurred within the first 2 weeks postsurgery. While alarming, most were resolved before the 2-month follow-up. The most common adverse events were temporary constipation, surgical wound pain, intercostal pain, hypoesthesia, paresthesia, and dysesthesia. Other less common adverse events were headache, vertigo, and insomnia.[40]

During the clinical follow-up, eight patients presented with progressive loss of the SSEPs. Nine patients showed some grade of psychological adaptation; however, no signs of clinical depression were detected. Interestingly, while patients exhibited progressive reduction in quality of life up to the procedure, this trend was seen to stabilize postprocedure as measured by the EuroQol scale[40] (patient questionnaire-based analysis evaluating health outcomes; www.euroqol.org).

Investigators explained that all patients had extradural-extraspinal postsurgical "collections" that were near the site of infusion. Throughout the course of the study three patients developed a small "subacute hemorrhage" on the postsurgical MRI near the site of administration. Other histological assays reported an increasing number of motor neurons as they approached the areas of infusion. Areas superior and inferior to T4–T5 (area of infusion) had almost 75% fewer motor neuron counts compared to the T4–T5 region. Small spherical cells, absent of neural markers, surrounded the motor neurons forming a cellular nest and were found exclusively in the grafted areas of the spine.[50]

Overall this study reported successful achievement of its primary goal due to the lack of procedurally related severe adverse events.[40] Investigators reported no evidence of accelerated declines in any of the ALS scoring parameters described above. Abnormalities observed through MRI during this trial were deemed transient or clinically irrelevant. Cells that were surrounded by the CD90+ cells did not display evidence of degeneration. Their data suggest that these cells survive a long time after grafting. TDP-43 deposits, often associate with ALS pathology, were also reduced within the spinal motor neurons.[40]

A single center, double-blinded phase I trial was launched in 2014 by Red de Terapia Celular and conducted at Hospital Universitario Virgen de la Arrixaxa. This phase I trial (NCT02286011) aimed to enroll 20 ALS patients to evaluate the safety and efficacy of using autologous MSCs. These cells are infused through four injections in the tibialis anterior muscle of the lower limb. Total volume will be 2 mL (0.5 mL per injection) and administered with a microinjector with a controlled infusion device. The average dose is 5.5×10^6 cells (100–1200 million total cells) in 2 mL of saline. Enrolled patients will receive MSC injections in the experimental limb, and the nontreated control will consist of a saline injection (same volumes as experimental) in the contralateral limb of the same patient. Patients will be followed up for 24 months for the primary purpose of establishing the frequency of adverse events following this therapy. Secondary outcome measures include a quantification of motor units population, muscle fiber density, muscle force, maximum muscle force, and maximal transverse area of tibialis anterior out to 24 months when compared with patient's initial baseline. This study began in 2014 and is expected to be completed by 2018.

Poland: Pomeranian Medical University Szczecin

An open-label phase I study (NCT02193893) conducted at the Pomeranian Medical University Szczecin in Poland was launched in 2010 to test the safety and efficacy of autologous bone-marrow-derived MSCs (NeuStem-ALS) for the treatment of ALS. The MSCs used in this study were purported to be able to secrete bioactive neurotrophic factors that may support or even regenerate neuronal cells, although no specific reference to evidence of this claim was provided. This prospective pilot study intended to establish safety information as well as efficacy relating to the therapy. Investigators planned to monitor patients to determine the proregenerative and neuroprotective functions of these MSCs. The Polish investigators aimed to enroll 50 patients that were previously diagnosed with ALS depending on date of onset. The "early" group will have had their ALS diagnosis within the 6 months of study participation, whereas the "advanced" group will have been diagnosed between 6 and 12 months prior to the study. No results have been published and the NeuStem-ALS clinical trial is expected to be completed by the end of 2017.

Turkey

Thirteen ALS patients designated as sporadic received bone-marrow-derived hematopoietic progenitor stem cells. Patients that had advanced

ALS progression were included in the study. These operations began in 2006. The authority under which this trial was launched was Akay Hospital Ethics Board, Ankara, Turkey, and the Ministry of Health. Patients received a complete laminectomy at the C1–C2 level and the dura was incised and opened to expose the spinal cord. Areas that were considered avascular were used to inject stem cells. Cells were injected through the dorsal region of the cord into the anterior part. 0.1 mL was injected in multiple locations.[51] A foam storage material was placed "on the lower cranial nerve and surface of the brainstem." Three milliliters was injected into the gel foam (10 million cells). Next they describe injecting 1.5 mL of cells into the subarachnoid space and then another 1.5 mL was given intravenously, by an anesthesiologist. This report claims that re-innervation as determined by electroneuromyography (ENMG) was confirmed in 7/13 patients. Additionally they report this therapy to be well tolerated and absent of adverse events.[51]

ASIA

Israel

In 2010 a phase I/II clinical trial (NCT01051882) was conducted at the Hadassah Medical Organization in collaboration with Brainstorm-Cell Therapeutics to test the safety, tolerability, and efficacy of autologous MSC transplantation for treating ALS. The trial planned to enroll 24 patients that would be divided into two treatment groups depending on disease progression. The first group, "A," would be ALS patients that were in a relatively early disease phase and they would receive the mesenchymal stem cells secreting neurotrophic factors (MSC-NTF) therapy through intramuscular injection into the patient's upper-arm biceps and triceps muscles. These patients were injected using a pre-determined grid map at 24 upper arm sites with 1×10^6 cells per site for a total of 2.4×10^7 cells. Group "B" included ALS patients characterized as progressive and they would receive an intrathecal injection in the lumbar region with a quantity of 6×10^7 cells.

Prior to bone marrow aspiration, patients were observed every 2 weeks during a run-in period consisting of 3 months. Six weeks after enrollment, patients underwent a bone marrow aspiration procedure to harvest the autologous MSCs. The marrow aspirate was then processed through a Brainstorm Cell Therapeutics LTD proprietary method to produce the cell product (MSC-NTF, or NurOwn™) for therapeutic administration. During the last run-in visit, patients received the MSC-NTFs through their designated administration route.

For six months following transplantation, patients were monitored monthly for signs of adverse reactions, ALS clinical progression, respiratory function, electrophysiology, and muscle volume changes. After the six-month follow-up period there were no observed adverse events associated with the procedure, and ALS progression had slowed when compared to the 3-month follow-up.[41]

In late 2012 the Hadassah Medical Organization initiated another clinical trial (NCT01777646) in collaboration with Brainstorm-Cell Therapeutics to test the safety, tolerability, and efficacy of autologous MSC transplantation for treating ALS. This phase II trial aimed to test the MSC-NTF treatment for early-stage ALS. This phase II pilot study was conducted at a single trial center, the Department of Neurology and Laboratory of Neuroimmunology at the Hadassah Hebrew University Medical Center in Jerusalem. All participants enrolled in the open-label, dose-escalating clinical study were required to have a "documented history of ALS disease prior to enrollment." A total of 12 participants, both male and female, were enrolled for the trial.

Patients enrolled in the trial were to be observed once a month over a 3-month run-in period to establish the baseline rate of the disease progression. At 6 weeks following their enrollment, patients underwent a bone marrow aspiration procedure to extract bone marrow for stem cell derivation. The MSC-NTF (NurOwn™) were cultured from the bone marrow aspirate using a proprietary method performed by Brainstorm Cell Therapeutics LTD. Study participants were injected with these MSC-NTF cells through intramuscular or intrathecal delivery. The number of cells included in each dosage was: "low," 9.4×10^7 cells; "medium," 1.41×10^8; and "high," 1.88×10^8. The lowest dosage was tested first, and subsequently the medium and high doses were to be conducted only after safety using lower doses had been established. Following intrathecal and intramuscular transplantation of the MSC-NTF cells, patients were clinically observed monthly over the course of 6 months. During this time, the safety and efficacy of MSC-NTF treatment was established. Although ClincalTrials.gov lists this trial as ongoing and active, in January of 2014 Brainstorm published a nonscientific press release describing this clinical trial's results (http://www.brainstorm-cell.com/index.php/news-events/331-january-5-2015). Fourteen patients were included in this trial and showed that the MSC-NTF therapy was safely tolerated in this cohort. According to Brainstorm, clinical benefit was experienced by "nearly" all the patients. Patients received up to 2×10^6 cells/kg through intrathecal administration and an additional 48×10^6 cells/kg intramuscularly. Eleven of 12 patients who participated in 3 or more months of follow-up reported a reduction in the rate of disease progression. The clinical outcome was unreported for two patients that underwent the procedure but did not participate in follow-up up to 3 months postprocedure.

A lead investigator in the trial stated, "... almost every subject experienced clinical benefit, either on ALS functional rating score (50%)

(ALSFRS), FVC (67%, FVC), or both measures." NurOwn™ demonstrated a 45% decrease in the progression of ALS during the 3-month follow-up period, and a 57% decrease during the 6-month follow-up when compared to the baseline that was established during the run-in period. The expected rate of decline in the FVC scores for these patients was reduced by 73%. The rate of decline during the run-in period was 2.6% per month which was lowered to just 0.7% per month up to 3 months following treatment. The reduced decline in lung function up to 6 months was reported as a 67% improvement when compared to the run-in period. This was reported via a press release from the company, but without a peer-reviewed publication detailing the methods and results of the trial, it is difficult to assess the true impact of this study.

NurOwn™ has been administered to over 30 patients with ALS in clinical trials conducted in Israel. Following the self-reported success of the phase II open-label trial, a subsequent randomized, double-blind, placebo-controlled phase II clinical trial is currently recruiting for enrollment of approximately 48 patients at multiple study centers in the United States (NCT02017912). This intervention with NuOwn™ (MSC-NTF) will again be administered through intrathecal and intramuscular injection. This trial's primary outcome is to establish the safety of NuOwn™ by monitoring patients over the course of 10 clinical follow-ups and quantifying adverse events in this cohort. Secondary outcomes are to measure the changes seen in patients' ALSFRS scores as well as slow vital capacity from the pretransplant period to the posttransplantation period between treatment and control groups through 24 weeks after NuOwn™ transplantation. Recruiting for this study is ongoing and the study's primary objectives are estimated to be completed by April of 2016.

Iran

The Royan institute launched a phase I clinical trial (NCT01759784) to determine whether autologous MSCs would be a safe and efficacious treatment for ALS by way of intraventricular injection. As previously described, the MSCs would be harvested through bone-marrow aspiration. Patients were monitored for acute adverse events including immune and neurological alterations for procedural safety validation. Additionally, patients were to be monitored at 1, 3, 6, and 12 months postprocedure. Secondary measures to establish efficacy included ALS-FRS, electromyogram-neve conduction velocity (EMG-NCV), and FVC. According to ClincalTrials.gov, participant recruitment has been suspended due the dangerous nature of stereotaxic surgery on ALS patients with compromised respiratory function.

In a similar study sponsored by Royan Institute, an open label, phase I clinical trial was initiated in 2012 (NCT01759797) to test the safety of intravenous injection using autologous MSCs to treat ALS. Primary outcome measures were aimed at monitoring prevalence of adverse outcomes such as fever, vomiting, or unconsciousness within 48 h posttransplantation. Secondary measures included monitoring for disease progression as determined by ALS-FRS, FVC, and EMG-NCV in the 6 months following the procedure. There are no published reports on the outcome of this trial. This study began in late summer of 2012 and was completed by the following summer.

A phase I/II trial (NCT02116634) has been planned to be conducted at and sponsored by the Alzahra Hospital in Iran. This trial is intended to test the efficacy of autologous MSCs for the treatment of ALS. Enrolled patients receiving the infusion will be monitored every 6 months for 2 years for ALS progression. ALS-FRS, EMG scale, FVC, and DWSE ± QoL score will be monitored as a secondary outcome measure. Delivery of the MSCs will be through intrathecal injection of 1×10^8 cells in 10 mL of saline. At the time of publication of this book, this trial had not yet begun recruiting patients.

India: Neurogen Brain and Spine Institute

This phase II study (NCT01984814) was conducted and sponsored by the Neurogen Brain and Spine institute in Mumbai, India. This trial was a nonrandomized endpoint classification for safety/efficacy. It aimed to improve the survival duration of ALS patients through the intervention of autologous MSC therapy compared to a retrospective control cohort. The experimental treatment group was compared to a retrospective cohort. This method of comparison is problematic in that the consistency of standards of care between a historical outpatient group and the experimental group cannot be confirmed. This study began in 2008, enrolled 57 patients, and was completed by August of 2013.

In Sharma et al. 2015[42] it is reported that all patients had an established clinical record for onset of ALS as was required for inclusion in the analysis. The cell harvest procedure removed 100–120 mL of aspirate from the superior iliac spine region. Following cell preparation the group reported recovery of 8×10^7 cells per aspiration with approximately 96.7% viability. These cells were "characterized" for the CD34+ expression. A total of 66.6% of the preparation was injected in the lumbar region between L4 and L5, while the remaining 33.3% was injected intramuscularly near the "motor-points of specific muscles." Patients were asked to return for a follow-up every 3 months following the procedure.

At the 2-year mark, there were no significant differences in survival between patients within the control group (50% mortality rate) when compared to the intervention group (48.64% mortality rate). However, when following-up patients to their death, this study reports that the intervention group survived an average of 87.76 months (range from 13 to 158 months in 37 patients) following ALS diagnosis. This is a 30-month improvement when compared to the control group who lived an average 57.38 months (range from 26 to 84 months in 20 patients).[42] A statistic demonstrating the astounding efficacy reported in this study is that the control group's longest survivor lived 84 months post-ALS diagnosis, a survival time that falls short of the mean survival time of the entire intervention group. While these results are very promising, it is of particular concern that the control group was a "retrospective control cohort," meaning that they were patients chosen from an outpatient population. Additionally, there is no description of how these 20 patients included in the control group were chosen.[42]

They did, however, compare their study's results with historical epidemiological populations, and showed a striking improvement in efficacy compared to the other studies. There were 12 epidemiological studies compared to this trial's results. Adverse events were reported in that "spinal headache" was present in nearly a third of the patients receiving treatment. Other minor adverse events, presenting in less than 10% of patients, included nausea, vomiting, injection site pain, aspiration site pain, and fatigue. No major adverse reactions were present after the procedure. Of the 37 patients that were enrolled in the study and underwent the procedure, only 6 patients were available for telephonic ALS-FRS scoring. It is reported that three of these patients showed improved ALS-FRS scores after the transplant during "most recent telephonic follow-up." This report states that impact of their findings were limited by the small size of the retrospective cohort, which is likely insufficient for conclusive clinical findings. They also state that the number of transplantations and the time between multiple transplantations could influence the outcome. Additionally the time to intervention after diagnosis should be taken into consideration, but this study did not have sufficient data to conduct these analyses.

Another clinical trial (NCT02242071) was launched in 2008 by the Neurogen Spine institute in India. This open-label phase I trial was intended to determine safety and efficacy of autologous MSC transplantation in ALS patients. This study aimed to enroll 200 patients, which is a large number of enrolled patients for a phase I trial when compared to others described here. There was no description found pertaining to either delivery method or dosage. ClincalTrials.gov lists this trial as currently recruiting patients. The study is expected to be completed by 2016.

South Korea

An open-label phase I/II clinical trial (NCT01363401) was launched in 2011 sponsored by Corestem, Inc. and conducted in Seoul, South Korea, at Hanyang University Center. This trial is believed to be the first to evaluate the safety of repeated intrathecal infusions, and aimed to enroll 71 patients to evaluate the efficacy of autologous bone marrow-derived MSCs, or "HYNR-CS," in the treatment of ALS.[43] It is the investigator's hypothesis that the stem cell treatment HYNR-CS will slow disease progression. The actual study enrolled only eight patients to participate in the trial. Prior to transplantation, patients were assessed during a lead-in period of 12 weeks to establish baseline ALSFRS-R scores. Patients who received transplantation underwent intrathecal infusion of 1×10^6 cells at a volume of 1 mL/kg of body weight to test the safe use of these cells after repetitive intrathecal infusions.

Seven of the eight patients enrolled underwent two intrathecal infusions of HYNR-CS with a 26-day interval in between. The remaining enrolled patient died after bone marrow aspiration but prior to receiving treatment and was withdrawn from the study.[43] Patients were assessed every 4 weeks for changes to ALSFRS-R. Also secondary outcome measures were monitored as alterations from the 12-week lead-in period when compared to scores immediately preceding surgery to 16 weeks postprocedure for both the Appel scale and FVC. SF-36, a measure of quality of life, were also monitored for changes. Other adverse events were reported to be minor and lasted less than 4 days following the transplantation. This study claims that repeated intrathecal infusion of autologous MSCs (HYNR-CS) is safe and feasible through the 12-month follow-up period.[43] An ongoing controlled, randomized, phase II trial is being conducted and its results are expected to be available at a later date.

Spinal Muscular Atrophy: Gene Therapy Clinical Trial

Jerry R. Mendell, in collaboration with Sophia's Cure Foundation and AveXis Inc, launched an open-label, phase I clinical trial (NCT02122952) in 2014 to evaluate viral-mediated gene therapy (scAAV9.CB.SMN) as a safe and efficacious method to treat SMA type 1. The estimated enrollment for this study was nine patients who have yet to reach 9 months of age. The primary purpose was to measure the safety and associated toxicity for 2 years following the administration of the virally mediated gene therapy.

The experimental design calls for the cohort to be broken up into three separate groups receiving scAAV9.CB.SMN in a dose escalation paradigm. Cohort "1" will receive 6.7×10^{13} vg/kg (vector genomes/kg); Cohort "2" will receive 2.0×10^{14} vg/kg; and Cohort "3" will receive 3.3×10^{14} vg/kg. After intravenous infusion, the subject will be monitored for side-effects or adverse events every week for 3 weeks following the procedure. Monthly visits for the remainder of the 2-year study period will also be conducted.

Hematology, serum chemistry, urinalysis, and immunologic response to scAAV9.CB.SMN therapy will be monitored to determine any significant changes attributed to the gene therapy. Additionally, researchers will monitor for alterations in projected life span, time until adverse event, and time from birth to medically prescribed respiratory assistance exceeding 16 hours a day. Nerve conduction and muscle assessment will also be recorded and reported. This trial is expected to be completed in the summer of 2017.

CONCLUSIONS

Currently available treatments have failed to offer adequate benefit in ALS and SMA. These diseases subject patients to compromised motor function, a reduced quality of life, and significantly shortened life spans. Riluzole is the only FDA-approved intervention for ALS, and it extends a patient's life span by approximately 2–3 months. Therefore, experimental therapeutics able to offer longer-term benefits for ALS patients must be explored and championed by doctors and scientists for rapid development in the coming years. Gene- and stem cell-mediated therapeutics have been touted for decades as the next big step in the treatment of a wide range of medically refractive conditions, and are currently being tested at a rapid pace in both preclinical and clinical trials.

These strategies are particularly attractive to medical researchers looking to develop cutting-edge therapeutics for diseases of the nervous system. Replacing damaged or dead cells with stem cells able to differentiate into a diverse set of functional cells is one strategy for the use of cell-mediated therapeutics. This is of particular importance in neurodegenerative diseases where postmitotic neuronal networks become dysfunctional leading to neurological deficit. The ability to replace lost or dysfunctional neurons in specific networks with stem cell therapies that would restore compromised neural function would be world changing for so many patients. Stem cells, such as iPSCs, also provide a valuable tool allowing doctors to harvest a patient's cells and then screen potential therapeutics in vitro for patient specific efficacy, possibly elevating the level of care.

Early enthusiasm for the potential of cell- and gene-based therapeutics has been tempered by relatively little efficacy in the decades since its rise to scientific and medical research prominence. Many studies using stem cell therapy to treat ALS initially showed promise in the open-label trials only to see the subsequent phase II trials fail to replicate, or have yet to report therapeutic effect.[52] Moreover, there exists little published data available for the rest of these trials.

Many patients with neurodegenerative disease who volunteer to participate in gene or stem cell clinical trials are usually suffering a severe, debilitating, and life-threatening disease process. This brings up the issue of how to adequately employ the necessary controls in order to scientifically establish efficacy. There are numerous ethical and health issues in having a large group of terminally sick ALS patients undergo invasive surgeries only to receive a placebo. Most stem cell clinical trials perform invasive surgeries on sick patients delivering stem cells to their target site, usually within the CNS. Potential surgical complications from invasive surgeries, leading to a further reduction in a patient's quality of life, are of questionable risk if a patient is only given the placebo. Could these patients and their families understand the potential hazards of the procedure and the probability of receiving a placebo as they hope for a cure? There is perpetual development of safer, less invasive surgeries through improved equipment and imagery to administer therapeutics.[48,53–57] Until clinical trials demonstrate strong efficacy in phase II it is unlikely that larger-cohort, randomized, double-blinded controlled trials will be supported.

Although widespread success of gene- and stem cell-therapy strategies have yet to become a reality, there remains great potential for progress in that many of these trials have successfully validated these biologics and surgical delivery procedures as safe and feasible interventions in the treatment of MND. As clinicians and investigators gain a better understanding of the specific mechanisms leading to MND, likewise new therapies will be adapted to exploit and alter these mechanisms. Extensive ground work has been laid in these early stem cell and gene therapy clinical trials and will provide invaluable knowledge for scientists and doctors combatting ALS and SMA in the decades to come.

References

1. Thomsen GM, Gowing G, Svendsen S, Svendsen CN. The past, present and future of stem cell clinical trials for ALS. *Exp Neurol* 2014;**262**(Pt B):127–37.
2. Thomsen GM, Gowing G, Latter J, et al. Delayed disease onset and extended survival in the SOD1G93A rat model of amyotrophic lateral sclerosis after suppression of mutant SOD1 in the motor cortex. *J Neurosci Off J Soc Neurosci* 2014;**34**(47):15587–600.
3. Simonato M, Bennett J, Boulis NM, et al. Progress in gene therapy for neurological disorders. *Nat Rev Neurol* 2013;**9**(5):277–91.
4. Donnelly EM, Lamanna J, Boulis NM. Stem cell therapy for the spinal cord. *Stem Cell Res Ther* 2012;**3**(4):24.
5. O'Connor DM, Boulis NM. Cellular and molecular approaches to motor neuron therapy in amyotrophic lateral sclerosis and spinal muscular atrophy. *Neurosci Lett* 2012;**527**(2):78–84.
6. O'Connor DM, Boulis NM. Gene therapy for neurodegenerative diseases. *Trends Mol Med* 2015
7. Federici T, Boulis NM. Gene therapy for amyotrophic lateral sclerosis. *Neurobiol Dis* 2012;**48**(2):236–42.
8. Borchelt DR, Lee MK, Slunt HS, et al. Superoxide dismutase 1 with mutations linked to familial amyotrophic lateral sclerosis possesses significant activity. *Proc Natl Acad Sci U S A* 1994;**91**(17):8292–6.

9. Saccon RA, Bunton-Stasyshyn RK, Fisher EM, Fratta P. Is SOD1 loss of function involved in amyotrophic lateral sclerosis? *Brain* 2013;**136**(Pt 8):2342–58.
10. Rotunno MS, Bosco DA. An emerging role for misfolded wild-type SOD1 in sporadic ALS pathogenesis. *Front Cell Neurosci* 2013;**7**:253.
11. Clermont O, Burlet P, Benit P, et al. Molecular analysis of SMA patients without homozygous SMN1 deletions using a new strategy for identification of SMN1 subtle mutations. *Hum Mutat* 2004;**24**(5):417–27.
12. Bevan AK, Hutchinson KR, Foust KD, et al. Early heart failure in the SMNDelta7 model of spinal muscular atrophy and correction by postnatal scAAV9-SMN delivery. *Hum Mol Genet* 2010;**19**(20):3895–905.
13. Duque SI, Arnold WD, Odermatt P, et al. A large animal model of spinal muscular atrophy and correction of phenotype. *Ann Neurol* 2015;**77**(3):399–414.
14. Foust KD, Salazar DL, Likhite S, et al. Therapeutic AAV9-mediated suppression of mutant SOD1 slows disease progression and extends survival in models of inherited ALS. *Mol Ther J Am Soc Gene Ther* 2013;**21**(12):2148–59.
15. Meyer K, Ferraiuolo L, Schmelzer L, et al. Improving single injection CSF delivery of AAV9-mediated gene therapy for SMA: a dose-response study in mice and nonhuman primates. *Mol Ther J Am Soc Gene Ther* 2015;**23**(3):477–87.
16. Marshall E. Gene therapy death prompts review of adenovirus vector. *Science* 1999;**286**(5448):2244–5.
17. LeWitt PA, Rezai AR, Leehey MA, et al. AAV2-GAD gene therapy for advanced Parkinson's disease: a double-blind, sham-surgery controlled, randomised trial. *Lancet Neurol* 2011;**10**(4):309–19.
18. Bevan AK, Duque S, Foust KD, et al. Systemic gene delivery in large species for targeting spinal cord, brain, and peripheral tissues for pediatric disorders. *Mol Ther J Am Soc Gene Ther* 2011;**19**(11):1971–80.
19. Foust KD, Nurre E, Montgomery CL, Hernandez A, Chan CM, Kaspar BK. Intravascular AAV9 preferentially targets neonatal neurons and adult astrocytes. *Nat Biotechnol* 2009;**27**(1):59–65.
20. Samaranch L, Salegio EA, San Sebastian W, et al. Adeno-associated virus serotype 9 transduction in the central nervous system of nonhuman primates. *Hum Gene Ther* 2012;**23**(4):382–9.
21. Ramalho-Santos M, Willenbring H. On the origin of the term "stem cell". *Cell Stem Cell* 2007;**1**(1):35–8.
22. Boulis NM, Federici T, Glass JD, Lunn JS, Sakowski SA, Feldman EL. Translational stem cell therapy for amyotrophic lateral sclerosis. *Nat Rev Neurol* 2011;**8**(3):172–6.
23. Murugan V. Embryonic stem cell research: a decade of debate from Bush to Obama. *Yale J Biol Med* 2009;**82**(3):101–3.
24. Noggle SA, James D, Brivanlou AH. A molecular basis for human embryonic stem cell pluripotency. *Stem Cell Rev* 2005;**1**(2):111–8.
25. Brack AS, Rando TA. Tissue-specific stem cells: lessons from the skeletal muscle satellite cell. *Cell Stem Cell* 2012;**10**(5):504–14.
26. Togel F, Westenfelder C. Adult bone marrow-derived stem cells for organ regeneration and repair. *Dev Dynam Off Publ Am Assoc Anatomists* 2007;**236**(12):3321–31.
27. Morshead CM, Reynolds BA, Craig CG, et al. Neural stem cells in the adult mammalian forebrain: a relatively quiescent subpopulation of subependymal cells. *Neuron* 1994;**13**(5):1071–82.
28. Muller FJ, Snyder EY, Loring JF. Gene therapy: can neural stem cells deliver? *Nat Rev Neurosci* 2006;**7**(1):75–84.
29. Richard JP, Maragakis NJ. Induced pluripotent stem cells from ALS patients for disease modeling. *Brain Res* 2015;**1607**:15–25.
30. Knoepfler PS. Deconstructing stem cell tumorigenicity: a roadmap to safe regenerative medicine. *Stem Cells* 2009;**27**(5):1050–6.

31. Mazzini L, Mareschi K, Ferrero I, et al. Stem cell treatment in Amyotrophic Lateral Sclerosis. *J Neurol Sci* 2008;**265**(1–2):78–83.
32. Martinez HR, Gonzalez-Garza MT, Moreno-Cuevas JE, Caro E, Gutierrez-Jimenez E, Segura JJ. Stem-cell transplantation into the frontal motor cortex in amyotrophic lateral sclerosis patients. *Cytotherapy* 2009;**11**(1):26–34.
33. Riley J, Glass J, Feldman EL, et al. Intraspinal stem cell transplantation in amyotrophic lateral sclerosis: a phase I trial, cervical microinjection, and final surgical safety outcomes. *Neurosurgery* 2014;**74**(1):77–87.
34. Riley J, Federici T, Polak M, et al. Intraspinal stem cell transplantation in ALS: a phase I safety trial, technical note & lumbar safety outcomes. *Neurosurgery* 2012;**71**(2):405–16.
35. Feldman EL, Boulis NM, Hur J, et al. Intraspinal neural stem cell transplantation in amyotrophic lateral sclerosis: phase 1 trial outcomes. *Ann Neurol* 2014;**75**(3):363–73.
36. Tadesse T, Gearing M, Senitzer D, et al. Analysis of graft survival in a trial of stem cell transplant in ALS. *Ann Clin Transl Neurol* 2014;**1**(11):900–8.
37. Mazzini L, Ferrero I, Luparello V, et al. Mesenchymal stem cell transplantation in amyotrophic lateral sclerosis: a phase I clinical trial. *Exp Neurol* 2010;**223**(1):229–37.
38. Mazzini L, Mareschi K, Ferrero I, et al. Mesenchymal stromal cell transplantation in amyotrophic lateral sclerosis: a long-term safety study. *Cytotherapy* 2012;**14**(1):56–60.
39. Mazzini L, Gelati M, Profico DC, et al. Human neural stem cell transplantation in ALS: initial results from a phase I trial. *J Transl Med* 2015;**13**:17.
40. Blanquer M, Moraleda JM, Iniesta F, et al. Neurotrophic bone marrow cellular nests prevent spinal motoneuron degeneration in amyotrophic lateral sclerosis patients: a pilot safety study. *Stem Cells* 2012;**30**(6):1277–85.
41. Karussis D, Karageorgiou C, Vaknin-Dembinsky A, et al. Safety and immunological effects of mesenchymal stem cell transplantation in patients with multiple sclerosis and amyotrophic lateral sclerosis. *Arch Neurol* 2010;**67**(10):1187–94.
42. Sharma AK, Sane HM, Paranjape AA, et al. The effect of autologous bone marrow mononuclear cell transplantation on the survival duration in Amyotrophic Lateral Sclerosis - a retrospective controlled study. *Am J Stem Cells* 2015;**4**(1):50–65.
43. Oh KW, Moon C, Kim HY, et al. Phase I trial of repeated intrathecal autologous bone marrow-derived mesenchymal stromal cells in amyotrophic lateral sclerosis. *Stem Cells transl Med* 2015;**4**(6):590–7.
44. Janson CG, Ramesh TM, During MJ, Leone P, Heywood J. Human intrathecal transplantation of peripheral blood stem cells in amyotrophic lateral sclerosis. *J Hematother Stem Cell Res* 2001;**10**(6):913–5.
45. Sen I, Nalini A, Joshi NB, Joshi PG. Cerebrospinal fluid from amyotrophic lateral sclerosis patients preferentially elevates intracellular calcium and toxicity in motor neurons via AMPA/kainate receptor. *J Neurol Sci* 2005;**235**(1–2):45–54.
46. Foran E, Trotti D. Glutamate transporters and the excitotoxic path to motor neuron degeneration in amyotrophic lateral sclerosis. *Antioxid Redox Signal* 2009;**11**(7):1587–602.
47. Riley J, Butler J, Park J, et al. Targeted spinal cord therapeutics delivery: stabilized platform and MER guidance validation. *Stereotact Funct Neurosurg* 2007;**86**(2):67–74.
48. Riley J, Federici T, Park J, et al. Cervical spinal cord therapeutics delivery: preclinical safety validation of a stabilized microinjection platform. *Neurosurgery* 2009;**65**(4):754–62. discussion 761–762.
49. Glass JD, Boulis NM, Johe K, et al. Lumbar intraspinal injection of neural stem cells in patients with ALS: results of a phase I trial in 12 patients. *Stem Cells* 2012;**30**(6):1144–51.
50. Blanquer M, Perez Espejo MA, Iniesta F, et al. Bone marrow stem cell transplantation in amyotrophic lateral sclerosis: technical aspects and preliminary results from a clinical trial. *Methods Find Exp Clin Pharmacol* 2010;(32 Suppl A):31–7.
51. Deda H, Inci MC, Kurekci AE, et al. Treatment of amyotrophic lateral sclerosis patients by autologous bone marrow-derived hematopoietic stem cell transplantation: a 1-year follow-up. *Cytotherapy* 2009;**11**(1):18–25.

52. Mitsumoto H, Brooks BR, Silani V. Clinical trials in amyotrophic lateral sclerosis: why so many negative trials and how can trials be improved? *Lancet Neurol* 2014;**13**(11):1127–38.

53. Boulis N, Federici T. Surgical approach and safety of spinal cord stem cell transplantation. *Neurosurgery* 2011;**68**(2):E599–600.

54. Riley JP, Raore B, Taub JS, Federici T, Boulis NM. Platform and cannula design improvements for spinal cord therapeutics delivery. *Neurosurgery* 2011;**69**(2 Suppl Operative) ons147-54.

55. Lamanna JJ, Miller JH, Riley JP, Hurtig CV, Boulis NM. Cellular therapeutics delivery to the spinal cord: technical considerations for clinical application. *Ther Deliv* 2013;**4**(11):1397–410.

56. Gutierrez J, Moreton CL, Lamanna JJ, et al. 203 Understanding cell migration after direct transplantation into the spinal cord: a tool to determine the optimal transplantation volume. *Neurosurgery* 2015(62 Suppl 1):234.

57. Federici T, Hurtig CV, Burks KL, et al. Surgical technique for spinal cord delivery of therapies: demonstration of procedure in gottingen minipigs. *J Vis Exp* 2012;**70**:e4371.

Index

K

Kinesin family member 5A (KIF5A), 54
Kugelberg–Welander disease, 106–107

L

Laminectomy, 290
Laminin, 10–11
LAP2 gene, 148–149
Lateral motor column (LMC), 7
Lateral motor column lateral (LMCl), 8
Lateral motor column medial (LMCm), 8
Lenti-SMN, 241
Lentivirus (LV), 143, 147–148
  vectors, 147–148
Lethal congenital contracture syndrome 1,
  121–122
LeX cells, 256–257
Lhx3. *See* LIM homeobox 3 (Lhx3)
LIM homeobox 3 (Lhx3), 5
Limb-innervating neurons, 8
LMC. *See* Lateral motor column (LMC)
LMCl. *See* Lateral motor column lateral
  (LMCl)
LMCm. *See* Lateral motor column medial
  (LMCm)
LMN signs. *See* Lower motor neuron signs
  (LMN signs)
Long ncRNAs (lncRNA), 65–66
Long terminal repeat (LTR), 147–148
Louisiana, stem cell clinical trials in ALS,
  293
Lower motor neuron signs (LMN signs),
  29, 30t
LTR. *See* Long terminal repeat (LTR)
LV. *See* Lentivirus (LV)

M

Magnetic resonance imaging (MRI), 292
Malnutrition, 36
MapK. *See* Mitogen-activated protein
  kinase (MapK)
*MAPT. See Microtubule associated protein tau
  (MAPT)*
Marrow derived stem cells (MSCs), 293
Matrix metalloproteinases (MMPs), 86–87
Maximum voluntary isometric contraction-
  arm (MVIC-arm), 293
Mayo Clinic, 291–292
MCP-1. *See* Monocyte chemotactic
  protein-1 (MCP-1)
MCT1. *See* Monocarboxylate transporter 1
  (MCT1)

Medial motor column (MMC), 7–8
Mesenchymal stem cells (MSCs), 149–150,
  152–153, 281–282
Mesenchymal stem cells secreting
  neurotrophic factors (MSC-NTF),
  298
Metallothioneins (MTs), 175–176
2′-*O*-Methoxyethyl (2′-MOE), 132, 238
2′-*O*-Methyl (2′-OMe), 238
Mexico, stem cell clinical trials in ALS, 293
Microglia, 83–84
MicroRNAs (miRNAs), 66–67
Microtubule associated protein tau (MAPT),
  54
mIGF-1. *See* Muscle expression of local
  isoform of IGF-1 (mIGF-1)
Migration, 6
Minimally invasive strategies of cell
  administration, 263–265
Minnesota, stem cell clinical trials in ALS,
  291–292
miRNAs. *See* MicroRNAs (miRNAs)
Missense mutations, 128
Mitogen-activated protein kinase (MapK),
  177–178
MLPA. *See* Multiplex Ligation-dependent
  Probe Amplification (MLPA)
MMC. *See* Medial motor column (MMC)
MMPs. *See* Matrix metalloproteinases
  (MMPs)
MN progenitor (pMN), 5
MND. *See* Motor neuron disease (MND)
MNs. *See* Motor neurons (MNs)
MNX1. *See* Homeobox gene 9 (Hb9)
MO. *See* Morpholino oligonucleotides
  (MO)
2′-MOE. *See* 2′-*O*-Methoxyethyl (2′-MOE)
Monocarboxylate transporter 1 (MCT1),
  84–85
Monocyte chemotactic protein-1 (MCP-1),
  214
Morpholino oligonucleotides (MO), 132,
  238
Motor neuron disease (MND), 26, 141–142,
  278–279
Motor neurons (MNs), 2–3, 65–69, 76,
  121–122, 168, 209. *See also* Survival
  motor neuron (SMN)
  amyotrophic lateral sclerosis (ALS), 87f
  axon targeting, 9
  axons, 75
  cell death, 13